Information Systems Engineering and Management 52

The book series "Information Systems Engineering and Management" (ISEM) publishes innovative and original works in the various areas of planning, development, implementation, and management of information systems and technologies by enterprises, citizens, and society for the improvement of the socio-economic environment.

The series is multidisciplinary, focusing on technological, organizational, and social domains of information systems engineering and management. Manuscripts published in this book series focus on relevant problems and research in the planning, analysis, design, implementation, exploration, and management of all types of information systems and technologies. The series contains monographs, lecture notes, edited volumes, pedagogical and technical books as well as proceedings volumes.

Some topics/keywords to be considered in the ISEM book series are, but not limited to: Information Systems Planning; Information Systems Development; Exploration of Information Systems; Management of Information Systems; Blockchain Technology; Cloud Computing; Artificial Intelligence (AI) and Machine Learning; Big Data Analytics; Multimedia Systems; Computer Networks, Mobility and Pervasive Systems; IT Security, Ethics and Privacy; Cybersecurity; Digital Platforms and Services; Requirements Engineering; Software Engineering; Process and Knowledge Engineering; Security and Privacy Engineering, Autonomous Robotics; Human-Computer Interaction; Marketing and Information; Tourism and Information; Finance and Value; Decisions and Risk; Innovation and Projects; Strategy and People.

Indexed by Google Scholar. All books published in the series are submitted for consideration in the Web of Science.

For book or proceedings proposals please contact Alvaro Rocha (amrrocha@gmail.com).

Maria José Sousa
Editor

Human Resources Development in a Digital Age

Springer

Editor
Maria José Sousa
Avenida das Forças Armadas
Lisbon, Portugal

ISSN 3004-958X ISSN 3004-9598 (electronic)
Information Systems Engineering and Management
ISBN 978-3-031-92547-4 ISBN 978-3-031-92548-1 (eBook)
https://doi.org/10.1007/978-3-031-92548-1

This Springer imprint is published by the registered company Springer Nature Switzerland AG
The registered company address is: Gewerbestrasse 11, 6330 Cham, Switzerland

Preface

In the rapidly evolving panorama of the digital world, the field of Human Resources Development (HRD) faces unprecedented challenges and opportunities. This book, "Human Resources Development in a Digital Age," aims to provide a comprehensive exploration of the dynamic interplay between technology and HRD practices. As organizations cross the complexities of digital transformation, it is imperative to understand how strategic HR management, artificial intelligence, and innovative practices can be leveraged to enhance productivity, employee engagement, and organizational resilience.

This collection of chapters brings together insights from leading experts and researchers in the field, offering a diverse range of perspectives on the critical issues shaping HRD today. From the impact of AI on civil service work to the development of employee engagement strategies in hybrid work environments, each chapter investigates the shades of HRD in the context of digital advancements.

The book is structured to provide both theoretical foundations and practical applications, making it a valuable resource for HR professionals, scholars, and students. Each chapter is designed to offer deep insights into specific aspects of HRD, supported by empirical research and case studies. The aim is to equip readers with the knowledge and tools necessary to navigate the complexities of HRD in the digital age effectively.

Maria José Sousa

Organization of the Book

The organization section of this book, "Human Resources Development in a Digital Age," is inspired by the groundbreaking research presented at the 24th UFHRD Conference held in Lisbon in June 2024. Hosted at ISEC Lisboa, this conference was a collaborative effort between UFHRD and ISEC Lisboa, under the chairmanship of Maria José Sousa, Associate Professor at ISCTE, Portugal

The UFHRD 24 Conference brought together leading scholars, practitioners, and thought leaders from around the globe to explore the multifaceted dimensions of HRD in the digital era. The conference's theme, "Human Resources Development in a Digital Age," provided a rich tapestry of ideas and innovations that have significantly influenced the structure and content of this book.

As you explore the chapters, you will find that each one reflects the cutting-edge research and practical insights shared during the conference. From strategic HR management and AI integration to organizational culture and employee engagement, the topics covered in this book are a testament to the collaborative spirit and intellectual rigor that characterized the UFHRD 24 Conference.

An overview of the chapter is presented in the following text:

Chapter 1: Introduction to Human Resources Development in a Digital Age

This chapter sets the stage for the book, providing an overview of the key themes and concepts that will be explored. It introduces the reader to the challenges and opportunities presented by the digital age in the context of HRD.

Chapter 2: Strategic Human Resource Management and Dynamic Capabilities in Life Sciences: Leveraging Hybrid Team Management and Performance Monitoring in Fluid Working Spaces for Advanced Productivity

This chapter examines the strategic HR management practices in the life sciences sector, focusing on hybrid team management and performance monitoring to enhance productivity in dynamic work environments.

Chapter 3: Artificial Intelligence and Its Impact on the Work of Civil Servants: A Systematic Review

A comprehensive review of the impact of AI on the work of civil servants, this chapter synthesizes existing research to provide insights into how AI is transforming public administration.

Chapter 4: Harnessing the Power of AI for Human Resources Development in the Digital Age: An Exploratory Study

This chapter explores the potential of AI to revolutionize HRD practices, presenting findings from an exploratory study on the integration of AI in HR processes.

Chapter 5: Digital Age and Hospitality Human Resources Strategy of Development: The Case of the Japanese and Chinese Customers

Focusing on the hospitality industry, this chapter discusses HR strategies tailored to the needs of Japanese and Chinese customers in the digital age.

Chapter 6: In-Person and Remote Employees and Information Security Policy Compliance

This chapter addresses the challenges of ensuring information security policy compliance among in-person and remote employees, highlighting best practices and strategies.

Chapter 7: A History of Incivility, Workplace Bullying, Cyberbullying, and Hate Crimes: A Toxic Continuum of Behavior

An in-depth analysis of the continuum of toxic behaviors in the workplace, this chapter examines the history and impact of incivility, bullying, and hate crimes.

Chapter 8: Impact of Generative AI on Supply Chain and Human Resources

This chapter explores the transformative impact of generative AI on supply chain management and HR practices, discussing the implications for organizational efficiency and innovation.

Chapter 9: Organizational Culture and Turnover Intentions: The Role of Affective Commitment

Investigating the relationship between organizational culture and employee turnover intentions, this chapter emphasizes the importance of affective commitment in retaining talent.

Chapter 10: Psychological Capital as a Predictor of Perceived Performance: Study Applied to Portuguese Air Force Academy Students

This chapter presents a study on the role of psychological capital in predicting perceived performance, with a focus on students at the Portuguese Air Force Academy.

Chapter 11: Competencies Development and Turnover Intentions: The Mediating Effect of Affective Commitment

Exploring the link between competencies development and turnover intentions, this chapter highlights the mediating role of affective commitment in employee retention.

Chapter 12: A Well-being Culture Influencing the Retention of Talent Generation Y: Case Study of Top 50 Companies in Thailand 2023

A case study on the well-being culture in top companies in Thailand, this chapter examines how such a culture influences the retention of Generation Y talent.

Chapter 13: Employee Churn Prediction in Commercial Airlines in Portugal

This chapter discusses methods for predicting employee churn in the commercial airline industry in Portugal, offering insights into data-driven HR practices.

Chapter 14: Employer Branding in Public Organization

Focusing on employer branding in public organizations, this chapter explores strategies for attracting and retaining top talent in the public sector.

Chapter 15: Definitions of Organization Well-being in the New Normal Era: A Qualitative Research

This chapter presents qualitative research on the definitions and perceptions of organizational well-being in the new normal era, providing a nuanced understanding of this critical concept.

Chapter 16: Exploring Organizational Practices for Gender Equality: A Narrative Inquiry Approach

A narrative inquiry into organizational practices for gender equality, this chapter highlights the experiences and challenges faced by organizations striving for gender parity.

Chapter 17: Organizational Capability Building in Transformation with University-Company Collaboration

This chapter examines the role of university-company collaboration in building organizational capabilities, discussing the benefits and challenges of such partnerships.

Chapter 18: Gamifying Knowledge Management Systems in Small and Medium-Sized Enterprises: Development and Validation of a Data Collection Tool

This chapter explores the gamification of knowledge management systems in SMEs, presenting the development and validation of a data collection tool.

Chapter 19: Artificial Intelligence and New Pedagogies: Transforming Teaching and Learning through AI-Driven Innovations

This chapter discusses the transformative potential of AI-driven innovations in teaching and learning, highlighting new pedagogies enabled by AI.

Chapter 20: Support and Enhance Learning Communities in Entrepreneurship Education Service through the Focus Group

Focusing on entrepreneurship education, this chapter discusses the role of focus groups in supporting and enhancing learning communities.

Chapter 21: Becoming Change Agents at University: Challenges and Opportunities

This chapter explores the challenges and opportunities faced by university students in becoming change agents, emphasizing the importance of leadership and innovation.

Chapter 22: Educating the Digital Native Generation of Physicians in American Medicine: Generation Z in Graduate Medical Education Training Programs

This chapter examines the unique needs and characteristics of Generation Z in graduate medical education, discussing strategies for effectively educating digital native physicians.

Chapter 23: Tailoring and Transferability of Coaching and Coaching Supervision Models: Challenges and Considerations for Educational Application

This chapter addresses the challenges and considerations in tailoring and transferring coaching and coaching supervision models for educational applications.

Chapter 24: AI as the New Antecedents of Technostress at Work

The final chapter explores the role of AI as a new antecedent of technostress at work, discussing the implications for employee well-being and organizational performance.

Book Description

This book leads to a transformative journey through the world of Human Resources Development (HRD) in the Digital Age, where technology and innovation converge to redefine the workplace. It unveils the dynamic shifts driven by artificial intelligence (AI), data analytics, and immersive technologies, offering a roadmap to harness these advancements for organizational success. This comprehensive exploration examines the critical aspects of HRD, including learning and development, engagement, competencies, organizational culture, and other important topics. Through engaging insights and real-world examples, the book highlights how AI personalizes learning experiences and enhances decision-making processes. It also addresses the ethical considerations of adopting digital tools, emphasizing data privacy, fairness, and the human element in technology-driven strategies. From overcoming the digital divide to fostering a culture of continuous learning, this book equips readers with the knowledge to navigate challenges and seize opportunities in a rapidly evolving landscape. This book is an essential resource for HR professionals, organizational leaders, professors and students, and anyone seeking to future-proof their workforce and drive sustainable growth. Redefine your HR strategy and embrace the limitless potential of digital transformation with this groundbreaking book.

Contents

About the Editors

Maria José Sousa (PhD in Management, with Aggregation in Management and Public Management) is a lecturer and researcher at ISCTE/University Institute of Lisbon. She was Pro-Rector between March 2022 and January 2024, and a member of the scientific committee of the PhD "in Law and Economics" (Cycle XXXIX), Università degli Studi del Molise. She is a member of the Advisory Board of Banco Português de Fomento for Public Policies. She worked as an expert in a European Commission project for the creation of a new category of digital skills to be integrated into the European Innovation Scoreboard (EIS). She was also a member of the Coordinating Committee of the PhD in Management at the European University, a Senior Researcher at the GEE (Strategy and Studies Office) of the Ministry of Economy Portuguese, responsible for Innovation, Research and Entrepreneurship Policies, and Knowledge and Skills Manager at AMA, IP, Agency for Administrative Modernization (Ministry of the Presidency and the Council of Ministers) and a Project Manager at the Ministry of Labor and Employment, responsible for Innovation Projects and Qualification Assessment and Development Projects. Her research interests are currently in the public policies of Innovation and Research. She is the co-author of more than 200 scientific articles and book chapters (European Planning Studies, Information Systems Frontiers, Systems Research, and Behavioral Science, Computational and Mathematical Organization Theory, and others). She is guest editor of more than 10 special editions of Elsevier and Springer. She was President of ISO/TC 260—Human Resources Management, representing Portugal in ISO—International Organization for Standardization. She has coordinated several European innovation projects (funded by programs such as Erasmus+, Horizon Europe, and ICMPD) and is also an External Expert of COST Association—European Cooperation in Science and Technology.

Introduction to Human Resources Development in a Digital Age

Maria José Sousa(✉)

Instituto Universitário de Lisboa, Lisbon, Portugal
maria.jose.sousa@iscte-iul.pt

Abstract. The digital age has revolutionized Human Resources Development (HRD), introducing advanced technologies that transform traditional practices and enhance workforce potential. This article explores the integration of Artificial Intelligence (AI) in HRD, highlighting its transformative role in talent acquisition, learning and development, performance management, and strategic workforce planning. AI-powered tools enable organizations to streamline recruitment, personalize learning experiences, and foster employee engagement, resulting in a more agile and efficient workforce. Additionally, the digital age offers unprecedented opportunities for inclusivity, scalability, and real-time data-driven decision-making, enabling organizations to adapt to rapid technological changes. However, challenges such as the digital divide, data privacy concerns, and the need for ethical AI implementation remain critical. By fostering a culture of continuous learning and adopting a balanced approach that combines digital innovation with human-centric strategies, organizations can effectively navigate these challenges and harness the potential of AI. The article underscores the importance of addressing HRD themes and promoting employee reskilling to prepare for a technology-driven future. As advancements in AI and digital tools continue, their integration in HRD will remain pivotal in shaping a resilient and future-ready workforce.

Keywords: Human Resources Development · Artificial Intelligence · Digital Age · Workforce Planning · Continuous Learning · UFHRD24

1 Introduction

In the modern era, characterized by rapid technological advancements, the field of Human Resources Development (HRD) has undergone a transformative evolution. The integration of digital technologies has reshaped how organizations approach talent development, workforce management, and employee engagement. This article explores the concept of HRD in the digital age, highlighting the opportunities and challenges posed by digitalization, and underscores the importance of fostering a culture of continuous learning and innovation, as a representation of the 24th Edition of the University Forum for Human Resource Development (UFHRD) Conference, held in Lisbon, on 11th to 14th June 2024.

M. J. Sousa (Ed.): UFHRD 2024, ISEM 52, pp. 1–5, 2025.
https://doi.org/10.1007/978-3-031-92548-1_1

1.1 Understanding Human Resources Development

Human Resources Development (HRD) focuses on enhancing the capabilities of employees within organizations through training, education, and career development programs. It aims to align individual growth with organizational objectives, ensuring that employees remain productive, motivated, and prepared for future challenges (Poba-Nzaou, et al., 2020). Traditionally, HRD relied on in-person workshops, classroom training, and manual performance evaluations. However, the digital age has revolutionized these practices, introducing a wide array of tools and methodologies that leverage technology to optimize human potential Huang, 2022).

1.2 The Impact of Digitalization on HRD

Digital technologies have transformed HRD in several significant ways. E-learning platforms, for instance, have democratized access to knowledge, enabling employees to acquire skills at their own pace and convenience. According to a report by Deloitte (2021), organizations that invest in digital learning technologies experience higher employee satisfaction and productivity levels. Virtual reality (VR) and augmented reality (AR) are also being utilized for immersive training experiences (Abdeldayem & Aldulaimi, 2020), particularly in industries such as healthcare, manufacturing, and aviation.

Another critical development is the use of artificial intelligence (AI) in talent development. AI-powered systems can personalize learning content based on individual needs, track employee progress, and predict future skill requirements using data analytics (Abdeldayem & Aldulaimi, 2020).

1.3 The Importance of AI in Human Resources Development

The integration of Artificial Intelligence (AI) in HRD has ushered in a transformative era for organizations worldwide. AI's capabilities in data analysis, personalization, and automation have redefined traditional HRD practices, offering unparalleled efficiency, accuracy, and strategic insight. This section explores the significance of AI in HRD, examining its role in talent acquisition, learning and development, performance management Halid et al., 2020), and the ethical considerations that accompany its implementation.

1.4 AI in Talent Acquisition

One of AI's most prominent applications in HRD is talent acquisition. AI-powered tools can streamline recruitment processes by automating resume screening, matching candidates with job requirements, and conducting initial assessments. For example, platforms like HireVue use AI algorithms to analyze video interviews, assessing candidates' facial expressions, tone, and word choice to predict job suitability (Smith, 2021).

By leveraging AI, organizations can significantly reduce hiring biases. Algorithms can be programmed to focus solely on skills, qualifications, and experience, minimizing the influence of unconscious human biases. However, it is crucial to monitor these algorithms to prevent the perpetuation of existing biases in training data.

1.5 AI in Learning and Development

AI has revolutionized learning and development (L&D) within HRD by enabling personalized, scalable, and efficient training solutions. Traditional training programs often adopt a uniform approach, which may not cater to individual employee needs. AI systems, such as Learning Management Systems (LMS) with embedded AI capabilities, analyze employee performance data to recommend tailored learning paths. For instance, platforms like Coursera and LinkedIn Learning use AI to suggest courses based on users' roles, career goals, and past learning behaviors (Amalia, 2024).

Moreover, AI enhances the learning experience through interactive technologies such as chatbots, virtual reality (VR), and augmented reality (AR). AI-driven chatbots serve as virtual mentors, providing instant feedback and support to learners. VR and AR offer immersive simulations for skills training, particularly in industries like healthcare, manufacturing, and aviation. These technologies make learning more engaging and effective, improving knowledge retention and skill application (Emran & Elhony, 2023).

1.6 AI in Performance Management and Employee Engagement

Performance management is another critical area where AI demonstrates its importance. AI tools enable continuous performance tracking, replacing traditional annual reviews with real-time feedback mechanisms (Halid et al., 2020). AI-powered platforms analyze performance metrics, identify skill gaps, and recommend targeted development initiatives.

For instance, AI systems like Workday and SAP SuccessFactors provide actionable insights into employee performance trends, enabling managers to make informed decisions. These tools also facilitate predictive analytics, forecasting future performance and potential leadership capabilities based on historical data (Deloitte, 2021).

In addition, AI fosters employee engagement by creating a more personalized and inclusive work environment. AI-driven sentiment analysis tools monitor employee feedback and morale, identifying areas of concern and suggesting strategies for improvement. Such tools empower HR professionals to proactively address issues, enhancing employee satisfaction and retention.

2 Opportunities in the Digital Age

The digital age offers numerous opportunities for HRD. First, the scalability of digital platforms allows organizations to train large, geographically dispersed teams efficiently. Second, the availability of real-time data empowers HR professionals to make informed decisions about talent development strategies. Third, digital tools enhance collaboration and communication, fostering a more connected and agile workforce (Huang, 2022).

Additionally, digital HRD initiatives promote inclusivity. Employees with disabilities or those from remote areas can access training resources and participate in development programs without physical barriers. This inclusivity strengthens organizational culture and broadens the talent pool.

3 Challenges and Ethical Considerations

Despite its advantages, the digitalization of HRD presents challenges. One significant concern is the digital divide, where employees lacking access to technology or digital literacy skills are at a disadvantage. Organizations must address this gap by providing necessary resources and training to ensure equitable participation (Amalia, M. R. (2024)).

Another challenge is data privacy. The extensive use of AI and analytics raises ethical questions about how employee data is collected, stored, and used. Companies must adhere to data protection regulations, such as the General Data Protection Regulation (GDPR), to maintain trust and compliance.

Moreover, the reliance on digital tools can sometimes lead to a lack of human connection in HRD processes (Abdeldayem & Aldulaimi, 2020). While technology enhances efficiency, it is crucial to balance digital solutions with personalized, human-centric approaches to maintain employee engagement and well-being.

4 The Role of Continuous Learning

In the digital age, continuous learning is vital for both employees and organizations. The World Economic Forum (2020) highlights that over 50% of all employees will need reskilling by 2025 due to the rapid pace of technological change. Organizations must foster a culture of lifelong learning by encouraging employees to embrace new skills and adapt to evolving industry demands.

Gamification, microlearning, and social learning are emerging as effective strategies to promote engagement in digital HRD programs. These methods leverage interactive and collaborative elements to enhance knowledge retention and application.

5 The Future of AI in HRD

The integration of AI in HRD is expected to grow exponentially, driven by advancements in machine learning, natural language processing, and robotics. Future applications may include hyper-personalized career development plans, AI-powered mentorship programs, and even greater use of immersive technologies for training (Abdeldayem & Aldulaimi, 2020).

Organizations that embrace AI as a strategic partner in HRD will be better positioned to adapt to evolving market demands and technological disruptions (Amalia, 2024). However, the successful implementation of AI requires a balanced approach that considers ethical implications, employee engagement, and the human touch in HR processes.

References

Abdeldayem, M.M., Aldulaimi, S.H.: Trends and opportunities of artificial intelligence in human resource management: aspirations for the public sector in bahrain. Int. J. Sci. Technol. Res. **9**(1), 3867–3871 (2020)

Amalia, M.R.: The impact of digital era 4.0 transformation on human resources management. Manage. Stud. Business J. **1**(1), 89–98 (2024)

Deloitte.: Global Human Capital Trends. https://www2.deloitte.com (2021)

Emran, H.A., Elhony, F.: The implications of digital transformation and its impact on human resource management strategies. East Asian J. Multi. Res. (2023)

Halid, H., Yusoff, Y.M., Somu, H.: The relationship between digital human resource management and organizational performance. In: First ASEAN Business, Environment, and Technology Symposium (ABEATS 2019), pp. 96–99. Atlantis Press (2020)

Huang, J.: Adaptive change of human resource management from the perspective of digital transformation: a case study of J company. BCP Business Manage. (2022)

McLagan, P.: Models for HRD Practice. ASTD Press, Alexandria, VA (1989)

Poba-Nzaou, P., Galani, M., Tchibozo, A.: Transforming human resources management in the age of industry 4.0: a matter of survival for HR professionals. Strategic HR Rev. **19**(6), 273–278 (2020)

Smith, J.: AI in recruitment: transforming the hiring process. J. Hum. Res. Innov. **34**(2), 45–58 (2021)

World Economic Forum. The Future of Jobs Report. www.weforum.org (2020)

Strategic Human Resource Management and Dynamic Capabilities in Life Sciences: Leveraging Hybrid Team Management and Performance Monitoring in Fluid Working Spaces for Advanced Productivity

Antonio Pesqueira[1](✉) and Andreia de Bem Machado[2]

[1] ISCTE, Lisbon, Portugal
antonio_pesqueira@iscte-iul.pt
[2] Universidade Federal de Santa Catarina, Florianópolis, Brazil
andreia.bem@ufsc.br

Abstract. Developing robust strategic human resource management (SHRM) and dynamic capabilities (DC) across contemporary organizations is considered one of the most critical factors for enhancing workforce performance and fostering organizational innovation. Hybrid team management (HTM) initiatives are frequently scrutinized from various perspectives like Work-From-Home (WFH) or Return-to-Office (RTO), encompassing how they are employed and how they contribute value to enterprises. Life Sciences organizational leaders were surveyed on the influence of DC and HTM strategies in fluid working environments through an online structured questionnaire that included closed questions such as monitoring performance metrics and evaluating productivity indicators. This chapter conducted assessments based on multiple assumptions regarding workforce adaptability, the significance of agile capabilities, and the impact on organizational performance, aiming to illuminate the complexity of these propositions and examine the collective value of strategic HR approaches and capacity building. The findings and supporting questionnaire data emphasize the importance of adopting comprehensive HR frameworks, DC, metric-driven insights, and innovative solutions that strengthen the ability of professionals to optimize task allocation, improve collaboration mechanisms, and ensure access to essential information. Statistical analyses indicate that these results pertain not only to the influence of strategic HR management on workforce development but also to the broader implications for driving innovative capacity within organizations.

Keywords: Strategic HR Management · Dynamic Capabilities · Hybrid Teams · Performance Monitoring · Advanced Productivity

1 Introduction

Managing geographically dispersed teams and establishing collaborative, dynamic working environments have become increasingly prevalent concerns in modern organizational contexts. Recent shifts toward flexible working arrangements have challenged human

M. J. Sousa (Ed.): UFHRD 2024, ISEM 52, pp. 6–34, 2025.
https://doi.org/10.1007/978-3-031-92548-1_2

resource (HR) practitioners and leaders to design, implement, and monitor innovative practices that address productivity and employee well-being. In many cases, these challenges are exacerbated by the need to maintain cohesive organizational cultures, align strategic goals across distributed teams, manage employees' preferences around Work-From-Home (WFH) or Return-to-Office (RTO), and ensure continuous knowledge-sharing. We also need to align flexible work options with our people-centric HR strategy to empower employees and drive organizational success (Jacob, Bourke, Heuss, 2022)

Amid this growing complexity, new organizational structures are emerging to cope with rapid changes in market conditions, the influx of technological tools, and the demand for heightened adaptability (Dillard-Wright, Shields-Haas, 2021). These evolving structures underscore the necessity of dynamic solutions to HR challenges in fluid working spaces. Unlike traditional, rigid office setups with fixed roles and schedules, fluid workplaces are designed to accommodate dynamic work styles, evolving technologies, and the changing demands of the business landscape (Junaid et al., 2023).

Solving these problems is essential for advancing knowledge and improving practices in strategic human resource management (SHRM). In particular, the dynamic capabilities (DC) framework offers a theoretical lens through which organizations can systematically develop processes for sensing market shifts, seizing emerging opportunities, and reconfiguring internal competencies (Kaplan & Hays, 2022). Such agility not only can support innovative work processes but can also empower the inclusion of neurodivergent talents, who thrive in environments that value flexibility and individuality. Consequently, enhanced productivity, higher job satisfaction, and more inclusive work cultures can emerge as by-products of adopting DC at all levels of HR decision-making (Gebhardtet al., 2022).

A growing body of literature has explored the implications of hybrid team management (HTM) and strategic HR practices on organizational performance, where different researchers argue that organizations must develop robust policies and systems for managing dispersed workforces, suggesting that proactive HR interventions can bolster employee engagement and long-term retention (Kendzerska et al., 2021). Additionally, studies on digital collaboration tools demonstrate how sophisticated performance monitoring systems can foster transparency, streamline project timelines, and mitigate communication bottlenecks. Despite this progress, unresolved issues remain (Bhattamisraet al.,2023). Notably, few studies offer comprehensive models that integrate the three stages of DC—sensing, seizing, and reconfiguring—directly into the design and management of hybrid teams. Furthermore, while current scholarship affirms the value of inclusive and neurodivergent-friendly workplaces, specific guidelines for merging such practices with fluid working spaces have yet to be thoroughly articulated (Kwok, Treiblmaier, 2023).

In addition, current literature does not fully address how performance monitoring should be adapted to dynamic, hybrid settings without compromising employee autonomy or stifling creativity and especially managing effectively WFH and RTO cases. Empirical evidence suggests that certain monitoring practices may inadvertently hinder innovation by imposing restrictive metrics and standardized protocols. At the same time, advanced productivity demands real-time analytics and consistent feedback loops that facilitate strategic reconfiguration (Ball, 2021). These tensions highlight the need for further research that investigates how best to balance the pursuit of high performance with the preservation of employee agency and diverse cognitive styles in fluid work environments (Ebrahimi, Bridgelall, 2021).

To address the complexities outlined above, this chapter aims to assess the influence of DC on strategic HR frameworks for managing hybrid teams and also identify best practices in performance monitoring tailored to fluid working spaces, with particular attention to neurodivergent individuals. By achieving these goals, the research endeavors to generate practical, societal, and technological insights. Practically, the findings could inform HR policies that optimize team performance and employee satisfaction. Societally, the chapter results could propose more inclusive workplace models that leverage diverse cognitive perspectives, reduce barriers faced by neurodivergent employees and enable more effective employee performance in the WFH or RTO context. Technologically, insights on real-time monitoring systems could guide firms in deploying analytics tools that enhance strategic decision-making without compromising workforce autonomy.

Despite increasing efforts to adopt hybrid and fluid working models, a critical gap persists in understanding how DC—particularly in the HR domain—can be effectively operationalized to bolster collaboration, inclusivity, and productivity. Consequently, this research focuses on the following research questions: RQ1: How does DC shape strategic HR approaches for managing hybrid teams in fluid working spaces? RQ2: Which performance monitoring methods most effectively balance advanced productivity with employee autonomy and inclusivity? RQ3: What roles do the three stages of DC (sensing, seizing, reconfiguring) play in facilitating collaborative practices that support neurodivergent employees and broader workforce adaptability?

Each question is designed to probe a specific dimension of the overarching research problem. By investigating these areas through both quantitative and qualitative measures, the chapter aims to elucidate evidence-based strategies that can be integrated into organizational practice. By employing a structured online survey instrument distributed to life sciences professionals and leaders, organizational leaders, and team members working in hybrid environments, it includes closed-ended items for quantifying attitudes and experiences, along with a limited number of open-ended questions to capture qualitative nuances.

By situating these findings within ongoing scholarly debates, the chapter aims to clarify theoretical discrepancies and propose new lines of inquiry. Furthermore, this work offers practical guidelines that may influence organizational policies and frameworks, potentially guiding life sciences managers in crafting strategies that promote regulatory compliance and organizational efficiency. Such guidelines could encompass ethical data usage, flexible performance metrics, and frameworks for fostering neurodivergent-inclusive cultures, each of which is increasingly essential for organizations operating in rapidly shifting, technology-driven environments.

2 Literature Review

2.1 Strategic Human Resource Management and Dynamic Capabilities

The evolution of SHRM as a distinct discipline has been closely tied to the growing recognition that human capital constitutes a critical driver of sustainable competitive advantage. Early conceptualizations of SHRM centered on aligning core HR functions—such as recruitment, training, and performance evaluation—with overarching organizational strategies (Patil et al., 2022). Recent work posited that firms stand to benefit significantly by embedding HR policies into higher-level corporate objectives, ensuring that personnel development and strategic growth proceed in tandem. Over time, research has revealed that this alignment requires continuous adaptation, especially as markets and technologies evolve at unprecedented rates. Modern SHRM frameworks, therefore, not only address transactional issues like benefits administration or regulatory compliance but also place a premium on cultivating a workforce that exhibits the agility necessary to navigate dynamic market landscapes (Jalilet al., 2023).

In parallel with this development, scholars have turned to the concept of DC to elucidate how organizations can remain competitive in environments characterized by frequent technological advancements and shifting consumer demands. At its core, the DC framework underscores an organization's aptitude for sensing emerging trends or opportunities, seizing those opportunities through decisive resource allocation, and reconfiguring internal competencies to sustain long-term relevance when applied to HR practices, DC emphasizes the cultivation of internal processes that enable rapid learning, strategic redeployment of talent, and continuous re-skilling (Vallentin, Thórisson, Latapie, 2023). This extends beyond incremental improvements in hiring or performance management, advocating for an overarching culture in which innovation, experimentation, and cross-functional collaboration become integral to daily operations. From a theoretical standpoint, merging SHRM with DC entails a shift from static workforce planning to proactive talent pipelines, where employee development is seen as an ongoing, strategic endeavor that mitigates skills obsolescence and bolsters organizational resilience (Pesqueira, Sousa, 2022).

Despite the compelling synergies highlighted by this body of research, several debates persist. One line of discourse questions whether an overemphasis on agility might inadvertently undermine employee stability, job satisfaction, or well-being. Proponents of DC argue that strategic reconfigurations of human capital resources, such as restructuring teams or rotating employees across projects, remain critical for spurring creativity and innovation (Pesqueira, Sousa, 2022). However, skeptics caution that frequent organizational restructuring may engender uncertainty and burnout among employees, thus diminishing the engagement required to actualize DC (Amini & Jahanbakhsh Javid, 2023). Moreover, the literature acknowledges the potential for inclusivity to be overlooked, raising questions about how to design adaptive HR systems that accommodate neurodivergent employees, who may thrive under flexible work conditions but also require consistent support structures. Consequently, a multifaceted approach that balances innovative reconfigurations with stable support networks has emerged as a central theme in recent studies (Jacob, Bourke, Heuss, 2022; Javaid, Haleem, Singh, 2023).

In synthesizing these debates, it becomes evident that the efficacy of integrating DC into SHRM hinges on a nuanced understanding of organizational culture, leadership styles, and technology adoption. Scholars assert that leadership engagement is paramount; top-level executives must champion both human-centric policies and the strategic agility that underpins DC (Wani et al., 2022). Equally important is the role of data analytics and real-time performance insights, which can inform decisions about redeploying talent or revising skill development initiatives. By identifying how best to leverage DC within the HR domain, organizations can simultaneously advance workforce productivity and ensure a culture conducive to continuous learning and neurodivergent inclusion. This theoretical nexus thus lays the groundwork for further empirical examination, highlighting the necessity of well-orchestrated practices that not only drive innovation but also maintain ethical, inclusive, and human-centered approaches to workforce management (Vallentin, Thórisson, Latapie, 2023).

2.2 Hybrid Team Management and Performance Monitoring in Fluid Working Spaces

The emergence of hybrid team management as a critical focus of organizational research mirrors the broader societal shift toward flexible, technology-mediated work environments. Early examinations of remote and distributed teams highlighted communication challenges, such as ensuring consistent information flow and trust-building in the absence of face-to-face interactions (Masih, 2022). Over time, these inquiries broadened to encompass the multifaceted dynamics of teams that operate partly on-site and partly remotely, coining the term "hybrid" to reflect a duality that combines the benefits of physical proximity with the freedoms afforded by remote work (Bouguerra et al., 2022). Recent scholarships have delved deeper into the strategic implications of hybrid team structures, noting that while flexible work arrangements can improve job satisfaction and widen the talent pool, they also demand more nuanced leadership models and advanced technological infrastructures (Shiwlani et al., 2023).

Central to effective hybrid team management is performance monitoring, which is a fluid working space that often relies on digital tools capable of tracking progress, productivity, and collaboration metrics in real-time. Fluid workplaces embody a contemporary approach to work environments, prioritizing flexibility, adaptability, and responsiveness to both employee needs and organizational objectives. Contrary to conventional, rigid office setups with fixed roles and schedules, fluid workplaces are designed to accommodate dynamic work styles, evolving technologies, and the shifting demands of the business landscape (Cifuentes-Faura, 2022). This concept is becoming increasingly pertinent in today's diverse and rapidly evolving work environments, encompassing remote, hybrid, and in-office settings (Masih, 2022; Mostafiz et al., 2022). Fluid workplaces are particularly pertinent in sectors like Life Sciences, where collaboration, innovation, and adaptability are crucial. The COVID-19 pandemic accelerated the adoption of remote and hybrid work models, highlighting both the potential and the challenges of fluid workplaces. Organizations that successfully implement fluid workplace strategies are better positioned to attract diverse talent, respond to changing industry demands, and foster a culture of continuous improvement and innovation (Pesqueira, Sousa, Pereira, 2023).

Fluid workplaces embody a shift towards more adaptable, employee-centric work environments that align with contemporary business needs and workforce expectations. By embracing flexibility in location, schedule, and team dynamics, and leveraging technology to facilitate seamless collaboration, organizations can enhance productivity, innovation, and employee satisfaction. However, achieving the benefits of fluid workplaces requires careful planning, effective communication, and a commitment to maintaining a strong organizational culture (Ullahet al., 2023).

In addition, the literature suggests that advanced analytics platforms and communication technologies can, in theory, provide unparalleled visibility into task completion rates, engagement levels, and interpersonal dynamics. However, scholars are divided on the optimal degree of monitoring required to maintain productivity without undermining employee autonomy and innovation. On one side of this debate, advocates argue that transparent metrics reduce ambiguity, facilitate timely feedback, and enable data-driven decision-making—an imperative in fast-paced markets (Wamba et al., 2023). On the other side, critics highlight the risk of creating surveillance cultures, where overly intrusive monitoring tools breed distrust and inhibit the creative risk-taking essential for innovation (Eager, Brunton, 2023). This tension is further complicated by the need to accommodate neurodivergent individuals, whose performance and collaboration preferences may diverge significantly from standardized monitoring protocols. While studies recognize the importance of adaptability in today's volatile markets, relatively few have explored how monitoring technologies and organizational practices can be reconfigured dynamically to respond to sudden shifts in team composition or project requirements (Mostafiz et al., 2022). Some scholars emphasize that a single, static dashboard of key performance indicators may become obsolete if the organization's strategic orientation changes rapidly. Instead, they advocate for agile tools that provide metrics capable of evolving alongside shifting priorities or team structures. Such flexibility can prove particularly beneficial for neurodivergent employees, who often excel when they can adapt work processes to fit their unique cognitive profiles (Yulianti, Sulistyawati, 2021).

Synthesizing these insights underscores the essential role that leadership, organizational culture, and thoughtful technological adoption play in hybrid team management. Leaders must navigate the delicate balance between enforcing accountability and fostering psychological safety, especially in fluid workplaces where the absence of physical co-location can heighten misunderstandings (ElBaih, 2023). Organizational cultures that champion inclusivity and experimentation may be better suited to implement adaptive performance monitoring, ensuring that employees view metrics not as punitive but as enablers of growth and strategic alignment (Kant, Anjali, 2021). Moreover, the selective application of DC could allow organizations to revise their performance indicators based on real-time feedback, thereby optimizing both team output and employee well-being (Alawidaet al., 2023). This expanded theoretical landscape highlights the critical need for empirical inquiry into how hybrid teams can leverage performance monitoring frameworks that remain ethically sound, agile, and attuned to diverse workforce needs. By addressing these intertwined challenges, the present study aims to provide actionable guidance on structuring fluid working spaces that maximize productivity, foster collaboration, and embody the principles of inclusive, human-centered design (Vlasta, 2023).

2.3 Governance, Leadership, and Mental Health in Hybrid Environments

An increasingly complex facet of modern organizations concerns governance structures and the corresponding leadership responsibilities in contexts marked by fluid work arrangements. Governance, in this regard, extends beyond the mere legal and regulatory framework to encompass how leaders design decision-making processes, implement accountability measures, and ensure ethical conduct across geographically dispersed teams (Koubaet al., 2023). Within hybrid environments, the decentralization of workflow often requires new models of coordination, calling for adaptive governance mechanisms that can align strategic objectives with local autonomy. Such adaptive governance strategies must also account for the systemic risks introduced by emerging technologies and remote work practices, which can make it difficult to enforce uniform standards across diverse operational units. Indeed, scholars have underscored the necessity of cohesive governance policies that integrate DC, enabling organizations to reconfigure oversight structures quickly in response to fluctuating market conditions or rapid technological disruptions (Haleem et al., 2022).

Leadership challenges associated with hybrid and fluid working models revolve predominantly around fostering trust, inclusivity, and engagement. When teams are partially or fully remote, leaders grapple with more pronounced communication gaps and cultural fragmentation (Breaux et al., 2023). Additionally, the diminished visibility of daily work routines can complicate efforts to provide timely feedback and maintain equitable treatment of team members (Tariqet al., 2022). Studies suggest that leaders in such contexts must exhibit a heightened level of emotional intelligence, adaptability, and empathy (Javaid, Haleem, Singh, 2023). These qualities are crucial for recognizing individual challenges, facilitating collaborative decision-making, and sustaining organizational culture in a setting where team members may seldom share the same physical space, in which the literature also points to the strategic use of technology—such as virtual collaboration platforms and advanced analytics tools—as central to effective leadership, but highlights the importance of governance protocols that regulate data usage, privacy, and security (Spiel et al., 2022). This interplay between governance and leadership is emblematic of a broader trend in which leaders must evolve from unilateral decision-makers to facilitators of distributed knowledge and innovation networks (Kasneciet al., 2023).

A related yet increasingly urgent dimension in hybrid work environments is the mental health of employees, where rapid organizational changes, social isolation, and blurred boundaries between personal and professional domains can exacerbate stress, burnout, and other forms of psychological distress. Leadership practices and governance structures can unintentionally contribute to or mitigate these challenges, depending on how they balance productivity targets with well-being support mechanisms. While performance monitoring systems offer efficiency gains, they can also heighten pressures if perceived as overly invasive. Moreover, the shift to remote or hybrid models can complicate the early detection of mental health risks, as managers may have fewer face-to-face interactions through which to observe changes in employees' behavior or emotional states. Scholars emphasize the importance of implementing comprehensive well-being frameworks that incorporate flexible working hours, mental health resources, and inclusive management practices. To operate such frameworks, leaders must cultivate a culture of psychological safety, ensuring employees feel empowered to communicate challenges and seek support without fear of negative repercussions (Kotcher et al., 2021).

Synthesizing these strands of research indicates that governance and leadership in the era of hybrid work are not merely administrative functions, but strategic imperatives that shape organizational resilience and cultural cohesion. A rapidly expanding literature base suggests that DC—particularly those involving reconfiguration—can offer a template for redefining governance protocols and leadership styles to accommodate evolving employee needs (Pereira, Bianchi, Araújo, 2021; Pesqueira, Sousa, 2022). For instance, sensing mechanisms can be employed to monitor workforce sentiment or detect emerging stressors; seizing involves translating these insights into actionable policies or resource allocations, while reconfiguring might include reorganizing leadership roles or redistributing decision-making authority to address mental health concerns more effectively (Leffa et al., 2022). Yet, empirical gaps remain regarding how precisely DC is best integrated into governance frameworks without overstretching managerial capacity or eroding team autonomy (Mele et al., 2023).

The scholarly discourse on governance concerns leadership challenges, and mental health in hybrid teams highlights the intricate balance organizations must strike to maintain rigorous oversight while fostering empowerment and well-being. It is crucial to acknowledge that these aspects governance, leadership, and mental health—are interconnected with broader discussions on hybrid work productivity. To ensure the long-term viability, fairness, and adaptability of hybrid work environments, comprehensive solutions are essential. These solutions should encompass formal policy changes, leadership training programs, and the adoption of ethical performance monitoring systems (Secinaro et al., 2021).

3 Methodological Approach

This chapter utilizes an online questionnaire to gather comprehensive insights from life sciences leaders and professionals across various organizational levels regarding different assessment parameters in connection with the defined research objectives and questions. The questionnaire was meticulously designed with a combination of closed and open-ended questions that explored key dimensions such as flexibility, technology utilization, team dynamics, employee well-being, and performance monitoring. The choice of an online questionnaire is justified by its ability to efficiently reach busy professionals in the life sciences field, ensure respondent anonymity, encourage honest feedback, and allow for scalable data collection. This approach was essential for capturing diverse viewpoints on fluid workplace practices, identifying best practices, assessing the influential levels from the studied parameters, quantifying participants' attitudes, and identifying key factors and their impact on productivity and innovation (Tariqet al., 2022).

The survey was disseminated to 114 organizational leaders and various organizational professionals to ascertain their perceptions of critical strategic HR practices and the most pertinent influential and impact factors in enhancing the value of hybrid team management, performance monitoring, and DC in dynamic working environments, from March 2nd to April 18th, 2024.

The questionnaire was administered in English and included demographic items, as well as questions about organizational culture, hybrid team structures, leadership approaches, and the practical application of DC. Respondents were asked to complete the questionnaire anonymously to safeguard their privacy, and they were not required to specify their affiliated organizations in the database.

To validate the coherence and reliability of the questionnaire, a pilot group of ten respondents—representing the broader study demographic—completed the survey before its widespread distribution. We selected participants from diverse global organizations with a confidence level of 95% (and $p = q = 0.5$), where the inclusion criteria targeted individuals with pronounced involvement in life sciences decision-making and leadership strategy, a working knowledge of technological enablers for hybrid teams, and a demonstrated interest in performance metrics and DC for advanced productivity (Wada et al., 2022).

4 Results and Analysis

4.1 Descriptive Analysis

Table 1 provides an overview of the organizations and participants surveyed, highlighting their backgrounds and professional roles. A significant majority of respondents, accounting for 78%, are affiliated with pharmaceutical or biotech companies, while 22% work in medical device companies. Most participants are engaged in executive management (40%) or marketing (36%), with smaller proportions in areas such as sales, operations, and compliance. Regarding seniority, 73% of respondents hold senior leadership positions, including VP, SVP, or director roles, indicating a high level of expertise within the sample. Experience in managing hybrid or fluid workforce structures varies, with 39% reporting large experience, 37% moderate experience, and 24% reduced experience. The geographic distribution reveals that the majority of organizations are located in North America (47%) and Europe (32%), with fewer participants from regions like Asia-Pacific, China, and other global markets. This figure captures the diversity and relevance of the study's sample.

Regarding the responses to whether the COVID-19 pandemic influenced organizations' approaches to hybrid team strategies and fluid work arrangements, an overwhelming majority, with 111 respondents, confirmed that the pandemic significantly impacted their strategies (Fig. 1). A minimal number of participants, only two, reported uncertainty, while just one respondent indicated that their organization was unaffected. This data underscores the substantial influence of the pandemic on workforce management practices, particularly in adapting to hybrid and fluid working models.

Table 1. Organizational and Workforce Characteristics

What best describes your organization?	N	%
Pharmaceutical/biotech company	89	78%
Medical device company	25	22%
	114	100%
In which area do you work?		
Executive Management	46	40%
Marketing	41	36%
Sales Management	12	11%
Operations	10	9%
Compliance	1	1%
Legal	1	1%
Regulatory	1	1%
Clinical Operations	1	1%
Medical Affairs	1	1%
	114	100%
What best describes your level of seniority?		
VP/SVP	42	37%
Assoc. Dir./Director/Exec.Dir	41	36%
Manager	16	14%
Executive	8	7%
Researcher	2	2%
Consultant	1	1%
Specialist	1	1%
Analyst	1	1%
Developer	1	1%
Junior	1	1%
	114	**100%**
Experience managing hybrid or fluid workforce structures?		
Large experience	44	39%
Moderate experience	42	37%
Reduced experience	28	25%
	114	**100%**
In what region or market is your organization primarily located?		

(continued)

Table 1. (*continued*)

What best describes your organization?	N	%
North America	54	47%
Europe - UK, France, Germany, Spain, Italy	37	32%
Broader APAC	7	6%
China	7	6%
Broader EMEA	4	4%
India	2	2%
Africa	1	1%
South-America	1	1%
Russia	1	1%
	114	**100%**

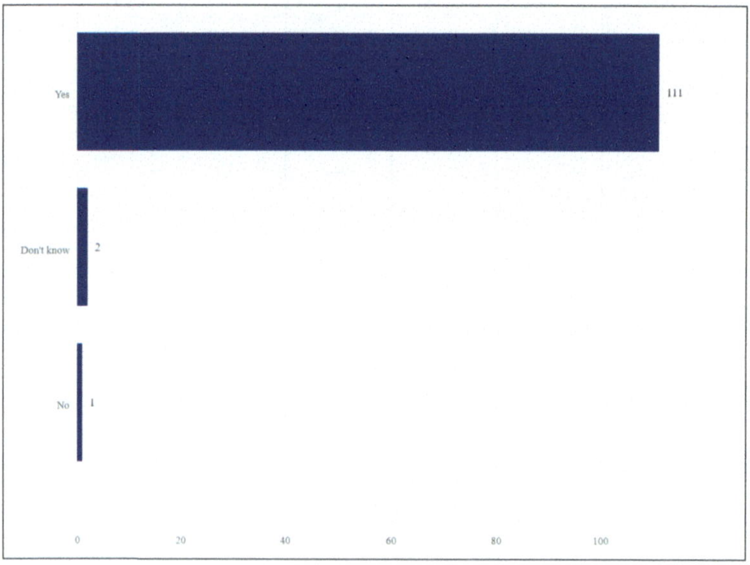

Fig. 1. Impact of the COVID-19 Pandemic on Hybrid Team Strategy

Respondents' evaluation of three key metrics for assessing hybrid team effectiveness and strategic HR initiatives were also collected where "Talent Retention and Engagement" is considered the most critical metric, with 46 participants rating it highly. "Inclusive Collaboration" follows closely, identified as important by 43 respondents. "Organizational Agility," while still significant, is rated as less critical, with 25 participants highlighting its importance. These results reflect a strong emphasis on workforce stability and collaboration within hybrid work models, alongside a recognition of agility as a contributing factor (Fig. 2).

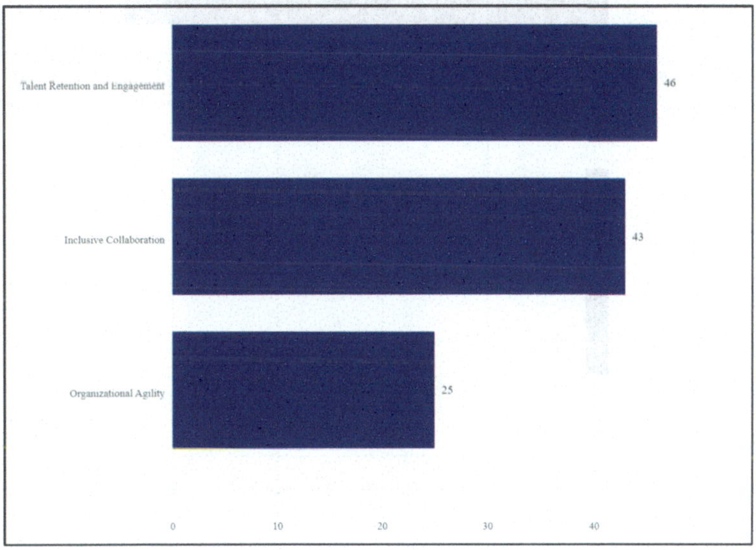

Fig. 2. Importance of Metrics for Evaluating Hybrid Team Effectiveness and Strategic HR Programs

Figure 3 outlines the perceived importance of various performance indicators for monitoring success in fluid working environments. "Project Turnaround Time" is identified as the most critical metric, with 81 respondents emphasizing its significance. "Knowledge Sharing" is the second most valued indicator, highlighted by 25 participants. Both "Cross-Functional Coordination" and "Return on Investment (ROI)" are rated considerably lower, with only four respondents each considering them important. This distribution underscores a predominant focus on efficiency and collaboration as key measures of success in dynamic workspaces.

The role of emerging technologies, such as AI-driven dashboards and collaborative platforms, in ensuring the effectiveness of hybrid teams in the future presents a majority of respondents, 96, "agree" that such technologies will be decisive, while 15 "strongly agree." Only one respondent selected "disagree," "strongly disagree," or "neither agree nor disagree." These responses indicate a strong consensus on the importance of advanced performance analytics for optimizing hybrid team management (Fig. 4).

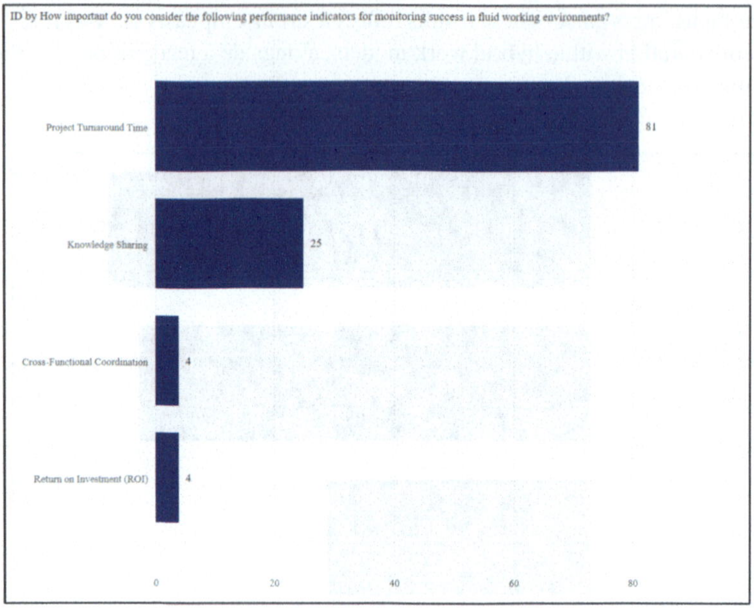

Fig. 3. Importance of Performance Indicators for Success in Fluid Working Environments

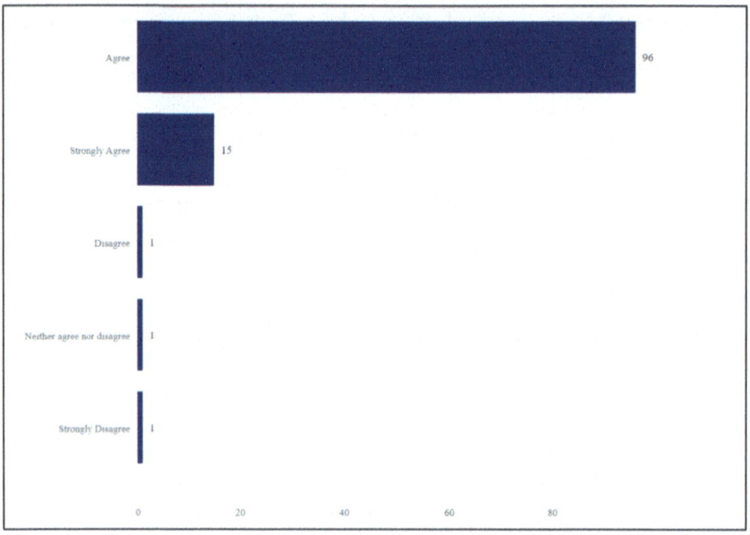

Fig. 4. Perceived Importance of Emerging Technologies in Performance Analytics for Hybrid Teams

Figure 5 illustrates the responses regarding whether organizations conducted feasibility assessments of decentralized or agile elements aimed at improving workforce flexibility and productivity in the past 12 months. A total of 57 respondents indicated

"Yes," confirming such assessments were performed, while 56 respondents selected "Don't know," suggesting uncertainty about these initiatives within their organizations. Only one respondent explicitly stated "No." These results highlight varying levels of engagement and awareness regarding decentralized or agile workforce practices.

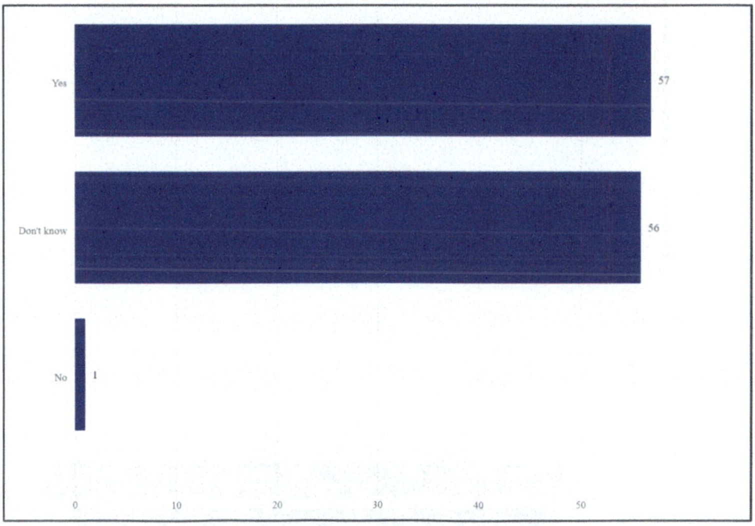

Fig. 5. Feasibility Assessments of Decentralized or Agile Elements for Workforce Flexibility and Productivity

Regarding the percentage of weekly work time respondents dedicated to coordinating or optimizing hybrid team practices, including remote collaboration or advanced performance metrics, a vast majority, of 111 respondents reported spending more than 60% of their work time on these activities. Only one respondent indicated spending less than 25%, 25–40%, or 41–60% of their weekly time. This highlights the significant focus on hybrid team practices and performance optimization in respondents' professional responsibilities (Fig. 6).

Figure 7 depicts the training formats respondents consider most effective for educating teams on innovative HR practices and DC. Badge or certificate programs (skill-based certifications), university/academy online education programs, and internal organizational masterclasses are equally preferred, with 33 respondents selecting each format. Live training sessions, including networking events, were chosen by five respondents, while eBooks, research papers, MOOCs (massive open online courses), self-paced courses, and topic-specific online videos each received two responses. This highlights a strong preference for structured, formal learning formats over less interactive or informal methods.

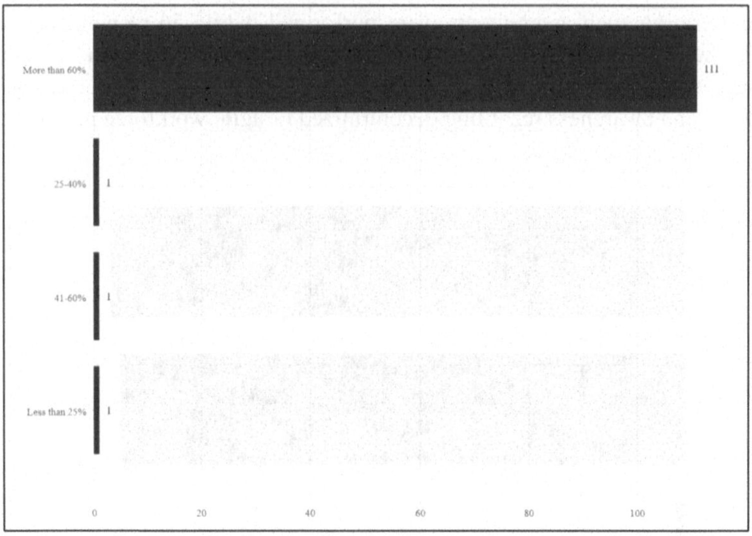

Fig. 6. Weekly Time Devoted to Hybrid Team Coordination and Optimization

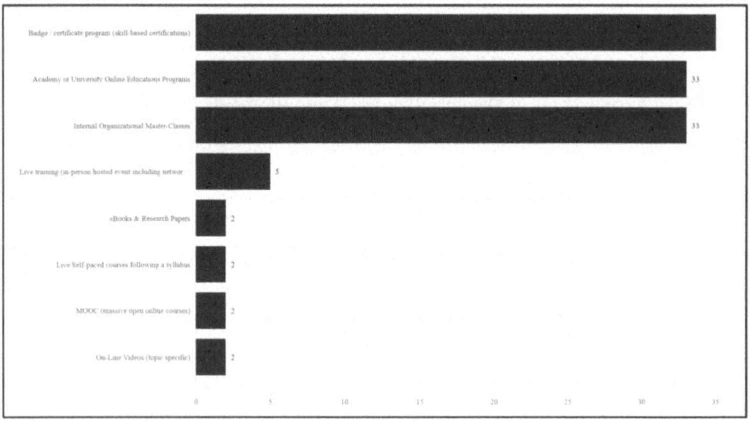

Fig. 7. Preferred Training Formats for Innovative HR Practices DC

Another key collected parameter was connected with the relationships between performance indicators for monitoring success in fluid working environments and metrics for evaluating hybrid team effectiveness and strategic HR programs. Key connections include "Project Turnaround Time" being strongly associated with "Inclusive Collaboration" (27 instances) and "Organizational Agility" (12 instances). "Knowledge Sharing" is linked to "Talent Retention and Engagement" (42 instances) and "Organizational Agility" (9 instances). Less frequent associations are seen for "Return on Investment (ROI)" and "Cross-Functional Coordination." These results highlight the interconnectedness of performance indicators, with a pronounced emphasis on collaboration, agility, and knowledge sharing as critical components for success in hybrid and fluid work settings (Fig. 8).

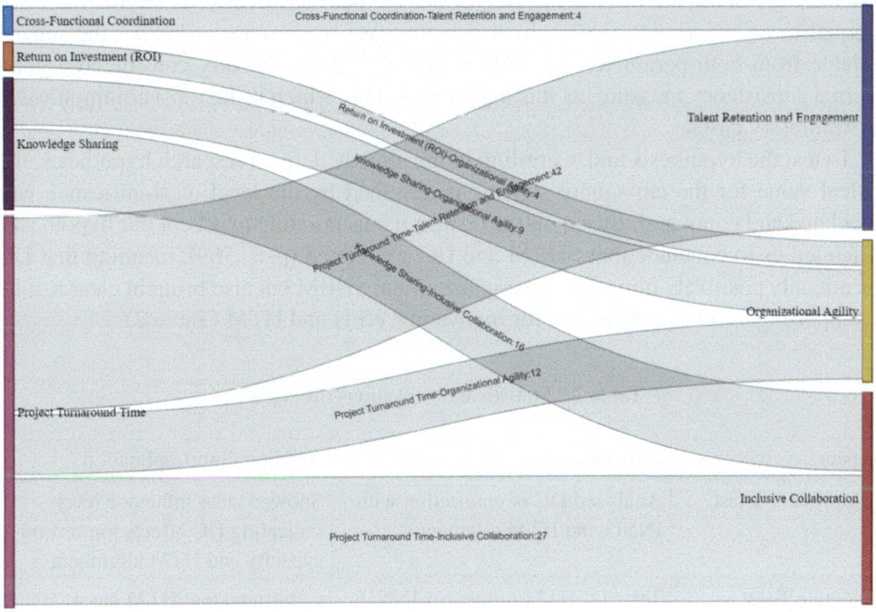

Fig. 8. Importance of Metrics and Indicators for Success in Fluid Work Environments and Hybrid Teams

4.2 Statistical Analysis

To determine whether there is a significant difference between the key variables and the controls between totally different treatments, several univariate calculations were conducted using univariate statistics. In the case of a classical hypothesis test, only one effect from the treatment group is considered to be responsible for the effects. By analyzing some of the statistical data from the dependent variables, the corresponding analysis could show that SHRM and DC impact were connected with the independent variables which were the WFH and HTM innovation parameters.

Analyzing DC in connection with the grouping variables of the organizational innovation capacity (INNO) and the alignment of effective HTM and performance monitoring in fluid working spaces, the independent T-test indicates clear influencing levels and through the student's t-test that the HTM can influence INNO. To get as much information as possible for a compressed analysis, Welch's t-test was also applied. This uses the square root of mean-variance to standardize differences in means. As a final step in defining the correct t-test, the tests were performed on all cases with valid data. When analyzing the p-value, it was needed to understand differences in means, i.e., the difference in the sample means.

Upon testing the reliability coefficient scale through Alpha Cronbach, both SHRM and DC impact exhibited positive reliability coefficient scores (Alpha Score = 0.721 and 0.915, respectively). As evidenced by the high score level obtained for both dependent variables, the internal consistency scale evaluates them positively. When this indicator is equal to or superior to 0.80, which was the case in the case at hand - the impact variable from both parameters of SHRM and DC - it is generally considered a good internal consistency measure, as it is higher than 0.60, which is the most common value in exploratory cases.

To test the hypothesis and according to the already defined research hypothesis, the critical value for the chi-square statistic determined by the level of significance was calculated and compared. Based on the results, it was possible to accept our hypothesis, which led us to conclude that SHRM and DC are related ($p = .569$), meaning that DC has not only positively impacted the organizational SHRM but also brought clear results in supporting a robust framework for innovating WFH and HTM (Table 2).

Table 2. Statistical observations overview

Aspect	Description	Interpretation/Implication
Independent T-test	Analyzed DC in connection with INNO and HTM alignment	Showed clear influence levels, indicating DC affects innovation capacity and HTM alignment
Student's T-test	Tested if HTM influences INNO	Confirmed that HTM has a significant impact on organizational innovation capacity
Welch's T-test	Applied to standardize differences in means using the square root of mean-variance	Ensured accurate mean differences despite variance disparities
P-value Analysis	Used to understand differences in sample means; $p = .569$ for SHRM and DC relationship	Determined significance of results, indicating SHRM and DC are related
Reliability Testing	Cronbach's Alpha: SHRM = 0.721, DC = 0.915	Both SHRM and DC scales are reliable, with DC showing excellent internal consistency
Internal Consistency	SHRM (0.721) and DC (0.915) both above acceptable thresholds (0.60 and 0.80 respectively)	Validated the reliability of the measurement scales for SHRM and DC
Hypothesis Testing	Calculated critical chi-square statistic at defined significance level	Hypothesis that SHRM and DC are related was accepted
Chi-square Result	$p = .569$	Indicates that SHRM and DC are significantly related, supporting the hypothesis

5 Discussions

5.1 Key Findings

The results provide valuable insights into the evolving dynamics of HTM, the application of DC, and the metrics deemed critical for evaluating workforce productivity and innovation in fluid working environments. These findings illuminate the increasing importance of emerging technologies, structured training formats, and adaptive governance frameworks in addressing challenges associated with hybrid work. The open-ended questions also revealed that organizations can create effective hybrid work environments that drive productivity and innovation by focusing on key factors such as communication, technology infrastructure, leadership practices, employee well-being, and performance monitoring. Additionally, understanding the essential characteristics and variables, such as flexibility, trust, inclusivity, and resilience, is crucial for successful implementation. The results also highlighted that allowing individual flexibility choices for employees in terms of selecting WFH or RTO was positively understood.

The open-text fields from the questionnaire further highlighted that SHRM aligns HR practices and initiatives with the overarching business strategy, ensuring that human capital contributes effectively to organizational goals.

In the context of WFH and hybrid models, SHRM can align workforce strategies by developing policies that support flexible work arrangements while maintaining productivity and engagement. The talent management dimensions highlight the attraction, retention, and development of talent capable of thriving in remote and hybrid settings. This can be achieved through targeted training and career development programs, which are connected with performance management to implement performance evaluation systems that accurately reflect remote work contributions and outcomes. Feedback from the DC has highlighted the organization's capacity to integrate, develop, and adapt internal and external competencies in response to rapidly evolving environments. In the HTM context, the ability to swiftly adjust workflows, communication channels, and collaboration tools in response to changing project needs or external factors and foster a culture that encourages experimentation with new technologies and processes to enhance remote collaboration and productivity, is crucial. In terms of resource allocation, the company deploys its resources, including human capital and technological infrastructure, to support flexible work arrangements. Combining SHRM with DC enables Life Sciences organizations to create resilient, adaptable, and innovative work environments. This synergy ensures that HR strategies are not only aligned with current operational needs but also flexible enough to evolve with future challenges and opportunities.

Moreover, the results revealed key priorities for organizations, including talent retention, project turnaround time, inclusive collaboration, and knowledge sharing. In this section, we interpret these results, linking them to existing literature and exploring their implications for theory and practice.

The pandemic has had a profound effect on organizations' adoption of hybrid and fluid work strategies, with an overwhelming majority (97%) of respondents affirming its influence. Organizations were compelled to adopt new technologies and frameworks to ensure continuity, flexibility, and productivity, underscoring the critical role of adaptive governance, where the results suggest that the pandemic acted as a catalyst for experimentation with hybrid team structures, paving the way for more permanent changes in workforce management practices.

The implications of these changes are significant, and it was highlighted that organizations must balance the benefits of flexibility with the challenges of maintaining engagement and alignment across dispersed teams. The emphasis on hybrid models reflects a shift toward employee-centric policies that prioritize well-being and autonomy, consistent with the DC framework whereby sensing and seizing opportunities for innovation, organizations have enhanced their agility in managing workforce dynamics, a trend that warrants further exploration in future research.

The results underscore the importance of specific metrics in evaluating hybrid team effectiveness and strategic HR programs. "Talent Retention and Engagement" emerged as the most critical metric, followed closely by "Inclusive Collaboration" and "Organizational Agility." These findings align with the literature on human capital management, which identifies retention and engagement as foundational to organizational success. The emphasis on collaboration and agility further underscores the need for HR practices that facilitate seamless interaction and adaptability, particularly in hybrid settings. Interestingly, "Project Turnaround Time" was consistently ranked as a high-priority performance indicator for success in fluid working environments. This focus on efficiency reflects organizational priorities in meeting project deadlines while navigating the complexities of hybrid work. However, lower rankings for metrics such as "Return on Investment (ROI)" and "Cross-Functional Coordination" suggest that organizations are placing greater emphasis on people-centric outcomes over financial metrics or broader organizational structures. This shift may reflect a recognition of the long-term value of investing in workforce well-being and collaboration.

The interconnectedness of these metrics highlights the need for integrated performance evaluation frameworks, where for example, the strong association between "Project Turnaround Time" and "Inclusive Collaboration" suggests that fostering a collaborative environment can directly enhance operational efficiency. Similarly, the link between "Knowledge Sharing" and "Talent Retention and Engagement" underscores the importance of creating a culture of continuous learning to retain top talent. These insights support the argument that DC—particularly the ability to reconfigure resources—is critical in aligning HR strategies with organizational goals.

Relevance of Fluid Workplaces, SHRM, and Enhanced WFH Options for Neurodivergent Employees

The open-ended responses underscore the critical role of fluid workplaces and SHRM in fostering an inclusive environment for neurodivergent employees. Respondents highlighted that fluid workplaces, characterized by flexibility in work location and adaptable schedules, significantly benefit neurodivergent individuals by accommodating diverse working styles and reducing environmental stressors. Several collected comments reported enhanced productivity and job satisfaction when allowed to work in environments tailored to their sensory and cognitive needs, such as quiet home offices or customizable workspaces.

SHRM emerged as a pivotal framework in developing and implementing inclusive HR policies, and respondents emphasized that strategic alignment of HR practices with organizational goals must include specific provisions for neurodiversity. This includes tailored onboarding processes, continuous professional development, and personalized support systems. Effective SHRM ensures that neurodivergent employees receive the necessary accommodation, such as flexible hours, noise-canceling headphones, and clear, structured communication channels, which are essential for their success and well-being. Enhanced WFH options were frequently mentioned as vital for supporting neurodivergent employees. The ability to control one's work environment mitigates common challenges faced in traditional office settings, such as sensory overload and social anxiety. Respondents noted that remote work arrangements allow for better concentration and the management of energy levels, leading to improved performance and reduced burnout rates. Additionally, technology-enabled solutions, such as virtual collaboration tools and AI-driven assistive technologies, were praised for facilitating seamless communication and task management, further empowering neurodivergent professionals.

HR policies must evolve to address the unique needs of neurodivergent employees within fluid workplaces. Survey comments advocated comprehensive policies that promote inclusivity, such as mandatory training for managers on neurodiversity awareness and the implementation of Employee Resource Groups (ERGs) focused on neurodivergent support, where policies should also incorporate regular feedback mechanisms to continuously refine and adapt support structures based on employee input. Furthermore, respondents highlighted the importance of fostering a culture of acceptance and understanding, that involves proactive efforts to destigmatize neurodiversity and encourage open dialogue about individual needs and preferences. Organizations that prioritize mental health and provide resources for neurodivergent employees not only enhance their workforce's diversity but also drive innovation and resilience. The survey's free-text responses reveal that fluid workplaces, when integrated with strategic HR management and robust WFH options, create a supportive and empowering environment for neurodivergent employees. By embracing flexibility, tailored accommodations, and inclusive policies, life sciences organizations can harness the full potential of their diverse talent, leading to greater organizational success and a more equitable workplace.

5.2 Workforce Priorities: Retention, Engagement, and Collaboration

A significant finding from the survey is the strong consensus on the importance of emerging technologies in performance analytics, but also findings that effective HTM and performance monitoring in fluid working spaces hinge on several critical factors and dimensions like communication and collaboration, leadership and management practices, employee engagement and well-being and training and development.

Nearly all respondents agreed or strongly agreed that technologies such as AI-driven dashboards and collaborative platforms will be decisive for effective HTM in the future. This finding reflects the growing reliance on digital tools to address the complexities of hybrid work, where technology plays a central role in enabling remote collaboration and productivity but also leveraging analytics to track performance, identify trends, and inform decision-making processes.

The adoption of these technologies presents both opportunities and challenges, but implementing robust IT systems that support remote access, data security, and real-time collaboration and having communication protocols in establishing clear guidelines for communication frequency, channels, and responsiveness to ensure consistency and clarity can have positive results in leveraging HTM and performance monitoring according to the open questions collected feedback.

On the one hand, advanced analytics and AI can provide real-time insights into team performance, enabling more informed decision-making, and training leaders to manage remote teams effectively, focusing on trust-building, delegation, and virtual team dynamics was considered as good practices. On the other hand, the ethical implications of performance monitoring must be carefully considered, where overly invasive monitoring practices could erode trust and stifle creativity. To address these challenges, organizations must implement governance frameworks that balance transparency with employee autonomy, ensuring that technology serves as an enabler rather than a barrier.

The emphasis on emerging technologies also underscores the importance of digital literacy and training, where respondents identified structured training formats, such as badge or certificate programs and internal masterclasses, as the most effective methods for educating teams on innovative HR practices and DC. These findings suggest that organizations recognize the need to upskill their workforce to fully leverage the potential of new technologies. However, the relatively low preference for informal learning formats, such as MOOCs or research papers, indicates a need for more hands-on, interactive approaches to training.

The results reveal mixed levels of engagement with decentralized or agile elements aimed at improving workforce flexibility and productivity. While 50% of respondents confirmed that their organizations had conducted feasibility assessments for such practices, an equal proportion expressed uncertainty. This ambiguity highlights a potential gap in organizational awareness or communication regarding the adoption of decentralized models.

Decentralized and agile practices align closely with the DC framework, particularly the principles of sensing and seizing opportunities. By decentralizing decision-making and fostering agility, organizations can respond more effectively to changing market conditions and employee needs. However, the uncertainty expressed by many respondents suggests that these practices are not yet fully embedded in organizational cultures. Future efforts should focus on clarifying the value of decentralized models and providing leaders with the tools and frameworks needed to implement them effectively.

The findings emphasize the critical importance of talent retention, engagement, and inclusive collaboration in hybrid work settings. These priorities are consistent with existing research on workforce management, which identifies engagement as a key driver of performance and retention as a determinant of organizational stability. The strong focus on collaboration further underscores the importance of fostering a sense of community and connection among team members, particularly in hybrid environments where physical interaction is limited.

One notable finding is the high level of time commitment to hybrid team coordination. A majority of respondents reported spending more than 60% of their weekly work time on activities related to HTM, highlighting the operational demands of these models. The results underscore the need for efficient tools and processes to streamline hybrid work practices, enabling leaders to focus on strategic priorities, and emphasis on metrics such as collaboration, agility, and retention aligns with the principles of sensing, seizing, and reconfiguring resources.

Practically, the results highlight several areas for organizational improvement. First, the strong emphasis on emerging technologies underscores the need for investments in digital tools and training programs. Organizations must prioritize the development of digital literacy among their workforce to fully leverage the potential of AI-driven analytics and collaborative platforms. Second, the focus on people-centric metrics suggests that HR strategies should prioritize employee well-being and collaboration over purely financial outcomes. This shift requires a rethinking of traditional performance evaluation frameworks to incorporate more holistic measures of success.

Finally, the findings point to the need for stronger governance frameworks to guide the adoption of decentralized and agile practices. Organizations must ensure that these practices are communicated and effectively implemented, with a focus on balancing flexibility with accountability. By addressing these challenges, organizations can create more resilient, adaptive workforces capable of thriving in dynamic and fluid environments (Table 3).

Table 3. Significant survey findings related to hybrid team management (HTM) and performance monitoring

Key Finding	Description	Implications/Actions
Emerging Technologies in Performance Analytics	– Strong consensus on the importance of AI-driven dashboards and collaborative platforms for effective HTM	– Invest in digital tools to enhance remote collaboration and leverage analytics for informed decision-making
Adoption of AI and Collaborative Platforms	– Nearly all respondents agree that technologies like AI-driven dashboards and collaborative platforms are decisive for future HTM	– Integrate AI and collaborative tools into existing workflows to support remote and hybrid team operations
Communication and Collaboration	– Effective HTM relies on clear communication protocols and advanced collaborative tools to facilitate seamless interaction	– Establish guidelines for communication frequency, channels, and responsiveness to ensure consistency and clarity

(*continued*)

Table 3. (*continued*)

Key Finding	Description	Implications/Actions
Leadership and Management Practices	– Training leaders in trust-building, delegation, and managing virtual teams is crucial for successful HTM	– Develop leadership training programs focused on remote management skills and virtual team dynamics
Employee Engagement and Well-being	– High importance placed on employee retention, engagement, and inclusive collaboration to maintain organizational stability and performance	– Implement strategies to foster community, support well-being, and ensure inclusive practices in hybrid work environments
Training and Development	– Preference for structured training formats like certificate programs and internal masterclasses over informal learning methods	– Offer interactive, hands-on training programs to improve digital literacy and effectively utilize new technologies
Technology Infrastructure	– Need for robust IT systems that support remote access, data security, and real-time collaboration to enable effective HTM	– Invest in secure, scalable IT infrastructure to support hybrid work models and ensure data protection
Ethical Performance Monitoring	– Concerns about overly invasive monitoring practices potentially eroding trust and stifling creativity among employees	– Develop governance frameworks that balance transparency with employee autonomy, ensuring technology serves as an enabler rather than a barrier
Digital Literacy and Training	– Emphasis on the importance of upskilling the workforce to leverage new technologies effectively, with a preference for structured training formats	– Prioritize digital literacy through structured training programs to fully leverage AI-driven analytics and collaborative platforms
Decentralized and Agile Practices	– Mixed engagement with decentralized/agile practices; some organizations have conducted feasibility assessments, while others remain uncertain about adoption	– Clarify the benefits of decentralized models and provide tools and frameworks to support their effective implementation

(*continued*)

Table 3. (*continued*)

Key Finding	Description	Implications/Actions
Alignment with Dynamic Capabilities	– Decentralized and agile practices support the principles of sensing and seizing opportunities, enhancing organizational responsiveness	– Foster agility and responsiveness within the organizational culture to better adapt to market changes and employee needs
Talent Retention and Engagement	– Critical for performance and organizational stability, with a strong focus on fostering a sense of community and connection among team members in hybrid environments	– Focus HR strategies on retention and engagement initiatives, emphasizing collaborative and inclusive practices to maintain high morale and productivity
Time Commitment to HTM Coordination	– Majority of respondents spend over 60% of their weekly work time on activities related to HTM, highlighting the operational demands of hybrid models	– Streamline HTM processes and utilize efficient tools to reduce the time burden on leaders, allowing them to focus on strategic priorities
People-Centric Metrics in HR Strategies	– HR strategies should prioritize employee well-being and collaboration over purely financial outcomes, necessitating a shift in traditional performance evaluation frameworks	– Redefine performance metrics to include holistic measures of success, such as employee satisfaction and collaboration effectiveness
Governance Frameworks for Decentralized Practices	– Need for clear governance to guide the adoption of decentralized and agile practices, ensuring flexibility is balanced with accountability	– Implement governance frameworks that clearly communicate and effectively manage decentralized practices, maintaining a balance between flexibility and accountability
Resilience and Cultural Cohesion	– Essential characteristics for managing hybrid teams, promoting a unified organizational culture that transcends physical locations and fosters team cohesion	– Promote unified culture initiatives and develop resilience strategies to ensure teams can withstand and quickly recover from disruptions

(*continued*)

Table 3. *(continued)*

Key Finding	Description	Implications/Actions
Performance Outcomes and Technology Adoption	– Importance of tracking the impact of hybrid management on project completion rates, quality of work, and innovation, alongside assessing technology adoption rates	– Monitor and evaluate technology adoption and its effects on performance outcomes to ensure positive impacts on project success and overall organizational innovation

5.3 Statistical Analysis Findings

The results mean that the two subsets of the variables (SHRM and DC) allow us to understand statistically significant evidence that there exists a statistically significant association between the dependent variables of impact, meaning there is a significant impact relationship. Another analysis applied was to analyze the difference between the means of multiple groups through the ANOVAs, where our dependent continuous variables of impact allow us to test the DC variable group, where the level of the independent variables included in the analysis and the descriptive statistics for each combination of levels for the independent variables allowing to understand that leveraging SHRM and DC can drive significant innovation and develop effective solutions for managing remote and hybrid teams. The results showed that there is an association among the variables of interest, specifically between the variables of interest and the independent variable. The final step of our analysis was to apply a linear mixed model, allowing us to explore the relationship between the variables of interest, including the interaction between the variables of interest. We found that projects related to DC have an impact on ensuring that human capital contributes effectively to organizational goals, and in the context of WFH and hybrid models. Here, there was a clear association ($df = 1$, 0.26, $F = 0.114$, and $p = 0.864$) between DC and impact on outcomes connected with an organization's ability to integrate, build, and reconfigure internal and external competencies to address rapidly changing environments, where in HTM, DC can Foster a culture that encourages experimentation with new technologies and processes to enhance remote collaboration and productivity, and including human capital and technological resources management, to support flexible work arrangements. The analysis used sum contrast coding for categorical predictors and allowed for better interpretability of models with interactions and resulted mainly from the shape of the p-value distribution. To understand the complexity of these assumptions and evaluate the overall value of HR empowerment and knowledge increase around flexibility for HTML models.

6 Conclusions

In light of the results and findings presented, the results highlight the critical importance of HTM, the application of DC, and the integration of emerging technologies in addressing the challenges of fluid working environments. The findings emphasize the need for organizations to adopt innovative governance frameworks, implement advanced performance metrics, and prioritize inclusive and collaborative practices to ensure workforce adaptability and productivity. Additionally, the results revealed that hybrid work environments require DC—specifically, the ability to sense emerging opportunities, seize these opportunities through flexible work structures, and reconfigure resources to sustain organizational performance. These observations align with the principles of modern HR strategies, which prioritize workforce agility, knowledge sharing, and employee engagement. The descriptive analysis indicates that most respondents are based in North America and Europe, holding senior leadership positions, which underscores the relevance of the study's focus to a global audience of life sciences professionals and decision-makers.

Furthermore, this chapter identifies "Talent Retention and Engagement," "Inclusive Collaboration," and "Organizational Agility" as key metrics for evaluating hybrid team effectiveness. These findings underscore the growing emphasis on people-centric outcomes in workforce management, with a clear priority placed on fostering collaboration and supporting employee well-being. "Project Turnaround Time" emerged as a critical indicator of success in fluid workspaces, highlighting the operational efficiency expected from hybrid teams. The chapter also underscores the value of knowledge-sharing frameworks, which are essential for retaining talent and building a culture of innovation. Also, the findings provide valuable insights into the role of emerging technologies, such as AI-driven dashboards and collaborative platforms, in enabling effective HTM, where a strong consensus among respondents affirmed the transformative potential of these technologies, which can provide real-time performance insights and facilitate seamless communication across dispersed teams. However, the study also highlights the importance of ethical considerations in performance monitoring, emphasizing the need for governance frameworks that protect employee autonomy while leveraging data to drive productivity and innovation.

Further findings reveal that structured training formats, such as badge programs, online education, and internal organizational masterclasses, are also effective methods for educating teams on innovative HR practices and DC, underscoring the necessity of continuous upskilling to ensure that teams are equipped to navigate the complexities of hybrid and fluid work environments. The emphasis on formal training also indicates that organizations value interactive and structured learning experiences over informal or self-paced formats. Additional key insights are the mixed level of engagement with decentralized and agile practices, while some organizations have actively conducted feasibility assessments to explore these approaches, many respondents expressed uncertainty about their adoption. This highlights a potential gap in organizational communication and awareness, suggesting a need for a clearer articulation of the value and application of decentralized models.

In terms of strategic implications, results demonstrate that HTM, when effectively supported by metrics and technology, can drive significant improvements in organizational performance. By focusing on collaboration, inclusivity, and agility, organizations can better align their workforce strategies with evolving market demands. Moreover, the adoption of DC offers a framework for organizations to remain competitive in an increasingly complex and unpredictable business environment. While metrics like project turnaround time and ROI remain important, the strong emphasis on engagement and collaboration reflects a broader shift toward creating more human-centered workplaces and also better-balanced discussions around employees' preferences for WFH or RTO. This balance is particularly important in hybrid settings, where the absence of physical interaction can present challenges to maintaining a cohesive and inclusive team culture.

This chapter contributes to the growing discourse on hybrid work by providing actionable insights and evidence-based recommendations for organizations. It highlights the transformative potential of DC and emerging technologies in optimizing HTM, while also emphasizing the need for ethical governance and inclusive practices. By addressing these challenges and opportunities, organizations can create resilient, adaptive, and innovative workforces capable of thriving in a rapidly changing world. Future research should build on these findings by exploring the long-term impact of hybrid work strategies and the role of emerging technologies in shaping the future of workforce management.

References

Alawida, M., Mejri, S., Mehmood, A., Chikhaoui, B., Isaac Abiodun, O.: A comprehensive study of ChatGPT: advancements, limitations, and ethical considerations in natural language processing and cybersecurity. Information **14**(8), 462 (2023)

Amini, M., Jahanbakhsh Javid, N.: A multi-perspective framework established on diffusion of innovation (DOI) theory and technology, organization and environment (TOE) framework toward supply chain management system based on cloud computing technology for small and medium enterprises. Organization and environment (TOE) framework toward supply chain management system based on cloud computing technology for small and medium enterprises (January 2023). Int. J. Inf. Technol. Innov. Adoption **11**, 1217–1234 (2023)

Ball, P.: What the COVID-19 pandemic reveals about science, policy and society. Interface Focus **11**(6), 20210022 (2021)

Bhattamisra et al.: Artificial intelligence in pharmaceutical and healthcare research. Big Data Cogn. Comput. (2023)

Bouguerra, A., Hughes, M., Cakir, M.S., Tatoglu, E.: Linking entrepreneurial orientation to environmental collaboration: a stakeholder theory and evidence from multinational companies in an emerging market. Br. J. Manage. (2022)

Breaux, R., et al.: Telehealth delivery of the RELAX intervention for families of adolescents diagnosed with ADHD: Preliminary treatment outcomes and evidence of acceptability and feasibility. Evid. Based Pract. Child Adolesc. Mental Health **8**(1), 24–38 (2023)

Cifuentes-Faura, J.: European union policies and their role in combating climate change over the years. Air Qual. Atmos. Health, 1–8 (2022)

Dillard-Wright, J., Shields-Haas, V.: Nursing with the people: Reimagining futures for nursing. Adv. Nurs. Sci. **44**(3), 195–209 (2021)

Eager, B., Brunton, R.: Prompting higher education towards AI-augmented teaching and learning practice. J. Univ. Teach. Learn. Pract. **20**(5), 02 (2023)

Ebrahimi, S., Bridgelall, R.: A fuzzy Delphi analytic hierarchy model to rank factors influencing public transit mode choice: a case study. Res. Transp. Bus. Manag. **39**, 100496 (2021)

ElBaih, M.: The role of privacy regulations in AI development (a discussion of the ways in which privacy regulations can shape the development of AI). SSRN 4589207 (2023)

Gebhardt, M., Kopyto, M., Birkel, H., Hartmann, E.: Industry 4.0 technologies as enablers of collaboration in circular supply chains: a systematic literature review. Int. J. Produ. Res. **60**(23), 6967–6995 (2022)

Haleem, A., Javaid, M., Singh, R.P., Suman, R.: Medical 4.0 technologies for healthcare: Features, capabilities, and applications. Internet Things Cyber-Phys. Syst. **2**, 12–30 (2022)

Jacob, C., Bourke, S., Heuss, S.: From testers to cocreators—the value of and approaches to successful patient engagement in the development of ehealth solutions: qualitative expert interview study. JMIR Hum. Factors **9**(4), e41481 (2022)

Jalil, S., Rafi, S., LaToza, T.D., Moran, K., Lam, W.: Chatgpt and software testing education: Promises & perils. In: 2023 IEEE International Conference on Software Testing, Verification and Validation Workshops (ICSTW), pp. 4130–4137. IEEE (2023)

Javaid, M., Haleem, A., Singh, R.P.: ChatGPT for healthcare services: an emerging stage for an innovative perspective. BenchCouncil Transa. Benchmarks, Stan. Evaluations **3**(1), 100105 (2023)

Junaid, M., Zhang, Q., Cao, M., Luqman, A.: Nexus between technology enabled supply chain dynamic capabilities, integration, resilience, and sustainable performance: an empirical examination of healthcare organizations. Technol. Forecast. Soc. Chang. **196**, 122828 (2023)

Kant, N., Anjali, K.: Blockchain technology: a strategic resource. In: Blockchain for Healthcare Systems, pp. 1–15. CRC Press. 13 (2021)

Kaplan, R.M., Hays, R.D.: Health-related quality of life measurement in public health. Annu. Rev. Public Health **43**, 355–373 (2022)

Kasneci, E., et al.: ChatGPT for good? on opportunities and challenges of large language models for education. Learn. Individ. Differ. **103**, 102274 (2023)

Kendzerska, T., et al.: The effects of the health system response to the COVID-19 pandemic on chronic disease management: a narrative review. Risk Manage. Healthc. Policy 575–584 (2021)

Kotcher, J., et al.: Views of health professionals on climate change and health: a multinational survey study. Lancet Planet. Health **5**(5), e316–e323 (2021)

Kouba, P., et al.: Machine learning-guided protein engineering. ACS Catal. **13**(21), 13863–13895 (2023)

Kwok, A.O., Treiblmaier, H.: Blockchain technology as a driver of economic development in small economies: a dynamic capabilities framework. J. Decis. Syst. 1–26 (2023)

Leffa, D.T., et al.: Transcranial direct current stimulation vs sham for the treatment of inattention in adults with attention-deficit/hyperactivity disorder: the TUNED randomized clinical trial. JAMA Psychiat. **79**(9), 847–856 (2022)

Masih, E.A.: Feasibility of using QR code for registration & evaluation of training and its ability to increase response rate–the learners' perception. Nurse Educ. Today **111**, 105305 (2022)

Mele, G., Capaldo, G., Secundo, G., Corvello, V.: Revisiting the idea of knowledge-based dynamic capabilities for digital transformation. J. Knowl. Manage. (2023)

Mostafiz, M.I., Musteen, M., Saiyed, A., Ahsan, M.: COVID-19 and the global value chain: Immediate dynamics and long-term restructuring in the garment industry. J. Bus. Res. **139**, 1588–1603 (2022)

Patil, et al.: Utilization of artificial intelligence in healthcare system, pharmaceutical industry & education. Int. J. Adv. Res. Sci. Commun. Technol. (2022)

Pereira, R., Bianchi, I.S., Araújo, V.M.D.S.: Multi-framework implementation of the problem management process. Форсайт, **15**(3 (eng)), 94–104 (2021)

Pesqueira, A.M., Sousa, M., Pereira, R.: Individual dynamic capabilities influential factors in blockchain technology innovation from hospital settings (2023)

Pesqueira, A., Sousa, M.J.: 12 Social analytics influencing KOLs profiling in the dissemination of innovations and scientific knowledge in the life sciences industry. Innovations Soc. Media Anal. Digit. Soc. (2022)

Pesqueira, A., Sousa, M.J.: Life sciences industry blockchain value management, investments, and finance impact assessment: global questionnaire survey of 1524 industry professionals and executives. In: Incorporating AI Technology in the Service Sector, pp. 67–107. Apple Academic Press (2022)

Secinaro et al.: The role of artificial intelligence in healthcare: a structured literature review. BMC Med. Inf. Deci. Mak. (2021)

Shiwlani, A., Khan, M., Sherani, A.M.K., Qayyum, M.U.: Synergies of AI and smart technology: revolutionizing cancer medicine, vaccine development, and patient care. Int. J. Soc. Hum. Life Sci. **1**(1), 10–18 (2023)

Spiel, K., Hornecker, E., Williams, R. M., Good, J.: ADHD and technology research–investigated by neurodivergent readers. In: Proceedings of the 2022 CHI Conference on Human Factors in Computing Systems, pp. 1–21 (2022)

Tariq, et al.: Software measurement by using artificial intelligence. J. Nanomat. (2022)

Ullah, H., Manickam, S., Obaidat, M., Laghari, S.U.A., Uddin, M.: Exploring the potential of metaverse technology in healthcare: applications, challenges, and future directions. IEEE Access (2023)

Vallentin, A., Thórisson, K.R., Latapie, H.: Addressing the unsustainability of deep neural networks with next-gen AI. In: International Conference on Artificial General Intelligence, pp. 296–306. Cham: Springer Nature Switzerland (2023)

Vlasta, M.: Methodological framework for European regulation of (artificial) intelligence and its limits (2023)

Wada, M., et al.: Dopaminergic dysfunction and excitatory/inhibitory imbalance in treatment-resistant schizophrenia and novel neuromodulatory treatment. Mol. Psychiatry **27**(7), 2950–2967 (2022)

Wamba, S.F., Queiroz, M.M., Jabbour, C.J.C., Shi, C.V.: Are both generative AI and ChatGPT game changers for 21st-century operations and supply chain excellence? Int. J. Prod. Econ. **265**, 109015 (2023)

Wani, et al.: Utilization of artificial intelligence in disease prevention: diagnosis, treatment, and implications for the healthcare workforce. Healthcare (2022)

Yulianti, T., Sulistyawati, A.: Enhancing public speaking ability through focus group discussion. J. Pajar (Pendidikan Dan Pengajaran) **5**(2), 287–295 (2021)

Artificial Intelligence and Its Impact on the Work of Civil Servants. A Systematic Review

José Dias Lopes[1](✉), Paulo Almeida Gonçalves[1], Ricardo Raimundo[2], and João Rodrigues dos Santos[3]

[1] ISEC-Lisboa and ISEG/ULisboa, Lisbon, Portugal
diaslopes@iseg.ulisboa.pt
[2] ISEC-Lisboa and IADE/Universidade Europeia, Lisbon, Portugal
[3] ISEC-Lisboa and Universidade Europeia and CETRAD/UTAD, Lisbon, Portugal

Abstract. Artificial intelligence could potentially have a major impact on government practice and the way citizens relate to state institutions. These impacts will affect the way public administration operates and the way civil servants carry out their duties. It is therefore foreseeable that, with artificial intelligence, work in public administration will be profoundly changed in terms of its content, but also in terms of the way public servants relate to citizens, elected representatives, and other entities.

This research studies the impacts of artificial intelligence on public administration, using a systematic analysis of literature.

Because the set of documents available is very limited, it was decided to compare the content of the articles identified with a more general sample of articles on artificial intelligence and Industry 4.0.

The study shows that the subject is highly topical, with no articles on the subject more than five years old. On the other hand, the study shows that most of the articles focus on the impact of artificial research on the general framework of the work of public servants, rather than on its content and the specific ways in which this work is carried out.

We, therefore, propose a research agenda for the study of the impacts of artificial intelligence on public administration work, focusing on the content of the work and its framework, without neglecting the more structural questions associated with issues such as rights, freedoms, and guarantees.

Keywords: artificial intelligence · human resources · public administration · civil servants · systematic literature review

1 Introduction

García-Orosa (2021) reflects on the ecology of e-democracy, concluding that a fourth wave of this phenomenon has been established since 2016 with the Brexit campaign and the Cambridge Analytica scandal, marked, among other things, by the increased

M. J. Sousa (Ed.): UFHRD 2024, ISEM 52, pp. 35–50, 2025.
https://doi.org/10.1007/978-3-031-92548-1_3

use of Artificial Intelligence and big data; the use of falsehoods as a political strategy, as well as other fake news and deep fake phenomena, associated with an element of technological determinism. The author concludes that this is a combination of factors that threaten democracy. In similar terms, Hernández Peña (2022) and Gutiérrez and Cruz (2024) both warn of the dangers that the use of these technologies can pose to individual freedoms. This leads Manheim and Kaplan (2019) to conclude that we may not need to stop AI, but we certainly need to pay attention to it. Stern et al. (2018) issue a similar warning, but regarding the whole process of digitalization, considering that while it offers many growth opportunities, it can also endanger the safety of citizens and the integrity of public infrastructures, concluding that a wide range of new challenges are emerging.

The digitalization process mentioned above should not be seen simply as the addition of a set of technologies which, for simplification, we will call Technologies 4.0 (T4.0). What makes the process new compared to previous technological developments is the potential that results from bringing these technologies together (Javaid et al. 2022). The combination of these technologies is the origin of the concept of Industry 4.0 (Laffi & Boschma, 2022)).

As Lopes, Estevão and Toth-Peter (2023) point out, although the concept of Industry 4.0 originally developed from a technological base - a set of very transformative technologies (T4.0) - and in manufacturing, it quickly evolved into other sectors, namely the public administration sector (Government 4.0).

Therefore, the effects of implementing artificial intelligence solutions in public administration should not be seen in isolation, but in the context of the combination of T4.0, i.e. in the context of the application of what is commonly referred to as Industry 4.0.

On the other hand, we will try to assess the impacts of implementation not in terms of their impact on society, as mentioned above, but in terms of their impact on the work carried out in public administration.

As aforementioned, the expectations are high regarding the impact of artificial intelligence on jobs, be it at general employment, or tasks design and flow, the ensuing content and demanded skills. This research aims to ascertain the importance of artificial intelligence regardin its impact on employment and work.

2 Literature Review

2.1 The Fourth Technological Revolution or Industry 4.0

The concept of Industry 4.0 (I4.0) developed in the early 2010s (Hazra et al., 2021; Lopes de Sousa Jabbour et al., 2018; Parida, Sunand Dash & Sharma, 2021; Strandhagen et al., 2017) revealed paramount. The changes introduced by the concept are huge, in such a way that Almada-Lobo (2015) and Yang and Gu (2021) considered it to be a revolution, namely the fourth industrial revolution.

As Kim, Wang and Boon (2021) point out, the introduction of new technologies has a major impact on jobs, which equals to say that the greater the technological innovation, the greater its impact on the workflow, as well. In the same vein, Matyushenko and

Redko (2019) consider that this Fourth Industrial Revolution settled new boundaries for economic development and ensuing future jobs. In particular, the impacts of this new industrial revolution on jobs must be assessed at both academic and practical levels (Kagermann, Wahlster, & Helbig, 2013).

This revolution could resemble what Carlota Perez (2010) defines as a new techno-economic paradigm. According to the author, the new paradigm is shaped by the mode new technologies are applied and extended, multiplying their impact on the economy and ultimately changing how socio-institutional structures are organized.

Freeman (1991) argues that the new techno-economic paradigm will redefine work both regarding content and the fashion it is socially considered. For the author, a new paradigm carries out, to some extent, a level of free will to reshape what is socially acceptable concerning the work loading, social climate, and working relationships.

Nübler (2016) describes the process of adjusting the workload issue along with the emergence of the new paradigm. According to the author, its initial phase is made up of process innovations and is linked to increasing productivity and job suppression. In this case, both employees and companies learn to apply the new technologies along a process of "unlearning the old and learning the new".

This view is also posited by Perez (2010) as he contends that as current paradigms become obsolete are replaced by new ones, whilst either procedures or management capabilities, of worth in the past become outdated and inefficient, requiring further unlearning, learning, and learning to learn new processes. As I4.0 is a new techno-economic paradigm, we are now in the stage of unlearning, learning, and learning to learn new processes.

Bécue, Praça and Gama (2021) and Peres et al. (2020), in turn, highlight the central role of artificial intelligence in implementing I4.0 solutions. In this way, the transformational role of work and employment associated with I4.0 can also be associated with Artificial Intelligence.

In this vein, the World Economic Forum which publishes the Future of Jobs Report every two years has underscored in its last four reports explicit references to the importance of artificial intelligence in jobs changes (WEF 2016, p.7; 2018, p.vii; 2020, p.8; 2023, p.4). In particular, the latest report (WEF 2023, p.6) states that artificial intelligence is expected to be adopted by around 75% of the companies that took part in the survey and that it will further lead to high staff turnover - with half of the organizations anticipating job creation and 25% expecting job losses, respectively.

While the impact on employment is expected to be significant, the costs regarding job content and skills are also expected to be noteworthy. According to the Ellingrud et al. report (2023), by 2030, activities that currently account for 30 percent of the workload could be automated in the USA. This trend could be further accelerated by artificial intelligence. Likewise, Hupfer (2020) posits that AI will lead to mild to significant changes in roles and skills over the next three years.

Unfortunately, no similar discussions can be found regarding the implications of I4.0 for operational processes in public administration.

2.2 Artificial Intelligence and Industry 4.0

Muhuri, Shukla and Abraham (2019) interpret the idea of the fourth industrial revolution, considering that the arrival of Industry 4.0 has seen a global transformation based on digital integration and intelligent engineering. For the authors, as all devices are now equipped with some form of artificial intelligence (machine learning in the original), automating processes has become a priority. In a similar vein, Matt et al. (2020) consider that Industry 4.0 enables revolutionary new design, manufacturing, operation, and production systems thanks to intelligent connectivity between manufacturing systems, logistics systems, and products/services.

Also, Lemos, Gaspar and Lima (2022) point out that integration is a determining element of I4.0. Likewise, Ahi et al. (2022) argue that the integration of all advanced digital technologies (e.g. Internet of Things) indicates the beginning of this fourth industrial revolution. The importance of combining I4.0 technologies is particularly evident when it comes to manufacturing (Balubaid & Alsaadi, 2023; Muhuri, Shukla & Abraham, 2019), where the concept was originally developed, forming the inception of smart manufacturing. Nonetheless, as Mühl and de Oliveira (2022) state in a paper on Agriculture 4.0, the issue of integrating technologies goes well beyond manufacturing. Artificial Intelligence plays a key role in integrating technologies Vignesh Saravanan et al. (2023, p. 285).

Meziane et al. (2000) also consider that intelligent systems are decisive for the integration of technologies in the context of new production systems, concluding that this integration effort is the biggest challenge in the design of these systems.

Zhong et al. (2017) describe General Electric's approach to the Factory of the Future. The authors believe that this approach is based on the idea that intelligent machines, advanced analysis systems, and interconnected people are the key elements of the Factory of the Future. The authors also believe that the maximum potential of this type of system will be achieved through the holistic integration of three components: intelligent equipment, intelligent systems, and intelligent decision-making (see also Posada et al., 2015, which lists intelligent machines, advanced analysis systems and people at work). Zhong et al. (2017) conclude that given the deeply integrative nature of Industry 4.0, a generic framework for smart manufacturing is important, as manufacturing science and technology, ICT, and sensor technology will be highly interconnected in the future.

Discussing the direct benefits of applying artificial intelligence in smart factories, Yang and Gu (2021) consider that these will mainly occur in reduced downtime, optimized production, and better energy management.

2.3 Human Resources in Industry 4.0 Context

Mallinguh, Wasike and Zoltan (2020) posit that capital, and human resources are the main constraints to SMEs' efforts in implementing new technologies [I4.0]. These limitations turn companies increasingly dependent on external sources of knowledge to strengthen their technological skills. The complexity in terms of human resources is also mentioned by Ingaldi and Ulewicz (2019) in their study on the implementation of I4.0 in SMEs. The authors concluded that a total of 80% of the companies analyzed have problems with human resources and hardly any specialists are available on the market, preferring to work for large companies.

Badri, Boudreau-Trudel and Souissi (2018) analyze occupational health and safety (OHS) issues in the context of Industry 4.0, concluding that artificial intelligence plays an important role in this context, especially in two categories: Learning and rapid recognition of hazards, and timely decision-making. In the same vein, Meziane et al. (2000) add that when implementing the production systems of the future, OSH, particularly when establishing organizational models that comply with the relevant legislation, requires the use of intelligent systems.

For Badri, Boudreau-Trudel and Souissi (2018), more research is needed, and it must be interdisciplinary, to improve the integration of human work with intelligent equipment. They also believe that more effort is needed to integrate different practices into OSH initiatives, such as virtual task analysis, dynamic occupational risk assessment, cognitive workload analysis, and skills management tools.

Castelo-Branco, Cruz-Jesus and Oliveira (2019) consider that the development of Industry 4.0 requires super-skilled personnel, which poses challenges for technical schools and companies, which must join forces to prepare the new operators of the "factory of the future". The issue of skills in the context of Industry 4.0 is also highlighted by Belinski et al. (2020); Felsberger and Reiner (2020); Matt et al. (2020); Mkhize and Parumasur (2022) always with the same underlying idea - skills are decisive in the implementation of T4.0.

2.4 Public Administration, Trust, and Legitimacy

Trust is a relevant dimension in organizational behavior. Tyler (2003) considers that managers are concerned with promoting desirable behavior and limiting negative and undesirable behavior. Trust is a determining factor in this process and plays an important role in institutions.

Public administration and governance serve the interests of citizens and therefore, although they are organizations, the issue of trust is framed differently in public administration and government compared to other organizations. Administration involves public servants, citizens, and elected representatives, and these three parties must establish relationships of trust. Irastorza, Crespo and Bouckaert (2022) acknowledge that there is little research on intra-organizational trust (civil servants and between politicians) and inter-organizational trust (between the former and society's organizations). However, how the public administration carries out its tasks is not irrelevant to establishing this trust.

Woodrow Wilson's works are often referred to as the founders of public administration as a field of knowledge. In the work most often referred to in this context - Wilson (1987) - the author defends an efficiency-oriented administration with a specific legitimacy that derives from the technical mastery of matters.

In addition, some authors consider that it is the essence of the bureaucratic model of public administration, legitimized as an administrative function of the state, that enables this large organization to be managed, allowing it to be efficient (e.g. Fatile, Olojede, & Adejuwon (2015)). In this sense, Rutgers (2000) concludes that the public administration should be conceptualized as having its legitimacy as an administrative function of the state or, in other words, it is the very construction of the administrative function by the state, which is guaranteed by the public administration, that gives it its legitimacy.

3 Methodology

In recent decades, systematic literature reviews (SLRs) have established themselves as a very useful research technique. Normally, the situations that are assessed using SLRs are in areas of great novelty in terms of knowledge creation. New themes, new approaches, new concepts, whose development is still very recent, and which still need to be systematized. Choudhury and Asan (2020), Fernández-Ordóñez et al. (2019), and McKinnel et al. (2019) are examples of SLR applications in the field of artificial intelligence that fit the above description. This type of study often gives rise to research agendas (e.g. Belinski et al., 2020; Centobelli et al., 2020; Liao et al., 2017). SLRs are thus a type of study that seeks to systematize knowledge and are more common in areas that, because they are more recent, still lack systematization.

Lopes (2022) mentions that at the end of the 20th century, there was an effort to standardize protocols for performing SLRs. The QUOROM protocol, published in 1999, was followed in 2009 by the first version of the PRISMA protocol (see the description of the protocol in Liberati et al. (2009) or Moher et al. (2009).

The PRISMA protocol (http://www.prisma-statement.org/), although originally established in health sciences, has been increasingly used in other areas of knowledge, having become popular due to its very frequent use. The protocol was, however, revised in 2020 (see Page et al. (2021), for a description of the latest version).

3.1 Data

The articles in the study were selected using the bibliometric database Scopus. Many studies have used this database as a platform for article selection, namely Centobelli et al. (2020); Masae, Glock and Grosse (2020); Elnadi and Abdallah (2024) Agazu and Kero (2024).

The study was carried out by evaluating two samples of articles. The first sample consisted of a selection of articles obtained from the following query:

((TITLE-ABS-KEY("public administration" OR "civil servant") AND TITLE-ABS-KEY("artificial intelligence") AND ALL("human resource")) AND (LIMIT-TO (SRCTYPE, "j")).

i.e. documents that had,

- The words "public administration" OR "civil servant" in the Title, Abstract or Keywords, and.
- The words "artificial intelligence" in the Title, Summary or Keywords, and.
- The expression "Human Resource" is throughout the text.

- The search focused only on documents published in academic journals.

The search resulted in a set of 11 documents which make up Sample 1 of the study, its core sample; for the rest of the article, it will be identified as ***AP_AI_HRM***.

As this sample is small, it was assessed against a second sample obtained from the following query key:

((TITLE-ABS-KEY("Industry 4.0") AND TITLE-ABS-KEY("artificial intelligence") AND ALL("human resource")) AND (LIMIT-TO (SRCTYPE, "j")).

i.e. documents that had,

- In the Title, Abstract, or Keywords the words "Industry 4.0", and.
- "Artificial intelligence" in the Title, Abstract or Keywords, and.
- "Human Resource" throughout the text.

Given that this set of articles is much larger in number than the ***AP_AI_HRM*** sample, it was decided to consider a subset limited to the most cited articles. The h-index was used as a threshold for selecting the articles to be included in the sample. This index is established from a list of documents, ordering them in descending order by the number of citations. The index is established by the order number of the last document whose number of citations is greater than or equal to the order number. Coombes and Nicholson (2013) and Sánchez-Rebull, Niñerola and Hernández-Lara (2023) are examples of SLRs that used h-index as a selection and analysis criterion.

For the set of documents analyzed in this study, the h-index is 35 (thirty-five documents with thirty-five or more citations), so the sample for comparison with the AP_AI_HRM sample was made up of these 35 articles. This second sample is identified in this study as ***I4.0_AI_HRM*** as it is made up of articles whose specific selection factor was that they addressed the topic of Industry 4.0.

4 Empirical Study

4.1 Characterization of the Two Samples

Both samples show a very recent interest in the subject. This interest has given rise to a significant volume of annual production since 2019 for the I4.0_AI_HRM sample, while for ***AP_AI_HRM*** a more significant interest is only seen in 2023. As the ***I4.0_AI_HRM*** sample only includes the most cited articles, it tends to "forget" more recent articles that have not yet had enough time to impact the literature. For this reason, Table 1 includes information on the complete set of articles from which the ***I4.0_AI_HRM*** sample was drawn (the last two columns of the table). This confirms that the topic is very recent, with more than 50% of the articles being published after 2021.

Table 1. Number of articles by year in both samples.

YEAR	AP_AI_HRM		I4.0_AI_HRM		Complete set ofI4.0_AI_HRM articles	
2024				9%	19	12%
2023	5	45%	3	34%	44	29%
2022	2	18%	12	17%	40	26%
2021	1	9%	6	20%	19	12%
2020	2	18%	7	14%	16	10%
2019	1	9%	5	6%	13	8%
2018		0%	2		2	1%
Total	11		35		153	

The journal that published the most articles on the subject in the ***AP_AI_HRM*** sample was Revista Catalana de Dret Public, which included two articles. Nine other journals published one article each. The publications are mostly in the public administration and governance area, with six articles from five different publications: Revista Catalana de Dret Public, Administrative Sciences, Government Information Quarterly, Public Personnel Management, Transylvanian Review of Administrative Sciences. Apart from this area, no other stands out specifically. Table 2 shows these results.

Table 2. Number of articles by source year sample AP_AI_HRM.

Source title (AP_AI_HRM)	#
Revista Catalana de Dret Public	2
Administrative Sciences	1
Frontiers in Human Dynamics	1
Government Information Quarterly	1
Humanities and Social Sciences Communications	1
International Journal of Environmental Research and Public Health	1
Journal of Intelligent Information Systems	1
Public Personnel Management	1
TQM Journal	1
Transylvanian Review of Administrative Sciences	1

Regarding sample I4.0_AI_HRM, the sources are completely different, with only Administrative Sciences having one article in both samples. In this second sample, five journals repeated the publication of articles (Technological Forecasting and Social Change, Journal of Engineering United Kingdom, Journal of Industrial Integration and Management, Journal of Self Governance and Management Economics, and Operations

Management Research) and twenty-one other journals published just one article. In this set of publications, those related to technology, and production and operations stand out, a theme in which seventeen of the 26 journals considered can be classified. In a second relevant area, which includes articles on economics and management, we can find four journals (Journal of Self Governance and Management Economics, International Journal of Entrepreneurial Behavior and Research, Economics Management and Financial Markets, and Human Resource Management International Digest). These results can be seen in Table 3.

Table 3. Number of articles by source year sample AP_AI_HRM.

Source Title (I4.0_AI_HRM)	#
Technological Forecasting and Social Change	6
Journal of Engineering United Kingdom	2
Journal of Industrial Integration and Management	2
Journal of Self Governance and Management Economics	2
Operations Management Research	2
AI and Society	1
Administrative Sciences	1
Automation in Construction	1
Economics Management and Financial Markets	1
Human Resource Management International Digest	1
Information Processing and Management	1
International Journal of Entrepreneurial Behaviour and Research	1
International Journal of Information Management	1
International Journal of Manpower	1
International Journal of Operations and Production Management	1
International Journal of Production Research	1
Internet of Things and Cyber Physical Systems	1
Journal of Cleaner Production	1
Journal of Enterprise Information Management	1
Journal of Environmental Management	1
Journal of Food Engineering	1
Journal of Industrial Information Integration	1
Measurement Journal of the International Measurement Confederation	1
Sustainability Switzerland	1
Sustainable Operations and Computers	1
Technovation	1

Regarding the geographical origin of the articles, the authors of the AP_AI_HRM sample come from eight different countries, while the authors of the I4.0_AI_HRM sample come from 38 different countries. The two lists are also different in terms of country typology, so in the first list, the countries are mainly large and usually associated with the development of Industry 4.0. Mizukami and Nakano (2022) studied published literature on Industry 4.0 that related to AI, Big Data, and IoT, concluding from their study that it is mainly large, developed countries with a solid industrial base that dominates the literature.

If we exclude these countries from the two lists below, we are left with the ***AP_AI_HRM*** list:

- Finland, Romania, and the Russian Federation.

And in the I4.0_AI_HRM list:

- Chile, Croatia, Czech Republic, Ethiopia, Greece, Hong Kong, Indonesia, Ireland, Kenya, Malawi, Morocco, Norway, Portugal, Romania, Serbia, Singapore, Slovakia, South Africa, Uganda, Viet Nam, Zambia, and Zimbabwe.

As can be seen, the literature in the ***I4.0_AI_HRM*** sample is much more diverse than that in the ***AP_AI_HRM*** sample (Tables 4 and 5). Despite a dimension factor that differentiates the two samples and is certainly relevant, these results seem to indicate that the topic of human resources discussed in the context of Artificial Intelligence/Industry 4.0 has a production with much more diverse and distinct origins than its discussion in the context of Artificial Intelligence/Public Administration.

Table 4. Origin of the articles sample (AP_AI_HRM).

Country/Territory	#	Country/Territory	#
Spain	3	China	1
Italy	2	Finland	1
United States	2	Romania	1
		Russian Federation	1
		United Kingdom	1

4.2 Comparative Analysis

A comparative analysis of the content of the two samples was carried out by counting the incidence of words related to the theme of human resources, using MaxQDA software version 24.2. This software counted the words and the number of articles in which these words were present.

The results were exported to Excel and, using this software, the lexicon used was identified and the words related to the topic were grouped into five clusters:

Table 5. Origin of the articles (I4.0_AI_HRM)

Country/territory	#	Country/territory	#
India	12	Chile	1
United Kingdom	8	China	1
France	6	Croatia	1
Italy	5	Ethiopia	1
Malaysia	4	Greece	1
Romania	4	Hong Kong	1
Australia	3	Indonesia	1
United States	3	Ireland	1
Czech Republic	2	Malawi	1
Germany	2	Morocco	1
Iran	2	Pakistan	1
Kenya	2	Portugal	1
Norway	2	Serbia	1
Saudi Arabia	2	Singapore	1
Slovakia	2	South Africa	1
Uganda	2	South Korea	1
		Spain	1
		Taiwan	1
		Turkey	1
		Viet nam	1
		Zambia	1
		Zimbabwe	1

Source: Authors, with data from Scopus

- **Job and job environment**, which includes 16 terms: career; co-working; employee; employer; empowerment; Job; job-related; labor; labor-intensive; labour; sociotechnical; socio-technical; task; taylorism; teamwork.
- **Organizational behavior**, includes 9 terms: behavior; behavioral; behaviour; behavioural; demotivated; motivation; occupational; psychological; satisfaction.
- **Work, Worker, and workplace**, which includes 10 terms: work; worker; workflow; workforce; work-life; workload; workplace; work-related; workshop; workstation.
- **Skills, competencies, and talent**, which includes 16 terms: competence; competency; education; know-how; less-educated; lower-skilled, reskilling; skillful; skill; skilled; skillsets; talent; training; upskill; upskilling; vocational.
- **Government, citizens, and public administration,** which includes 7 terms: trust; citizen; citizenship; commitment; government; legitimacy; parliament.

For each of these terms, the relative presence in the lexicon was evaluated (the number of times the word was used divided by the total number of words used), and the percentage of articles in each sample in which each of the terms was used (Table 6).

Table 6. Comparing samples

	AP_AI_HRM > I4.0_AI_HRM	*I4.0_AI_HRM > AP_AI_HRM*
Job and job environment	7	23
Organizational behavior	6	12
Work, Worker, and workplace	2	18
Skills, competencies, and talent	6	26
Government, citizens, and public administration	10	4

5 Conclusions

Artificial intelligence is a significant challenge for governments and public administration. By asserting itself as a new reality, artificial intelligence will test the limits of states and their governments, and of public administration and the space for intervention and legitimacy in the actions of its officials. Predictably, their work will be altered both in terms of content and the limits of their intervention. But how public administrations relate to citizens and elected representatives is also likely to change. Issues such as trust in and of public servants, and how the actions of public administrations legitimize their actions or not are other relevant aspects.

All of this is very dependent on how the human resources of the public administration actually work, and this study is about to be carried out. This is the main conclusion of this study.

We, therefore, propose a research agenda for the study of the impacts of artificial intelligence on public administration work, focusing on the content of the work and its framework, without neglecting the more structural questions associated with issues such as rights, freedoms, and guarantees.

This work naturally has limitations, mainly those arising from the limited literature. This has led to methodological choices, which themselves limit the generalizability of the conclusions.

References

Agazu, B.G., Kero, C.A.: Innovation strategy and firm competitiveness: a systematic literature review. J. Innov. Entrepreneurship **13**(1), 24 (2024). https://doi.org/10.1186/s13731-024-00381-9

Ahi, A.A., Sinkovics, N., Shildibekov, Y., Sinkovics, R.R., Mehandjiev, N.: Advanced technologies and international business: a multidisciplinary analysis of the literature. Int. Bus. Rev. **31**(4), 101967 (2022)

Almada-Lobo, F.: The industry 4.0 revolution and the future of manufacturing execution systems (MES). J. Innov. Manage. **3**(4), 16–21 (2015)

Badri, A., Boudreau-Trudel, B., Souissi, A.S.: Occupational health and safety in the industry 4.0 era: a cause for major concern? Saf. Sci. **109**, 403–411 (2018)

Balubaid, M., Alsaadi, N.: Achieving sustainability in manufacturing through additive manufacturing: an analysis of its enablers. Sustainability **15**(12), 9504 (2023)

Bécue, A., Praça, I., Gama, J.: Artificial intelligence, cyber-threats and industry 4.0: challenges and opportunities. Artif. Intell. Rev. **54**(5), 3849–3886 (2021)

Belinski, R., Peixe, A.M., Frederico, G.F., Garza-Reyes, J.A.: Organizational learning and industry 4.0: findings from a systematic literature review and research agenda. Benchmarking Int. J. **27**(8), 2435–2457 (2020)

Castelo-Branco, I., Cruz-Jesus, F., Oliveira, T.: Assessing industry 4.0 readiness in manufacturing: evidence for the european union. Comput. Ind. **107**, 22–32 (2019)

Centobelli, P., Cerchione, R., Chiaroni, D., Del Vecchio, P., Urbinati, A.: Designing business models in circular economy: a systematic literature review and research agenda. Bus. Strateg. Environ. **29**(4), 1734–1749 (2020)

Choudhury, A., Asan, O.: Role of artificial intelligence in patient safety outcomes: systematic literature review. JMIR Med. Inform. **8**(7), e18599 (2020)

Coombes, P.H., Nicholson, J.D.: Business models and their relationship with marketing: a systematic literature review. Ind. Mark. Manage. **42**(5), 656–664 (2013)

Ellingrud, K., et al.: Generative AI and the future of work in America. McKinsey Center for Government (2023). https://www.mckinsey.com/mgi/our-research/generative-ai-and-the-future-of-work-in-america Accessed 27 Apr 2024

Elnadi, M., Abdallah, Y.O.: Industry 4.0: critical investigations and synthesis of key findings. Manage. Rev. Quart. **74**, 711–744 (2024). https://doi.org/10.1007/s11301-022-00314-4

Fatile, J.O., Olojede, I., Adejuwon, K.D.: Techno-bureaucratic governance and public service delivery: Indonesia and Nigeria in perspectives. Africa's Public Serv. Delivery Perform. Rev. **3**(3), 149–176 (2015)

Felsberger, A., Reiner, G.: Sustainable industry 4.0 in production and operations management: a systematic literature review. Sustainability **12**(19), 7982 (2020)

Fernández-Ordóñez, J.M., Jiménez, L.E.M., Torres-Carrión, P., Barba-Guamán, L., Rodríguez-Morales, G.: Experiencia afectiva usuario en ambientes con inteligencia artificial, sensores biométricos y/o recursos digitales accesibles: una revisión sistemática de literatura. Revista Ibérica de Sistemas e Tecnologias de Informação **35**, 35–53 (2019)

Freeman, C.: Innovation, changes of techno-economic paradigm and biological analogies in economics. Revue économique, 211–231 (1991)

García-Orosa, B.: Disinformation, social media, bots, and astroturfing: the fourth wave of digital democracy. Profesional de la información **30**(6), e300603 (2021). https://doi.org/10.3145/epi.2021.nov.03

Gutiérrez, O.Y.C., Cruz, Y.A.C.: Democracia en riesgo: la amenaza a las libertades políticas en la era de la inteligencia artificial. Justicia, 29(45) (2024)

Hazra, A., Adhikari, M., Amgoth, T., Srirama, S.N.: A comprehensive survey on interoperability for IIoT: taxonomy, standards, and future directions. ACM Comput. Surv. (CSUR) **55**(1), 1–35 (2021)

Hernández Peña, J.C.: Campañas electorales, big data y perfilado ideológico. Aproximación a su problemática desde el derecho fundamental a la protección de datos. Revista Española de Derecho Constitucional, 124, 41–73 (2022). https://doi.org/10.18042/cepc/redc.124.02

Hupfer, S.: Talent and workforce effects in the age of AI: insights from deloitte's state of AI in the Enterprise, 2nd Edition survey. Deloitte Development (2020). https://www2.deloitte.com/content/dam/insights/us/articles/6546_talent-and-workforce-effects-in-the-age-of-ai/DI_Talent-and-workforce-effects-in-the-age-of-AI.pdf. Accessed 27 Apr 2024

Ingaldi, M., Ulewicz, R.: Problems with the implementation of industry 4.0 in enterprises from the SME sector. Sustainability, 12(1), 217 (2019)

Irastorza, X.B., Crespo, M.J.C., Bouckaert, G.: ¿Qué mueve a confiar en la gobernanza colaborativa? análisis de un programa gubernamental en el País Vasco. Revista española de ciencia política **60**, 251–275 (2022). https://doi.org/10.21308/recp.60.09

Javaid, M., Haleem, A., Pratap Singh, R., Khan, S., Suman, R.: Sustainability 4.0 and its applications in the field of manufacturing. Internet Things Cyber-Phys. Syst. **2**, 82–90 (2022)

Kagermann, H., Wahlster, W., Helbig, J.: Recommendations for Implementing the Strategic Initiative Industrie 4.0. Acatech, Frankfurt (2013). https://www.din.de/blob/76902/e8cac883f42bf28536e7e8165993f1fd/recommendations-for-implementing-industry-4-0-data.pdf Access 23 Apr 2024

Kim, S., Wang, Y., Boon, C.: Sixty years of research on technology and human resource management: looking back and looking forward. Hum. Resour. Manage. **60**(1), 229–247 (2021)

Laffi, M., Boschma, R.: (2022). Does a local knowledge base in Industry 3.0 foster diversification in Industry 4.0 technologies? Evidence from European regions. Papers in Regional Science, 101(1), 5–35

Lemos, J., Gaspar, P.D., Lima, T.M.: Individual environmental risk assessment and management in industry 4.0: An IoT-based model. Appl. Syst. Inno. **5**(5), 88 (2022)

Liao, Y., Deschamps, F., Loures, E.D.F.R., Ramos, L.F.P.: Past, present and future of industry 4.0-a systematic literature review and research agenda proposal. Int. J. Produ. Res. **55**(12), 3609–3629 (2017)

Liberati, A., et al.: The PRISMA statement for reporting systematic reviews and meta-analyses of studies that evaluate health care interventions: explanation and elaboration. Annals Internal Med. **151**(4), W-65 (2009)

Lopes de Sousa Jabbour, A.B., Jabbour, C.J.C., Godinho Filho, M., Roubaud, D.: Industry 4.0 and the circular economy: a proposed research agenda and original roadmap for sustainable operations. Ann. Oper. Res. **270**(1–2), 273–286 (2018). https://doi.org/10.1007/s10479-018-2772-8

Lopes, J.D.: Industry 4.0 and the small business something behind the technology - a literature review. Serbian J. Manage. **17**(1), 161–178 (2022). https://doi.org/10.5937/sjm17-36382

Lopes, J.D., Estevão, J., Toth-Peter, A.: Industry 4.0, multinationals, and sustainable development: a bibliometric analysis. J. Cleaner Produ. **413**, 137381 (2023). https://doi.org/10.1016/j.jclepro.2023.137381

Mallinguh, E., Wasike, C., Zoltan, Z.: Technology acquisition and smes performance, the role of innovation, export and the perception of owner-managers. J. Risk Financ. Manage. **13**(11), 258 (2020)

Manheim, K., Kaplan, L.: Artificial intelligence: risks to privacy and democracy. Yale J. Law Technol. **21**, 106–188 (2019)

Masae, M., Glock, C.H., Grosse, E.H.: Order picker routing in warehouses: a systematic literature review. Int. J. Prod. Econ. **224**, 107564 (2020)

Matt, D.T., Orzes, G., Rauch, E., Dallasega, P.: Urban production–a socially sustainable factory concept to overcome shortcomings of qualified workers in smart SMEs. Comput. Ind. Eng. **139**, 105384 (2020)

Matyushenko, I.Y., Redko, N.: The assessment of Ukraine's readiness for innovations in the conditions of the spread of technologies of the new industrial revolution. Acta Innovations **33**, 5–19 (2019)

McKinnel, D.R., Dargahi, T., Dehghantanha, A., Choo, K.K.R.: A systematic literature review and meta-analysis on artificial intelligence in penetration testing and vulnerability assessment. Comput. Electr. Eng. **75**, 175–188 (2019)

Meziane, F., Vadera, S., Kobbacy, K., Proudlove, N.: Intelligent systems in manufacturing: current developments and future prospects. Integr. Manuf. Syst. **11**(4), 218–238 (2000)

Mizukami, Y., Nakano, J.: International comparison of cross-disciplinary integration in industry 4.0: a co-authorship analysis using academic literature databases. Plos one **17**(10), e0275306 (2022)

Mkhize, L.V.B., Parumasur, S.B.: Strategically aligning human resources skills in preparation for the fourth industrial revolution. Corp. Governance Organ. Behav. Rev. **6**(4), 233–245 (2022)

Moher, D., Liberati, A., Tetzlaff, J., Altman, D.G.: Prisma group preferred reporting items for systematic reviews and meta-analyses: the PRISMA statement. PLoS Med. **6**(7), 1000097 (2009)

Mühl, D.D., de Oliveira, L.: A bibliometric and thematic approach to agriculture 4.0. Heliyon **8**(5) (2022)

Muhuri, P.K., Shukla, A.K., Abraham, A.: Industry 4.0: a bibliometric analysis and detailed overview. Eng. Appl. Artif. Intell. **78**, 218–235 (2019)

Nübler, I.: New technologies: a jobless future or golden age of job creation?. Research Department working paper, No. 13. Geneva: International Labour Office. ISSN: 2413–4589 (2016)

Page, M.J., Mckenzie, J.E., Bossuyt, P.M., Boutron, I., Hoffmann, T.C., Mulrow, C.D., et al.: The PRISMA 2020 statement: an updated guideline for reporting systematic reviews. PLoS Med. **18**, 1003583 (2021)

Parida, B., Sunand Dash, S., Sharma, D.: Role of culture-specific rights, responsibilities and duties in industry 4.0: comparing Indic and Western perspectives. Benchmarking Int. J. **28**(5), 1543–1557 (2021)

Peres, R.S., Jia, X., Lee, J., Sun, K., Colombo, A.W., Barata, J.: Industrial artificial intelligence in industry 4.0-systematic review, challenges and outlook. IEEE Access **8**, 220121–220139 (2020)

Perez, C.: Technological revolutions and techno-economic paradigms. Camb. J. Econ. **34**(1), 185–202 (2010)

Posada, J., et al.: Visual computing as a key enabling technology for industrie 4.0 and industrial internet. IEEE Comput. Graph. Appl. **35**(2), 26–40 (2015)

Rutgers, M.R.: Public administration and the separation of powers in a cross-Atlantic perspective. Adm. Theory Praxis **22**(2), 287–308 (2000)

Sánchez-Rebull, M.V., Niñerola, A., Hernández-Lara, A.B.: After 30 years, what has happened to activity-based costing? a systematic literature review. SAGE Open **13**(2), 21582440231178784 (2023)

Stern, S., Daub, M., Klier, J., Wiesinger, A., Domeyer, A.: Public services, government 4.0 – the public sector in the Digital age. McKinsey & Company (2018). https://www.mckinsey.de/~/media/mckinsey/locations/europe%20and%20middle%20east/deutschland/publikationen/2018%20compendium/government%2040%20the%20public%20sector%20in%20the%20digital%20age/kompendium_04_ps.pdf

Strandhagen, J.O., Vallandingham, L.R., Fragapane, G., Strandhagen, J.W., Stangeland, A.B.H., Sharma, N.: Logistics 4.0 and emerging sustainable business models. Adv. Manufact. **5**, 359–369 (2017)

Tyler, T.R.: Trust within organisations. Pers. Rev. **32**(5), 556–568 (2003). https://doi.org/10.1108/00483480310488333

Vignesh Saravanan, K., Jothi Thilaga, P., Kavipriya, S., Vijayalakshmi, K.: Data protection and security enhancement in cyber-physical systems using AI and blockchain. In: Bharat, B., Arun Kumar, S., Tu N.N., (ed.) AI Models for Blockchain-Based Intelligent Networks in IoT

Systems: Concepts, Methodologies, Tools, and Applications, pp. 285–325. Cham: Springer International Publishing (2023)

WEF (World Economic Forum) (2016). Future of Jobs Report. Geneva, Switzerland: World Economic Forum. https://www3.weforum.org/docs/WEF_Future_of_Jobs.pdf. Accessed 27 Apr 2024

WEF (World Economic Forum) (2018). Future of Jobs Report. Geneva, Switzerland: World Economic Forum. ISBN 978–1–944835–18–7

WEF (World Economic Forum) (2020). Future of Jobs Report. Geneva, Switzerland: World Economic Forum. https://www3.weforum.org/docs/WEF_Future_of_Jobs_2020.pdf. Accessed 27 Apr 2024

WEF (World Economic Forum) (2023). Future of Jobs Report. Geneva, Switzerland: World Economic Forum. ISBN-13: 978–2–940631–96–4

Wilson, W.: The study of administration. Polit. Sci. Q. **2**(2), 197–222 (1887)

Yang, F., Gu, S.: Industry 4.0, a revolution that requires technology and national strategies. Complex Intell. Syst. **7**, 1311–1325 (2021)

Zhong, R.Y., Xu, X., Klotz, E., Newman, S.T.: Intelligent manufacturing in the context of industry 4.0: a review. Engineering **3**(5), 616–630 (2017)

Harnessing the Power of AI for Human Resources Development in the Digital Age: An Exploratory Study

Diogo da Silva(✉)

Associação Visionarium, Paredes, Portugal
contact@diogodasilva.eu

Abstract. Nowadays, in the world undergoing technological revolution, the emergence of Artificial Intelligence has presented companies with new opportunities to employ innovative strategies to enhance productivity. In order to improve Human Resources Development, application of Artificial Intelligence presents a vast array of its possible uses, but also raises certain concerns (Wilkens, 2020). There has been a need to establish the fears and opportunities that workers associate with the use of AI in order to consider its applications in a workplace (Su et. al, 2022).

This study aims to determine the current state of AI in public consciousness as and its possible uses. An online survey was performed on 100 participants representing different levels of company hierarchy. Furthermore, a focus group was formed in order to perform tasks without the usage of AI and repeat similar tasks (after training) with appropriate AI tools. The survey on the public's consciousness revealed that a considerable portion (68%) of work-age population is unaware of the possibilities offered by AI, and that a high percentage (79%) consider that their career will be affected by it. The focus group results reveal that after the initial training, most participants (82%) were comfortable employing AI in their daily tasks.

Keywords: Artificial Intelligence · Human Resource Development · technology · training · productivity · ethics and regulations

1 Introduction

In the current technology-driven and rapidly evolving workplace, the integration of artificial intelligence (AI) is pivotal for the development of human resources (HR). The transformative force of AI cannot be overstated, as it is imperative that companies employ active HR development strategies. In this new landscape the success of organizations and individuals lies on the intersection between human expertise and AI-enabled innovation. In this article we explore different applications of AI in HR development and their possible short- and long-term benefits.

This article aims to provide an overview of the current state of the art of Artificial Intelligence (AI) in Human Resource development (HRD) as well as the operationalization and effectiveness of five possible scenarios of AI application towards it.

M. J. Sousa (Ed.): UFHRD 2024, ISEM 52, pp. 51–70, 2025.
https://doi.org/10.1007/978-3-031-92548-1_4

Although the emergence of modern AI into the public consciousness is recent, the roots for its implementation in the business world have long been planted. Nevertheless, only recently have AI tools become widely available. The fast-paced product deployment from the industry, combined with a general lack of training, knowledge and protocols create a pathway for AI misuse, which can lead to security risks and abuse. In response to this, the European Union (EU) recently developed a set of rules governing the application of AI and established the European AI Office. These seek to guarantee, amongst other goals, the protection of employees and the way data is used. Therefore, this article will also focus on the limitations of AI usage in the workplace, how it can comply with EU law and its capacity for positive or negative social impact. As with any new technological revolution, society has yet to catch up to the potential of AI as an HR development tool, and as history tells us, new is often met with fear and distrust, specifically the fear of being replaced by new technology.

In order to determine the current state of AI in the public's consciousness, anonymous online survey data was collected. This resulted in the development of five AI usage scenarios. These scenarios were transformed into training solutions and then implemented within a focus group. The results were obtained via interview with the participants.

2 Literature Review

The following is a comprehensive literature review on the topics determined to be the most relevant when addressing the usage of Artificial Intelligence (AI) in Human Resource Development (HRD). The study of AI in an operable work tool is a recent one, as only starting in 2022 have Language Learning Models (LLM) become widely available for public use, and only now is the impact of such technology becoming observable. As we move into what can be called the "adoption phase" of AI in the workplace, it becomes fundamental to understand its mechanisms and how it can be used to further develop HRD strategies and methodologies.

2.1 Artificial Intelligence

Understanding AI

Over the last two years, Artificial Intelligence (AI) has been one of the most ubiquitous terms in tech. As of the time of writing this chapter, the term "AI" reached its highest popularity index on Google Trends (see chart below), which is not surprising, as Generative AI reached a fervour peak with the announcement of such tools as ChatGPT 5, Gemini and Sora.Subsequent paragraphs, however, are indented (Fig. 1) .

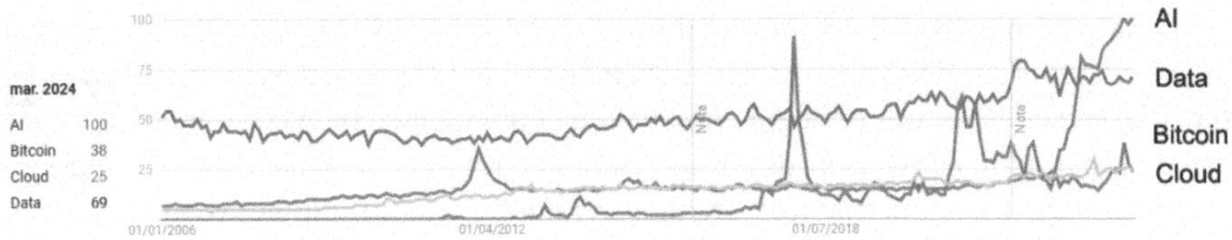

Fig. 1. *Interest over time (via Google Searches) from 2006 to 2024.* This graph represents the search interest in following terms: AI, XR, Bitcoin, Cloud and Data. The numbers represent the search interest relative to the highest point on the graph for the specified region and time range. A value of 100 is the peak popularity of the term. A value of 50 means that the term has had half the popularity. A score of 0 means that there was not enough data for this term. Google Trends (n.d.). " AI, Bitcoin, Cloud, Data". Retrieved May 9th, 2024, from https://trends.google.com/trends/explore?date=2006-01-01%202024-05-09&q=AI,Bitcoin,Cloud,Data&hl=pt-PT.

In order to validly approach the usage of AI in the context of Human Resource Development, it's imperative to find an understanding of the technology and its applicability. As such, the question "What is AI?" becomes relevant.

It is undeniable that AI has become fully present in the communal mindset. Its encounter is almost unavoidable and its implementation has become transversal to many products and services. It can be observed in nearly every subject of professional and personal life, either by allowing people to generate content or by working in the background and allowing daily life to continue to run its course. By definition, it comprises a series of processes, logical and mathematical, that aim to mimic the nature of the human brain, allowing connections to be formed between content, subject matter and data (Chowdhary, 2020). Just before the turn of the millennium, when there had been still a discussion between calling it Computational Intelligence or just Artificial Intelligence, it was already defined as the methods through which software or hardware would think, act and learn as humans do. Unlike simple automation, it operates through (machine) learning, as it obtains new data through the errors of previous iterations and establishes neural pathways between subjects, as humans do (Poole et. al, 1998).

AI and the Workplace

As artificial intelligence is lauded as a tool for facilitating and expediting tasks, experimentation of its application in the workplace has been obviously increasing. Companies are always looking for faster productivity, and AI promises to deliver on that matter. That being said, AI should present, as Wilkens (2020) states, "complementarity to individual intelligence, which can lead to an advancement, especially in quality, accuracy and precision. Moreover, AI has a potential to support individual competence development as long as AI-based approaches are thoughtfully combined with non-AI approaches."

There is, of course, fear that AI will end jobs and some manifestations of this have already taken place. Some companies have begun laying-off workers or implementing hiring freezes, and although many shy away from linking these actions to AI, other openly state this (Fortune, 2024).

Replacing some workers with AI chatbots might seem like a tempting financial decision, but users have long been drawn away from endless loop chatbots and human interaction is still a key factor in solving customer service problems.

Instead of replacing the current workforce and downsizing, AI should be used as an empowering tool. In 'Human + Robot', Daugherty (2018) poses the question "what are the workforce implications of being empowered or augmented by AI?", which they summarize by stating that "AI tools don't just automate routine workplace tasks—although they can do that, too—but that they create a symbiotic relationship between people and machines that upends the standard work flow".

Through AI tools workers are able to obtain data-driven insights as well as new baselines where to start their work from. An AI model can provide, for example, the basis for a social media copy and a generated image, which the worker can use in a company post, but the image will require work, personalization and the copy will need to be tailored to current company goals (HubSpot, 2023b).

These tools will eventually evolve to use more and more company data to become almost self-sufficient writers and designers, but that's not the current state of the art. AI allows humans to focus on tasks that are creative and require emotional intelligence, rather than mechanized and repetitive procedures. That implies a shift in company practices and a reimagining of procedures, where menial tasks are performed by AI and humans, working alongside it, shift to more originality and creative based processes (Daughtery, 2018; Wilkens, 2020).

Furthermore, while AI is seen as having the potential to replace some jobs, it is also important to note that it generates new opportunities and roles, specifically in areas that require decision making, creativity and handling of human interactions (HubSpot, 2023a). And when that time comes, companies need to be prepared with new roles, such as the AI trainer. The person in charge of predicting AI evolution and designing training solutions that will enable their peers to go beyond with new and more advanced tasks. They are already fundamental, though, to prevent users from being rogue AI users (Liebowitz, 1989).

The training process of users in the usage of AI is of the upmost importance, as the user must become the one in control, that is, AI needs to be driven by the users' intent and inputs. There are many traps that untrained workers will fall into, especially because of their lack of guidance and expectation of performance. Furthermore, companies must provide training in both the usage of AI and the application of AI delivered content in the company's ecosystem. Transforming a report into a presentation is easy with simple tools, but delivering engaging presentations and a comprehensive message is also fundamental. Workers need to be trained on both (Wilkens, 2020).

Fundamentally, the implementation of AI in a workplace setting should be complimentary to current and future internalized procedures. AI, no matter the tool used, is characterized by a non-human approach, which over time can erode the public perception on the company's commitment to community development and end-user satisfaction. On the onset of AI bots, some companies opted to disable their customer-service lines, replacing them with chatbots that often leave users with frustration, having no option to interact with a human in other way than an email which might take a few days to be replied to.

In order to mitigate the fear of being replaced that comes with all new technological revolutions, an investment has to be made into preparing workers for the usage of AI, which will both increase their productivity and provide them with a sense of security, a bolstering of their confidence and a sense that they belong. This, of course, reduces turnover rates and the need to train replacement employees.

"An essential element of a truly intelligent type of future of work, however, means that we do expand the workforce where both humans and machine will be part of, but with the aim to improve humanity and well-being while also being more efficient in the execution of our jobs." (De Cremer and Kasparov, 2021).

AI Ethics and Regulations

The capabilities of AI and its potential for misuse have become enough of a concern that international bodies are taking action to regulate how it can be used. There are specific cases of regulation that concern the context of this study, being the most prominent the fact that interviewees showed concern about their performance being evaluated using AI tools and the possibility of emotional surveillance.

In their narrative review of incidents and ethics with AI, Nasim et. al (2022) identified the matters of concern, as shown on Fig. 2, amongst 150 incidents from around the world.

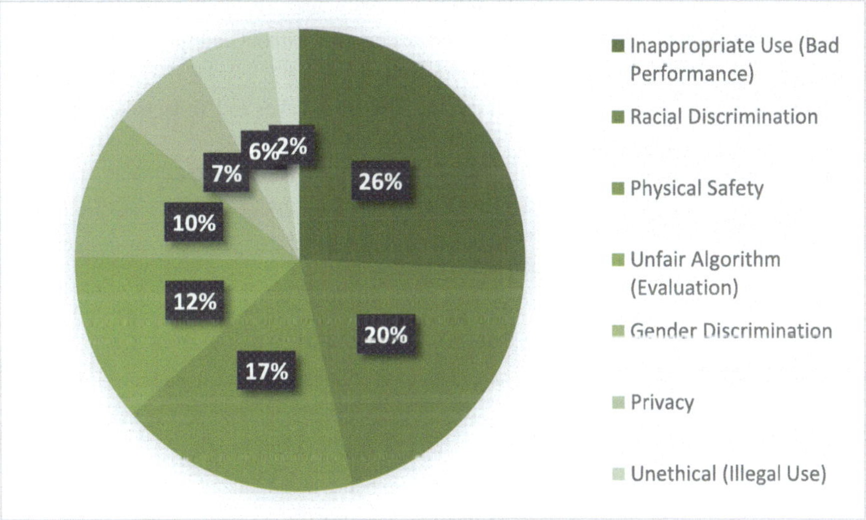

Fig. 2. AI ethical issues incidence between 2010 and 2021. This graph represents the frequency of AI ethics incidents from 2010 to 2021. The numbers the incidence percentage of each incident. Adapted from Nasim et. al (2022).

As observed in Fig. 2, the highest number of incidents concerns the inappropriate use of AI, which can stem from a number of events, but most concernedly for the context of this article, an example can be the usage of sensitive or protected data, which can impact a company's future or racial and gender discrimination, which can have an unintentional impact on the public's perception of the company, leading to erroneous conclusions.

In his 'Unite Paper: A Framework for Ethical AI at the United Nations', Hogenhout (2021) comprises a list of ethical problems. The following Table 1 offers a shortened overview of them.

Table 1. Shortened overview of Hogenhout (2021) ethical concerns.

Concern	Description
Incompetence	AI system failure resulting in physical or monetary damages
Loss of privacy	AI used to collect personal data from online sources
Discrimination	AI can perpetuate systemic group discrimination, affecting job opportunities or people's freedom
Bias	Models can use data that is biased and perpetuate that same bias
Erosion of Society	AI can be used to keep the user in their thought bubble, by offering a continuous stream of tailored content
Lack of transparency	AI processes running without auditing or "in the shadows" can erode trust
Deception	AI can produce Deep Fakes and Voice Fakes which can be used for deceptive objectives
Unintended consequences	An unregulated AI can take decisions that are not in conduct with social norms
Manipulation	AI can be used to gather data from public sources and target individuals
Lethal Autonomous Weapons (LAW)	AI can be used with autonomous weaponry and determine targets on its own
Malicious use of AI	AI can be used to support malware and hacking
Loss of Autonomy	Using AI to decide on personal freedoms and autonomy can erode mental health
Exclusion	AI can end up being monopolized, which can bring serious societal consequences

Note. This table shows a list of concerns put forward by the author of the article. The description has been abbreviated to facilitate reading. Adapted from Hogenhout, L. (2021)

Many models of AI have been trained on readily available online data, so their algorithms must be developed in order to avoid biases. As Hogenhout (2021) states: "The AI will only be as good as the data it is trained with. If the data contains bias (and much data does), then the AI will manifest that bias, too. If the AI is trained with data from a particular group only, it will start making the wrong assumptions".

But while it is expected that those protections are put in place, an employee that has been trained in ethics can spot when the model has produced biased content. Image generation models, for example, are very susceptible to scrutiny when generating content. To illustrate the data training on these and some security measures put in place, the following prompt has been requested to ChatGPT, Microsoft's CoPilot and to ideogram.ai (Fig. 3):

Fig. 3. Results of "Generate an image of an unmasked thief running from the police". This figure shows the results of the prompt "Generate an image of an unmasked thief running from the police" on both ChatGPT 4 and ideogram.ai. The images were generated on May 11th, 2024.

Microsoft's Copilot rejects the generation of images uncovered faces yet will allow the generation of images if the prompt is altered to "Generate an image of a thief running from the police". As it can observed from the images, a lot of the data gathered is set in scenarios alluring to streets in the United States of America. There is some racial variety with each model, but there are no female characters in any of the images, except for bystanders.

This anecdotal scenario shows how selective the "imagination" of an AI image generator can be. The AI algorithm resulted in the following:

- All thieves were male;
- All police officers were male;
- All streets depicted seem to be placed in the US;
- In one image, ChatGPT ignored the "unmasked" prompt.

With such a small pool of examples, it's unwarranted to get into other types of bias, such as racial depiction. That would be a topic for another article. That being said, this example denotes how two models produced scenarios with a single gender and where their data pool depicts criminality. This can occur because of the gender distribution in the United States of America when it comes to criminal/law enforcement agents (Statista, 2024, DataUsa, 2024).

There is, of course, a distinction to be made between AI ethics and AI regulations and this can be a muddled field. As Robles Carrillo (2020) states "there is widespread agreement on the need to endow AI with ethical principles. There is not the same concern or consensus on the importance of legal rules". As different regions of the world entered the foray of AI, it was observed that ethical concerns can vary. Ethics aren't rules, they are guidelines and there isn't as the author states, an "universal uniform ethical code". To determine what AI can be used for, laws and regulations are required.

"Ethics is necessary, even indispensable, but not sufficient to meet the challenge of AI. Ethics is specially needed when regulation is lacking. Law, however, is essential. Law implies a binding legal commitment, including for instance those ethical contents that are common and/or shared and therefore reach the statute of obligatory norms." (Robles Carrillo, 2020).

Such with the case of ethics, regulations on AI are not the same worldwide and while some regions/countries take more liberties with its usage in monitoring and surveillance of the population, the European Commission has taken steps to in order to prevent such situations. These scenarios are also behind some of the fears registered by interviewees, as an overreaching government or company administration has the potential to misuse Artificial Intelligence for control. Aware of the possible dangers that an unsupervised AI market presents, the European Commission, through the AI Act (EC, 2022), established the following as prohibitive practices:

The usage of AI systems to:

a) manipulate and impair the decision-making process of individuals;
b) exploit individual vulnerabilities based on age, disability or social-economical situation;
c) attribute a social score based on their behaviour;
d) profile an unsuspected individual to assess the risk of them committing a criminal offense;
e) create databases of untargeted facial images scraped from CCTV or the internet for purposes of facial recognition;
f) infer emotions of a person at work or in education institutions, unless for medical or safety reasons;
g) categorize individuals based on their biometric data, with exceptions related to law enforcement;
h) remotely identify individuals in public spaces, unless for safety and law enforcement concerns.

From the above list, the one that is directly related to this article is (f), which prohibits the usage of AI to gauge employee's emotional states and correlate them to their performance. This is done through computer vision and can be performed with rudimentary tools, which makes it an alluring for less scrupulous companies. Sinha et. al (2019) showcases a system that is able to, according to the article, detect emotions on students via CCTV with an accuracy of 83%. The employment of such a service, even if in benefit of the student population, carries some worrying privacy violations. The same could be said if this is applied to a company, where their employees' emotional status is being monitored and that information affects internal procedures such as contract renewal or promotions (EC, 2022).

Artificial Intelligence is bound to have a positive effect when applied consciously and when following ethical and regulatory guidelines. Its impact towards employee and company benefit is yet to be determined on a large scale, and without the possibility of backtracking to a place where it was not present, the best course of action is to make an appropriate use of it.

2.2 AI for Human Resource Development

Even after two years since its inception into the public's mindset, Artificial Intelligence is yet to become a tool that is accessible and regularly used by everyone. Although this statement holds itself on its own, it isn't exactly truthful, as many of the software and applications used daily are already operating with AI, "shadow AI", "invisible AI" or "background AI". These control the content people see, what route they should take to work and even tell them what to buy. These "background AI" processes are embedded in office productivitysoftware, language learning apps and, of course, in learning software. Before discussing the importance that AI HRD development can have for both companies and individuals, it is important to establish its current state of the art and remark on the profile that an employee must be encouraged to develop.

Profile of the AI Ready Employee

It would be a bold statement to say that there is a widespread lacking in qualified labour. If that same statement is addressed towards the field of Artificial Intelligence, then it might be easier to intake. This is supported by Thite (2020) who addresses it when discussing the external demands companies face, by stating: "The ageing population and lack of e-skills and digital fluency have contributed to serious skill shortages in the labour market. While many firms may therefore want to adopt global, rather than local or regional talent-sourcing policy to attract the digital workforce, government restrictions, political pressures and local employee resentment may restrict their ability to recruit foreign talent".

The reference to e-skills is perhaps the most relevant in the author's statement. It is quite difficult for an individual to alone keep up with a high-deployment technology such as Artificial Intelligence and guidance is required, not only so that the individual's company time is optimized, but also so that it complies with company and law guidelines.

When establishing the profile for an AI ready employee, it's essential that they undergo training that might not be directly correlated with Artificial Intelligence in the first place. This might sound contradictory, but the ever-evolving landscape requires, perhaps, more soft skills that previously imagined. Nowadays, there is a necessity for digital polyvalent and polyglot employees, individuals who are able to adapt to the aforementioned rapidly changing landscape and are able to intake new insights (Li, 2013). This, for many, doesn't come "out of the box". It has to be trained and it requires a healthy ecosystem for it to happen. Tools can be learned by following tutorials or in group sessions with a trainer, but these social and individual competences require specific scenarios (for example, team building activities), which mindful companies will happily invest in, as it provides a long-term return and increases retainability.

AI Usage in HRD, the State of the Art

To illustrate the state of the art in HRD, three studies were selected, each one pertaining to a different "field" of AI in HRD. These were selected on the basis of relevance to the current article.

1) Graßmann and Schermuly (2021)

In the exploration of AI's potential within organizational development, the research by Graßmann and Schermuly (2021) emerges as a pivotal contribution, providing deep insights into the application of AI in coaching roles and its significance for human resource development processes. Through the usage of AI mechanisms (machine learning, for example), the training tools can achieve a level of autonomy that enables the development of personalized approaches for each individual.

The authors are also quick to warn that while AI can automate replying procedures, for example through chatbots, "it [AI] currently cannot understand the clients' intention and explore what a client's core problem could be. There are too many options that are not well-defined, and even if it were possible, AI would demand large data sets for every option. Therefore, AI does not capture what clients did not know before. AI coaching may therefore only work well if the client is already aware of the core problem and simply needs to be prompted with questions to reflect upon it" (Graßmann and Schermuly, 2021). In a way, while AI can resolve some issues during the coaching and training process, it requires a training manager to determine the user's needs from the get to.

The integration of AI allows also for the into HRD programs allows also for the usage of diagnostic tools, as they would "fuel insights into the client's situation in a certain coaching process but would also professionalize coaching in general" (Graßmann and Schermuly, 2021).

2) Maity (2019)

The author illustrates a significant model aimed at forecasting the evolution of AI within HR training and development. The model underscores the growing significance of AI in enhancing knowledge management and employee training for organizations aiming to maintain a competitive edge. They defend that personalized training offers, which are easily deployable under the AI umbrella, have a more significant impact on developing the individuals' needs, unlike more traditional methods of learner grouping. And although training platforms such as these might incur in costs, they far outweigh the costs required to develop and deploy individualized training packages in the traditional sense. AI can determine the learning profile of a user, the fields they require investment in and develop training packages that are in line with their profile.

As an experimental example, a prompt was made to ChatGPT 4, for the purpose of illustrating the above statements (Table 2):

Table 2. Personalized AI training using ChatGPT 4

Prompt Create a personalized learning path for an entry-level employee. The subject is Artificial Intelligence and you must use readily available courses on Microsoft Learn. Create a roadmap with links to each course and a small justification of why the trainee should engage with each one
Result
Creating a learning path in Artificial Intelligence (AI) using Microsoft Learn resources is a great way for an entry-level employee to gain fundamental knowledge and practical skills. Here's a structured roadmap with courses that provide a comprehensive introduction to AI concepts, machine learning, and Azure AI services. Each course includes hands-on modules that build foundational skills relevant to modern AI applications
1. Microsoft Azure AI Fundamentals: AI Overview Link:https://learn.microsoft.com/en-us/training/modules/azure-ai-fundamentals/ Justification: This course introduces the core concepts of AI and how Azure services can be used to build AI solutions. It's designed for beginners and offers a broad understanding, making it an excellent starting point
2. Fundamentals of Machine Learning Link:https://learn.microsoft.com/en-us/training/modules/fundamentals-of-machine-learning/ Justification: Essential for understanding the underlying principles of AI. This course covers the types of machine learning—supervised, unsupervised, and reinforcement learning—which are crucial for grasping how algorithms improve
3. Introduction to Machine Learning for Beginners Link:https://learn.microsoft.com/en-us/training/modules/introduction-to-machine-learning-for-beginners/ Justification: Provides a hands-on introduction to classical machine learning. It's suitable for those new to the field and explores real-world applications
4. Microsoft Azure AI Fundamentals: Computer Vision Link:https://learn.microsoft.com/en-us/training/modules/azure-ai-fundamentals-computer-vision/ Justification: Focuses on computer vision, a key application of AI that enables machines to interpret the visual world. This course will be useful for understanding AI-driven image processing
5. Fundamentals of Azure OpenAI Service Link:https://learn.microsoft.com/en-us/training/modules/fundamentals-of-azure-openai-service/ Justification: Explores the capabilities of Azure OpenAI, including language and image generation models. This course is vital for understanding advanced AI applications and ethical AI practices
6. Machine Learning and AI with.NET for Beginners Link: Machine Learning and AI with.NET Justification: Tailored for.NET developers, this series integrates AI into.NET applications. It's perfect for understanding how AI can enhance software solutions

This table shows a list of training courses obtained via a prompt on ChatGPT 4. Requested and retrieved on May 11th, 2024.

As observed, a simple prompt without any specific platform other than an LLM resulted in developing a training pathway for a new employee to go through on their initial training. As no specific topic was given in the prompt, the results produced area generalized, with a focus on Azure and Machine Learning. Through the usage of proprietary tools or even in-house developed ones, these training packages can be personalized to the user's needs.

3) Chang Et. Al (2022)

It is not breaking news that AI can be used to determine a candidate's adequacy for a job. Whether the AI filtering practice of CVs is something is seen as more mechanical than human, the amount of information parsing that HR departments have to go through is enough of a justification. As such, new mechanisms for parsing information and finding the best candidates for positions is also something to be considered. Chang et. al (2022) propose a system using Machine Learning and Natural Language Processing that extracts data from different sources, which can, for example, be applied to CVs, university courses listed on them and online/internal job postings. The authors' algorithm performed a correlation analysis of these data sets and determined the skills lacking on potential future candidates, particularly in the area of soft skills. Soft skills have been a big talking point in various educational conferences over the past decades and perhaps a bigger investment on these might be required depending on the region and educational policy applied.

The Argument for Employees Learning About AI

Although HRD can benefit from AI when it comes to analytics, course development and individual-tailored content deployment, this article aims primarily to enlighten how AI can facilitate and improve employees' productivity, well-being and sense of accomplishment.

While that statement would work on its own, it is necessary to understand that AI training and usage can bring the following benefits:

- Employees spend less time on menial tasks, improving their overall productivity and becoming available to focus on more creative tasks.
- Employees become more technologically literate, being able to use transversal skills between platforms and software.
- Employees acquire skills in groundbreaking technologies, allowing them to quickly adapt to forthcoming updates and new generations.
- Employees sense of well-being and achievement can considerably increase, as they become more confident with the usage of tools that boos their productivity and, consequently, reduce the fear of becoming "dispensable".
- Employees can fact-check themselves and analyse their own productions for biases, providing them with the neutrality often expected.
- The company sees an increase in productivity, as their employees spend less time "reinventing the wheel".

- The company sees a return on training investment in both short and long term, as AI tools usually have small learning curves and as they become part of daily routine, the knowledge is easily retained.
- The company which implements ethical and regulatory training on AI application can expect their employees to safely and effectively use tools, avoiding "rogue use".
- The company sees a return on training investment in the form of employee retention.

3 Methodology

Data collected for this paper was gathered in the period between December 2023 and March 2024. An initial questionnaire survey, open online and limited to the first 100 form submissions, was used to gather information on the public's consciousness on the usage of AI and to define five scenarios of AI application in the workplace. The survey was delivered via email, social networks and posted on different social applications. To prevent the participants selection to be inside the author's interest bubble, which would lead to tendencies in the data obtained, the participation links were distributed by volunteers from different work fields, regions and ages. The criterion for the selection of these submissions was answering 0 or 1 on the scale of AI knowledge, 0 being "no experience" and 1 being "little to almost no experience". The link was deactivated when 100 valid submissions were obtained, which included the correct criterion and valid responses in the remaining form fields. From these, the 5 AI application scenarios were determined.

After the five scenarios were established, a focus group was put together, comprising of 17 individuals, chosen from a pool of online candidates with diverse backgrounds. The selection of the focus group was performed using the same randomization methodology as in the first survey (via social networks and through volunteer spreading) and was limited until all focus groups positions were filled. A second survey was performed because it was deemed essential that the new candidates were not part of the initial group of 100 form submissions. The second survey included the selection of one of the five scenarios determined by the first survey and each scenario had a limitation of either 3 or 4 candidates. The two scenarios which had the most responses in the first survey had 4 candidates, while the remaining ones had 3 candidates (see the scenarios section in 4. Data distribution).

The 17 individuals from the focus group were interviewed with a standardized series of questions about their opinion on AI and specific details about the task they selected (see 4). Particularly, the participants were asked to estimate the time they would take to perform the task. After the first interview, they were asked to participate in a MOOC style training with self-pace learning (two weeks) and perform the same task, now with the usage of AI. This time, they were asked to clock the time they took to perform the same task to the same standards.

All the surveys were answered after the participants agreed with the consent agreement and were made aware that they could stop at any time. It was also guaranteed that all data was confidential protected, and their identities would not be exposed.

All data gathered was stored on the author's personal computer, protected and encrypted, and will be destroyed after 5 years.

4 Data Distribution

In the initial survey, used to determine the public's consciousness about Artificial Intelligence, the questions present in Table 3 were presented to participants:

Table 3. Preliminary Questionnaire questions

Question	Answer format
Gender	Male/Female/Other
Age group	18 to 20, 21 to 30, 31 to 40, 41 to 50, Above 50
Current/previous job level	1 – Worker, 2 – Manager, 3 – Lower Admin, 4 – General Admin, 5 – Owner/President
Experience with AI	0 – No experience (never used) 1 – Little to almost no experience (tried a few times for entertainment purposes or curiosity) 2 – Some experience (have used it to produce some form of content) 3 – Good experience (use it regularly in work or other situations) 4 – Expert (familiar with the mechanics of language learning models, machine learning, etc.)
Can you imagine a specific way in which AI can benefit your job?	Yes No
Do you believe that AI will affect your career in the future, in any way?	Yes No
Please provide a task that you find menial in your career (previous, current or future job)	Open-ended question

Note. This table shows the questions provided to participants in an online survey, in order to gauge the public consciousness about Artificial Intelligence in the workplace. Data collected between December 2023 and January 2024

From the 100 questionnaire submissions, the following five AI scenarios were selected (highest number of occurrences and possibility cross-referenced with the possibility of training execution during the allocated time window):

(a) Productivity boosting with AI powered tools (32 occurrences),
(b) (re)(up)skilling - enhancing workplace learning (19 occurrences),

(c) AI driven time-consuming task automation (12 occurrences),
(d) AI for personal and professional schedule and time management (7 occurrences),
(e) AI for mental wellbeing and morale (4 occurrences).

From this original data set, the gender distribution can be observed in Table 4, with 51 male participants, 43 female participants and 6 participants identifying as other gender.

Table 4. Gender distribution in the preliminary survey (100 valid applicants)

Gender	Applicants
Male	51
Female	43
Other	6

Note. This table shows the gender distribution of participants in the preliminary survey. The numbers on the right column correspond to the number of individuals that participated

This is followed by an age distribution in Table 5 with five different age groups considered:

Table 5. Age distribution in the preliminary survey (100 valid applicants)

Age group	Applicants
18–20	12
21–30	27
31–40	33
41–50	19
> 50	9

Note. This figure shows the age distribution of participants in the preliminary survey, the age groups are listed on the left column. The numbers on the right column (applicants) correspond to the number of individuals that participated

The following Table 6 and Fig. 4 show the distribution of the hierarchy and the level of AI experience, combining the data of questions 3 and 4 from Table 3.

The following Table 7 shows the distribution in regard to the questions "Can you imagine a specific way in which AI can benefit your job?" and "Do you believe that AI will affect your career in the future, in any way?":

As previously mentioned, the focus group was comprised of 17 individuals who responded to a secondary survey, similar to the first one, which determined their viability to participate in the study.

The following Table 8 is an example of the first interview, with the answers collected before the focus group training implementation:

Table 6. Cross-reference of hierarchical level with perceived AI experience

Level	Worker	Manager	Lower Admin	General Admin	President Owner
Level 0 - No experience	19	4	2	2	1
Level 1 - Almost no experience	22	4	2	3	2
Level 2 - Some experience	10	1	1	4	1
Level 3 - Good experience	8	2	2	1	2
Level 4 - Expert	3	2	1	1	0

Note. This table shows a cross-reference of the AI experience levels with the hierarchy level in their current or previous position. Each row depicts the number of participants on each level of AI expertise

The participants were given two weeks and the training packages, as well as a communication line in case of any doubt that might arise. Afterwards, a secondary interview was performed. The questions for which are shown below (Table 9).

The results of the post-training interviews are as described on Table 10.

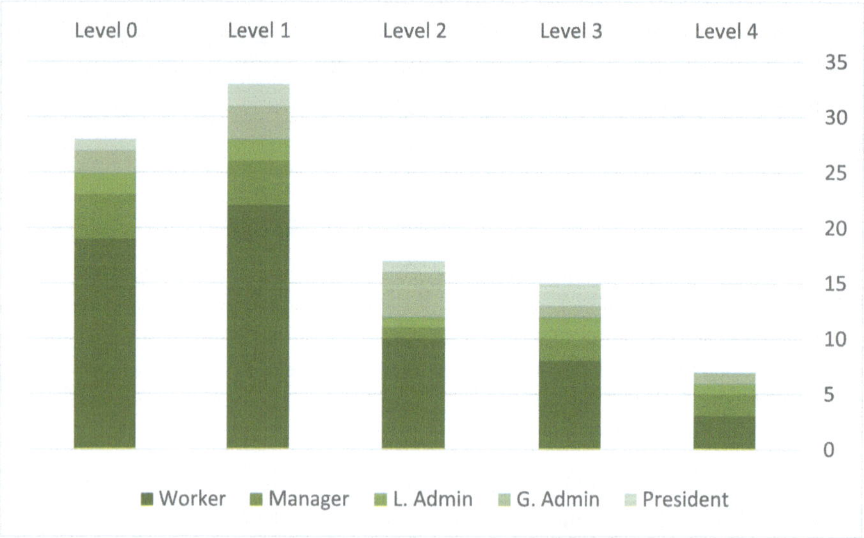

Fig. 4. Cross-reference of hierarchical level with perceived AI experience. Note. This figure shows a cross-reference of the AI experience levels with the hierarchy level in their current or previous position. Column represents the number of participants on each level (see Table 6), while the colour represents the position on the hierarchy (see Table 6).

Table 7. Opinion on AI influence and ability to change the workplace

Question	YES	NO
Can you imagine a specific way in which AI can benefit your job?	32	68
Do you believe that AI will affect your career in the future in any way?	79	21

Note. This shows the responses of participants to two questions regarding AI implementation in the workplace. All participants were required to answer either YES or NO

Table 8. Example of focus group pre-training interview

Question	Answer
What is your name?	Redacted
What is your current or latest job position?	"I am responsible for the production of social media content, producing training video content, preparing presentations, amongst other functions."
Are you familiar with the concept of AI?	"Yes"
Have you used AI during your practice? If so, how	"I haven't used it for work purposes. I used it to generate images and basically to see how it works."
Are you familiar with LLMs? If so, which ones	"Yes. I have used ChatGPT in the past."
Do you fear that your position might be replaced by AI in the future?	"If the software evolves to a point where it can be flawless, I think so. But I don't see anyone letting a computer decide what the company «says» online. It might get to a point that they'll just need someone to approve the content or make small changes."
State a normal task in your daily work routine. How long does it usually take you to complete it? Scenario (a) – Productivity boosting with AI powered tools	"Post creation. If it involves image collection, editing, the writing, etc., it usually takes me around 4 h to produce a post."

Note. This table shows the questions asked in the pre-training interview, as well as the answers from a participant. All participants answered these questions prior to receiving training

Table 9. Example of focus group post-training interview

Question	Answer
What is your name?	Redacted
Were you able to complete the training package content?	"Yes"
Did you find the contents of the training package to be useful for your practice?	"Yes"
Describe why, please	"It was easy to use and it allowed me to generate content very rapidly. If I was tasked with generating content that use generic images, this would be an ideal way. Sometimes, that isn't what I'm tasked with, but for this example it would fit perfectly."
Would you see yourself using the methodologies applied in the training package in the future?	"Yes."

Note. This table shows the questions asked in the post-training interview, as well as the answers from a participant. All participants answered these questions prior to receiving training

Table 10. Quantifiable results of the post-training interviews

Question	YES %	NO%
Were you able to complete the training package content?	100%	0%
Did you find the contents of the training package to be useful for your practice?	88%	12%
Would you see yourself using the methodologies applied in the training package in the future?	82%	18%

Note. This table shows the percentage of positive and negative answers asked in quantifiable post-training interview questions. All participants answered these questions after receiving and performing the tasks in the training package

5 Training Implementation

After the initial pre-training interviews, each member of the focus group was given a training package consisting of a roadmap document, multimedia tools and access (when required and not available freely) to necessary AI platforms.

The training packages were developed to be used in a MOOC style self-learning pace, with participants having a two-week period to complete them.

Participants were:

(a) instructed to follow the roadmap, which would guide them through a series of tutorials so that they could familiarize themselves with the AI tools to be used,
(b) asked to emulate training tasks which were described step-by-step in the training package,
(c) asked to create a task of their own accord, based on previously built task,
(d) challenged to replicate a task from their professional practice using their newly acquired knowledge.

Throughout the two-week training period, participants had access to a support email in case they needed support. The support was limited to access to platforms and other small issues, as to not compromise the self-paced nature of the learning process.

6 Discussion

As observed in 4. Data distribution, there was an overall positive effect when the focus group received appropriate guided training, in an area of their interest/need. As the training was delivered individually, participants were able to learn on their pace and focus only on the subjects of their concern, which greatly boosted their interest. From the focus group, when asked if the methodologies would be replicated in their future activity, 82% responded favourably. From the pool of participants, 12% found the training package not to be useful in their daily activities, but completed it successfully nonetheless.

There was also a significant reduction in the time required to perform the tasks, with all participants reporting shorter times (of at least 50% less) when using AI Tools to perform the tasks.

The two particular questions posed in the preliminary survey show that the public is not fully aware of the capabilities of AI, with 68 respondents stating they are not familiar with how the technology could benefit their work. When asked about the possibility of AI affecting their work in any way (positive or negative), 79 respondents addressed that possibility. Further studies on the positive or negative sentiment towards the technology could shed a light on how AI has been portrayed so far.

7 Conclusion

This study aimed to determine the effectiveness of AI training in a focus group with little to no experience in the usage of that technology, which resulted in an overall positive outcome. This supports the argument that HRD can benefit greatly from design and implementation of AI, not only for management purposes, but also for the development of the workforce. Amidst all the fears that come with this new technology, it is only through a symbiotic relationship between worker, company and HR that productivity can be boosted, wellbeing can be achieved and security can be assured.

References

Chang, X., Wang, B., Hui, B.: Towards an automatic approach for assessing program competencies, In: LAK22: 12th International Learning Analytics and Knowledge Conference. ACM (2022)
Chowdhary, K: Fundamentals of Artificial Intelligence. Springer India (2020)
DataUSA. Police Officer Gender and Age in the USA in 2021. DataUSA, Online (2024)
Daugherty, P.R.H., Wilson, J. (eds.): Human + Machine. Harvard Business Review Press, La Vergne (2018)
De Cremer, D., Kasparov, G: AI should augment human intelligence, not replace it. Harvard Business Rev. **18**(1) (2021)
Debabala, S., Prasant Kumar, P., Pradeep, K.G.: Machine learning and information processing. In: Proceedings of ICMLIP 2019. Springer Singapore (2020)
EC. 'The Artificial Intelligence Act', European Commission, Adapted from European Parliament legislative resolution of 13 March 2024 on the proposal for aregulation of the European Parliament and of the Council on laying down harmonisedrules on Artificial Intelligence (Artificial Intelligence Act) and amending certain UnionLegislative Acts (COM(2021)0206 – C9–0146/2021 – 2021/0106(COD)) (2024)
Fortune Over 4,000 workers have lost their jobs to AI since May, outplacement firm estimates—and that's 'certainly undercounting. Fortune, Article (2024)
Graßsmann, C., Schermuly, C.C.: Coaching with artificial intelligence: concepts and capabilities. Hum. Resour. Dev. Rev. **20**(1), 106–126 (2020)
Hogenhout, L.: 'Unite Paper | A Framework for Ethical AI at the United Nations', United Nations, Online (2021)
HubSpot. What Jobs Will AI Replace & Which are Safe in 2023. HubSpot, Online (2023a)
Hubspot. AI Trends for Marketers. Hubspot, Online (2023b)
Li, J.: Web-based technology and the changing landscape of HRD. Hum. Resour. Dev. Int. **16**(3), 247–250 (2013)
Liebowitz, J.: Artificial intelligence: new jobs from old. AI Soc. **3**(1), 61–70 (1989)
Maity, S.: Identifying opportunities for artificial intelligence in the evolution of training and development practices. J. Manage. Dev. **38**(8), 651–663 (2019)
Nasim Syeda Faiza, M.R.A., Kulsoom, U.: Artificial intelligence incidents & ethics - a narrative review. Int. J. Technol. Innov. Manage. (IJTIM) **2**(2) (2022)
Poole, D.L., Mackworth, A.K., Goebel, R.: Computational intelligence. Oxford Univ. Press, New York (1998)
Robles Carrillo, M.: Artificial intelligence: from ethics to law. Telecommun. Policy **44**(6), 101937 (2020)
Sinha, S., Mishra, S.K., Bilgaiyan, S.: Emotion analysis to provide counseling to students fighting from depression and anxiety by using CCTV surveillance. Springer Singapore, pp. 81–94 (2020)
Statista. Number of arrests for property crimes in the United States in 2021, by type and gender. Statista, Online (2024)
Thite, M.: Digital human resource development: where are we? where should we go and how do we go there? Hum. Resour. Dev. Int. **25**(1), 87–103 (2020)
Wilkens, U.: Artificial intelligence in the workplace – a double-edged sword. Int. J. Inf. Learn. Technol. **37**(5), 253–265 (2020)
Su, Z., Togay, G., Côté, A.-M.: Artificial intelligence: a destructive and yet creative force in the skilled labour market. Hum. Resour. Dev. Int.Resour. Dev. Int. **24**(3), 341–352 (2020)

Digital Age and Hospitality Human Resources Strategy of Development. The Case of the Japanese and Chinese Customers

Paula Tavares de Carvalho[1,2,3](✉)

[1] ISEC Lisboa, Lisboa, Portugal
paula.carvalho@iseclisboa.pt
[2] Department of Hospitality Management, ISCTE-BRU-Business Research Unit, Lisboa, Portugal
[3] Department of Data Analytics, University Institute of Lisbon, Lisboa, Portugal

Abstract. Tools of the digital age can be a useful tool to discuss and improve customer satisfaction, service quality and consequently the hotels' strategy. Analyzing a hotel´s digital tool though the online travel agency: booking.com, the study uses a qualitative approach of content analysis, identifying the Japanese and Chinese customers 'perceptions, satisfaction, and dissatisfaction about luxury hotels attributes. Findings of content reviews analysis (568 Japanese customers reviews and 968 Chinese customers reviews) reveal that Japanese and Chinese customers are pleased differently concerning hotel attributes and it falls into six dimensions: location, facilities, room, breakfast, staff-service, finance. This study has theoretical implications, confirming that these six dimensions are applied to Portugal but also to other countries, however this study also confirms that these dimensions are seen differently from Japanese and Chinese perspectives and can be grouped into the balanced scorecard dimensions. There are practical implications as Asian Customers are not all the same, hotels strategy should take in consideration the customer country of origin, this study contributes with highlights to develop Hotels Human Resources skills to meet Japanese and Chinese Hotel attributes´ expectations.

Keywords: Japanese customers · Chinese customers · hospitality strategy · hotel attributes · customer satisfaction

1 Introduction

The impact of digital transformation can be found to lead to changes in consumer behavior, evaluation of services and products, and expectations (Zhang and Chen, 2023). Digital transformation has not only had a profound impact on all aspects of our lives, but it has also had a significant impact on human resource management (HRM) processes and roles (Zhang and Chen, 2023; Schmid and Pscherer, 2021).

In a digital age, it is important to use the information available online for anyone. Online agencies are getting impact and influence in the customers choices when they

M. J. Sousa (Ed.): UFHRD 2024, ISEM 52, pp. 71–90, 2025.
https://doi.org/10.1007/978-3-031-92548-1_5

are reserving a room in a hotel. Concerning online travel agencies, booking.com is taking the lead. For these reasons, this study uses the online hotel reservations channel, booking.com, to source customer reviews of the luxury hotel chains operating in Lisbon Metropolitan Area and literature review to confirm the theoretical dimensions and to help construct the theoretical framework. Analysis of online reviews has become an indispensable marketing tool for managers because their content provides key information to help them allocate resources more appropriately to enhance service and thus increase profits. Customer generated media can be used in several different ways, to suit whatever purpose hoteliers and scholars might have (Oliveira et al., 2020). While positive reviews have been shown to improve hotel revenue (Phillips et al., 2017), negative reviews could lead to a reduction in hotel bookings.

Exploring open innovation in hotel service from online reviews, can be identified innovative opportunities from perspective of customer affective needs (Wu et al., 2023).

Several studies have already been based on text analysis of on-line reviews (Wang and Kirilenko, 2021; Li et al., 2020; Gunasekar and Sudhakar, 2019; Hu et al., 2019; Xu, 2019; Liu et al., 2017; Berezina et al., 2016; Ekiz et al., 2012; Magnini et al., 2011), these studies covered satisfaction with hotel products and services and their overall impact on customer satisfaction. Guests of luxury hotels rate the hotels significantly higher than guests of midrange and budget hotels.

In this study, the main research problem resides in evaluating the mismatch between Japanese and Chinese tourists' expectations and their perceptions about the attributes of luxury hotels (tangible and intangible) operating in Portugal. This mismatch can contribute towards the dissatisfaction of Japanese and Chinese tourists and, consequently, to their not returning or recommending that others come to Portugal, thus compromising sustainable growth in these markets, which are known to be very profitable.

The following research questions should be answered:

a) What do Japanese and Chinese travelers consider positive and negative hotels' attributes of the luxury hotels operating in Portugal? Are the preferences of the Japanese and Chinese similar or different?
b) Is there room for improvement to achieve a match between Japanese and Chinese tourists' hotels attributes expectations and hotels attributes of the Portuguese luxury hotels?

The main objective is to highlight the differences between Japanese c Chinese tourists, contributing with insights to improve hotels management strategy and, consequently, both tangible hotel attributes (such as: features) and intangible hotel attributes (such as: service), improving the customers 'experiences at the hotels and consequently their return to the country, passing a positive word-to-mouth to friends and relatives.

2 Review of Literature

2.1 Customer Satisfaction and Service Quality

Hotel guests consider not only the economic value of staying in a particular hotel (price and quality) but also seek other values of a more affective nature, such as self- gratification, aesthetic pleasure, prestige, transaction, and hedonism. Customer satisfaction has a significant direct positive effect on customer loyalty (El-Adly, 2019). As cited in Crotts and Erdmann (2000, p.410), "the role of customer satisfaction in influencing repeat patronage and positive word of mouth is well-documented (Crotts, 1999; Augustyn and Ho, 1998; Kotler et al., 1998; Oppermann, 1998; Heskett et al., 1997)". Asian tourists hold stricter standards when it comes to employee–customer relationships and satisfaction-based loyalty; thus, extensive and culturally customized services are required to establish conducive relationships with Asian customers to enhance long-term competitiveness. To attract Asian tourists, hotel management should create positive perceptions of service quality. For example, implementing excellent customer service strategies by engaging personalized interactions and understanding the needs of customers can enhance relationship marketing initiatives, satisfaction drives loyalty (Fam et al., 2021). As mentioned in Lu et al. (2015), numerous quantitative studies have shown that service quality is a predictor of customer satisfaction (Bitner, 1992; Cronin and Taylor, 1992; Gonzalez et al., 2007). Perceived service quality is part of a broader concept of customer satisfaction and behavioural intentions incorporating customer loyalty and WOM communication (Prasad et al., 2014 cited in Berezina et al., 2016). The importance of Customer satisfaction to the hospitality sector is mainly based on the following aspects: Physical attributes and commodities of the hotel room and performance of hotel employees (Berezina et al., 2016; Ekiz et al., 2012). Hotels with a high focus on sustainable HR practices are likely to have more satisfied customers than hotels that have little focus on sustainable HR practices (Wikhman, 2019, Chand, 2010; Choi and Chu, 2000; Dhar, 2015; Tsaur et al., 2005, Tsaur and Lin, 2004), contented, well-trained employees are a competitive advantage (Ottenbacher, 2007). There is a connection between service quality and customer satisfaction (Berezina et al., 2016, Bharwani & Jauhari, 2013; Buttle, 1996; Crick & Spencer, 2011; Prentice, 2013; Torres & Kline, 2013; Yee et al., 2009; Johnson & Grier, 2013), service quality is determined by customer perception of the level of service delivery (Zeithaml et al., 2006). Service quality, however, goes beyond just efficiently satisfying needs. Technical service is moderately significant, and the importance of expressive service in memorable guest experiences is highlighted (Sipe and Testa, 2018). However, managers need to understand the importance not only of customer satisfaction but also need to take delight in directly driving customer loyalty. Managers need to understand customers' emotions (Kim et al., 2014). For example, to achieve excellent service and customer satisfaction, it is crucial for the Hilton group (the most valuable hotels chain in the world) to know their customers' preferences in advance.

2.2 Cross-Cultural Differences and Market Approach

Characteristics of national cultures should be taken into consideration when planning the tourist offers and marketing (Ozdemir and Yolal, 2017). Understanding travellers' preferences and satisfaction is essential to better adapting the offer of products and services, as well as to developing different marketing strategies for each segment (Ahani et al., 2019; Xu, 2019; Xu et al. 2017; Zhou et al., 2014). Differences in cultural values and rules of behaviour determine satisfaction, and the differences in perceptions of service determine interaction. Tourism marketers cannot rely on perceptions of service quality alone (anticipating and understanding needs, and being attentive), to generate Asian tourist satisfaction in inter- personal contact with western hosts. Ultimately, to ensure overall holiday satisfaction it is important for tourism marketers to use their knowledge of cultural differences in values and rules of behaviour to create service quality that satisfies perceptions. Marketers can best directly influence Asian tourist satisfaction through issues associated with cultural values (intellectualism, security, self-control) and rules of behaviour (proper terms of address, respect, and avoidance of public criticism) (Reisinger and Turner, 1998). Lu and Zhu (2019) conducted a study on the Difficulties and Countermeasures of the Hilton Hotel Group's Cross-Cultural Management in China, cultural differences lead to cultural conflicts and increase management costs. Customers, owing to their cultural differences, require different types and forms of services. Communication difficulties, even in English since we cannot assume that the Chinese or Japanese will clearly understand everything said in English. With inhouse training results not being good, in order to ensure the scale of training and technology release, the Group has set up a professional hotel management training school. It also cooperates with many colleges and universities to joint improve the calibre of management and service level of hotel staff by combining theory with practice. The lack of unified corporate values and comprehensive management talent, need to be addressed, and a corporate employee culture needs to be strengthened. Depending on a country's stage of development the travel priorities are different. For less developed nations (which includes Portugal) the priorities are (from 1-the most important to 3- the least important): 1-service 2- food 3-location, and for more-developed nations (which includes Japan): 1-location 2- service 3-food. Cleanliness is the most important for all (Magnini et al., 2011). Complaints about Hospitality fall mainly under the following categories: "rooms", "arrogant and/or clueless staff" and "failure to respond" (Ekiz et al., 2012). Since it is evident that cultural differences determine customers' expectations of service and its required standard, tourism marketers and service providers should respond to the culturally accepted standards of the markets they serve (Resinger and Turner, 1998). 2.3. Japanese versus Chinese The know-how concerning the regionally/culturally differentiated travel motivations and tourism behaviour of customers is crucial to understanding tourists' perceptions about satisfaction with service quality (Weiermair, 2000). National culture plays a significant role in shaping customer expectations of service performance. Hospitality managers devise effective differentiation strategies that will best suit their customers' preferences and interests and thus increase customer satisfaction and intention to return (Zhang et al., 2015). To better meet the needs of their global customers, some hospitality firms already tailor their service offer and approach to communication to suit their customers' cultural background. For example, Hilton Hotels and Resorts (the most valuable hotel chain in the world) launched their "Huan Ying" program (which means welcome in Mandarin) in 2011. Chinese customers can complete their check-in using their native language and enjoy room amenities and breakfast items commonly found at home. There was also their approach to the Japanese market, using colours and pictures to create a unique and tailor-made website in Japanese. Hilton Worldwide launched a new website for Japanese travelers. Timothy Soper, vice president, operations for Japan, Korea and Micronesia, Hilton Worldwide, said, in an interview to Japan Today in 2014 "Japan remains equally important globally to Hilton Worldwide for both inbound and outbound markets, and we are committed to having the best possible services available to support our Japanese guests. The new website significantly improves usability, strengthens our connection with Japanese travelers and enriches their experience with us." The website has an intuitive and stunning, visually rich design, featuring enticing full screen images, simplifying the processes for ease of understanding. When choosing a destination, price is more important for the Chinese than for the Japanese (Assiouras et al., 2015). Japanese tourists were not perceived to bargain and mostly paid the asking price (Pizam and Sussmann, 1995). The Japanese prefer a more personalized and proactive attitude from service providers to service based solely on efficiency and standardization (Zemke & Pullman, 2008, Lee and Zhao, 2003). There are significant regional differences among Europeans and Japanese tourists routinely express less satisfaction with their visit (Wang and Kirilenko, 2021).

3 Research Method

The main research problem resides in evaluating the mismatch between Japanese and Chinese tourists' expectation and the actual attributes (perceptions) of luxury hotels (tangible and intangible) available in Portugal. This mismatch can contribute towards the dissatisfaction of Japanese and Chinese tourists and, consequently, to their not returning or recommending that others come to Portugal, thus compromising sustainable growth in the Japanese and Chinese markets, which are known to be very profitable. Relevant information concerning hotel attributes (tangible and intangible) preferences of Japanese and Chinese hospitality customers, type of trip (business or leisure), travel type (couple, group, individual), average days stayed, and room type preferences will be collected. Positive and negative key words will be grouped into relevant hotel´s attributes. Concluding which are the differences and similarities between them.

3.1 Sample and Data

Portugal registered a record number of tourists in 2019. Lisbon, capital of Portugal, was selected as the primary location of this study, with focus in luxury hotels (5 and 4 stars), as 144.007 Japanese visited Portugal, 59% (85.032) of whom visited Lisbon. In the same year, 378.561 Chinese visited Portugal, with 66% (250.567) visiting Lisbon (Travel BI, 2019). The type of accommodation that 78% of all Japanese tourists and 81% of all Chinese tourists preferred was 5 stars and 4 stars hotels (named as luxury hotels). In total in 2019, they were announced at booking.com: 152 hotels (5 stars and 4 stars) out of a total of 166 located in the Lisbon Metropolitan area (hotels over 10 km from Lisbon city center registered few or zero Japanese hotels reviews, for this reason, they were excluded from the sample analysis). The sample analyses, 47 hotels: 15 are 5 stars hotels and 32 are 4 stars hotels, from the total, 31 are in the top 10 hotel chains in Atlas da Hotelaria (Deloitte, 2019) and 16 are from other groups or independent hotels chains, representing more than 50% of total Lisbon´s total Hotel's accommodation and representing more than 53% of Japanese comments (598 comments of a total of 1.128) and 47% of Chinese comments (968 comments of a total of 2.071). Through the analysis of these 1.566 comments, it has reached a stage where an additional comment was not adding value to the research.

3.2 Internal Validity

In relation to internal validity, it is crucial for this research to divide the reviews' content correctly into positive and negative categories. The researcher assumed consistency in customer opinion about the hotel (expressed via rating and comments at booking.com). The reviews have name of hotel observed, username of the reviewer, reviewer´s residence location, trip and visit type include different categories: business, leisure, couple, family, group or individual, date and nights of stay. The reviews have positive and negative comments and an overall rating of the stay in the hotel. Ratings on Booking.com are presented on a scale from 1 to 10: 1 is bad and 10 wonderful, below 6 is no satisfactory, 6+ = pleasant, 7+ = good, 8+ = very good, 9+ = wonderful.

3.3 Data Collection and Word Categorization

The objective of content analysis is to convert recorded "raw" phenomena into data, which treat in essentially a scientific manner so that a body of knowledge may be built up. In fact, the researcher who wishes to undertake a study using content analysis must deal with four methodological issues: selection of units of analysis, developing categories, sampling appropriate content, and checking reliability of coding (Stempel, 1989 and Prasad, 2008). Booking.com Customer Reviews and Collection had been made using the qualitative data analysis software "NVivo" and manually. Every single review was content analyzed and counted for all hotels considered in the hotel sample. Manual analyses were important to confirm some wording contextualization. It was followed the open coding, axial coding and selective coding steps to group positive and negative key words into relevant hotel´s attributes. Open Coding refers to the initial and line-by-line conceptualization of data collected. Its purpose is to identify a problem and propose a possible solution by way of coding all the written data. This would often lead to many fragmented concepts and incidents, which need to be coded, compared, merged, modified, and re-named. Data is then turned into small, discrete components, with key words selected from the booking.com reviews. Each discrete piece of data is then coded with a descriptive label. Each word related to the same subject was labelled with the same code. Axial coding is used to find connections and relationships between codes, to thus establish a set of codes and identify connections between them. Causal conditions are sought, as well as the context behind observations, and the consequences of phenomena. Codes are aggregated and condensed into broader categories. The broader categories that make connections between codes are then determined.

Selective coding is used to bring everything together in one overarching, core category that captures the essence of the research being selected and makes the connections between this overarching category and the rest of codes and data being identified (Charmaz, 2006). The reviews were in Japanese and Chinese wording, automatic google translation to Portuguese on booking.com was used, and the content confirmed by translation into English, too. From a total of 1.566 reviews: 598 Japanese customers and 968 Chinese customers. SPSS26 analysis of means and standard deviation for the respondent scores was done. This study was undertaken between 2018, July 31st and 2022, January 31st.

4 Discussion and Results

4.1 Booking.Com Analysis: Japanese Versus Chinese Customers

4.1.1 Descriptive Statistics

The characteristics of the Asian travelers (598 Japanese and 968 Chinese) staying at 47 Hotels, are as follows (Table 1):

Table 1. Trip Type

	Frequency Japanese	Percent Japanese	Frequency Chinese	Percent Chinese
Business	85	14.2	162	16.7
Leisure	469	78.4	755	78.0
No info	44	7.4	51	5.3
Total	598	100.0	968	100.0

In the case of Japanese and Chinese customers, 78,4% and 78,0% were travelling for leisure, with only 14% and 17% of trips related to business. The percentage of Chinese travelling for business is higher than for the Japanese.

Both, the Japanese and Chinese prefer to travel with others, either in a Couple, as a Family or in a Group, representing 75% and 86% of the total respectively. Somewhat surprisingly, the percentage of Japanese travelling alone is higher than for the Chinese, representing 25% of total. The Chinese in the sample prefer to travel as a group much more than the Japanese do. As shown in Table 2, at almost 60%, travelling as a family (or in a couple) is the first preference among the Japanese. Almost 50% of both the Japanese and Chinese chose the Double/Twin room type:

Table 2. Travel Type

	Frequency Japanese	Percent Japanese	Frequency Chinese	Percent Chinese
Couple	179	29.9	253	26.1
Family	172	28.8	307	31.7
Group	99	16.6	277	28.6
Individual	148	24.7	131	13.6
Total	598	100.0	968	100.0

More than 90% of Japanese and Chinese stayed 4 nights or fewer. More than 30% stayed only one night in one of Lisbon's Luxury Hotels in the sample (47 hotels). However, on average, the Japanese stayed 2.6 nights, and the Chinese stayed 2.4 nights in one of Lisbon's Luxury Hotels.

On average, the Japanese scored the luxury hotels lower than the Chinese did, with 5-star hotels being rated higher than 4-star hotels, by both Japanese and Chinese customers. For both, however, reviews for 4-star hotels represent over 69% of the 47 hotels in the sample, as shown in Table 3.

Table 3. Number and Score of Japanese and Chinese reviews per hotel star rating.

	Hotel Stars	N	Perc	Mean	Max. Score	Min. Score
Japanese	4	415	69%	8.2	10.0	3.0
	5	183	31%	8.5	10.0	3.0
Total		*598*	100%			
Chinese	4	735	76%	8.3	10.0	1.0
	5	233	24%	8.7	10.0	3.3
Total		*968*	100%			

4.1.2 Wording Collection and Grouping by Hotel Attributes

For the word content analysis of the booking.com reviews, the following were used: Open Coding, Axial Coding and Selective Coding, 1.566 comments made in online reviews on booking.com were analyzed: 598 from Japanese hotel customers and 968 from Chinese hotel customers who stayed in one of the 47 hotels in the sample (4- and 5-star hotels operating in Lisbon). Each Japanese and Chinese booking.com review was fully content analyzed. First, words repeated in the Japanese and Chinese comments on booking.com were collected and classified in the Open Coding Dimensions. Second, the Open Coding dimensions with similar characteristics were gathered in six groups (Axial Coding): 1. Location, 2. Facilities, 3. Room (Bedroom, Bathroom and Room), 4. Breakfast, 5. Staff/Service, 6. Price. Third, some connections between these categories that capture the sense of the study were found (Selective Coding), the dimensions can be grouped as Hotel: Features, Processes and Procedures, People (Interaction/Service) and Finance, a connection to the Balanced Scorecard dimension could be made. As summarized in Table 4.

4.1.3 Customers Comments by Dimensions

Some examples of comments by dimension are described below, the sample of 47 hotels were listed and hotels numbered, for confidential reasons hotels are mentioned by the number in the list. The reference to the Hotels Top Ten is related to the Deloitte report "Atlas da Hotelaria, 2019", where all hotels' chains operating in Portugal are listed and ranked.

Table 4. Data Collection and wording categorization

Open Code Dimensions	Axial Code Dimensions	Selective Code Dimensions
Transport Shops, Restaurants, Supermarket Tourist Attractions Conference Quiet, Safe	Location	Features Processes/Procedures People Finance
View Elevator, Hardware, Software Design, Decoration, Building Cleaning Hotel Restaurant, bar, Lounge Lobby Free Drinks Swimming-pool, Gym, Spa Parking	Facilities	
Cleaning Comfort Size Relax, Calm, quite Kettle, Fridge, Amenities Design, Bright Free drinks, food Soundproofing Equipment working, how to use	Bedroom	
Comfort Mattress Clothes	Bed	
Design Size Cleaning Bathtub Hot Water, Water Flow Towels Equipment working, how to use	Bathroom	
Tasty Diverse Vegetables Availability Cleaning Calm	Breakfast	

(continued)

Table 4. (*continued*)

Open Code Dimensions	Axial Code	Selective Code
	Dimensions	Dimensions
Nice, Kind, Friendly Helpful, Pro-Active Enthusiastic, Smiling Attitude Front-office	Staff/Service	
Price Value for Money Additional payment Payment/invoicing	Price	

The differences found between the Japanese and Chinese concerning the most important Hotel Attributes:

a) Facilities (Features)

a.1.) Historical Design

The location or design of a hotel or room are hard to change, so these items were classified as features.

Japanese appreciate Hotel Historical Design, also new and renovated. Chinese see Hotel Historical Design as old and outdated. Modern and New facilities are appreciated.

A Japanese tourist's comment about hotel nº 33 - 5 stars (Palace) – Portuguese Hotel Chain included in the Top 10 of hotels operating in Portugal:

"It is a hotel that makes you want to come again. A wonderful hotel with a historical feel. The equipment is old, but well maintained and I could use it. The interior is beautiful, and La Belle Epoch style salon on the first floor where you drink afternoon tea is attractive. The breakfast restaurant is also beautiful. The staff were very friendly, and I was able to stay comfortably in general" (Review rating 9.2, made in March 2019).

Chinese Comments for the same hotel and in the same season:

"Old setting needs refurbishment and remodelling of room features-hotel outdated" (Review rating of 5.4, made in May 2019).

b) Room

b.1.) Bathtub and Hot Water

Japanese Comments

For the Japanese the hot water and bathtub existence in the bathroom is very important and related to their cultural habits with the spring hot waters practices. After the sightseeing, they like to relax in their room enjoying a relaxed hot water bathtub bath, if there is no bathtub in the bathroom some negative comments can be found, such as:

"No bathtub, only shower" (Review made in June 2019, hotel nº20 – Hotel Chain -included in the Top 10).

"No bathtub and it was requested in advance" (Review made in February 2020, Hotel nº 30 – included in the Top 10 and and Review in January 2019- Hotel nº 42- Independent Hotel Chain).

Bathroom integrated in the room is not recommended to be allocated to the Japanese as they feel it as a lack of privacy, as some negative comments below:

"Bathroom with glass (no privacy)" (Review made in August 2019, Hotel nº 43 – Hotel Chain included in the top 20).

"Bathtub´s location (middle of the room)" (Review made in June 2019, Hotel nº 13 – Hotel Chain- included in the Top 10).

"The toilet and bathroom in the room were very open and I felt uncomfortable. I don't understand why they used glass" (Review made in January 2022, Hotel nº1 - Hotel Chain- included in the Top 10).

Specifically written as positive: "There is a bathtub, and the water is really hot, good for relaxation" (Reviews made in February 2019, Hotel nº10 - Hotel Chain- included in the Top 10).

Chinese Comments

No specific comments about bathtub existence or hot water for relaxation.

b.2.) Rooms Equipment usage

Room´s Equipment easy to use and working well is also very important for the *Japanese.*

"Bath tap is difficult to use" (Reviews made in August 2019, Hotel nº1 - Hotel Chain- included in the Top 10).

"Shower not working properly, no employees to help" (Reviews made in April 2019, Hotel nº9 - Hotel Chain – included in the Top 10).

"Difficulty to find room number, didn´t know how to use the hot water". (Reviews made in November 2019, Hotel nº14 - Hotel Chain- included in the Top 10).

Concerning *Chinese,* there aren´t comments about the difficulty to understand of use the room´s equipment, but there are concerning mal function:

"Bad bath drainage" (Reviews made in June 2019, Hotel nº13 - Hotel Chain- included in the Top 10).

"Drainage sink. Shower (easy to splash)". (Reviews made in October 2019, Hotel nº19 - Hotel Chain- included in the Top 10).

b.3.) Room Amenities

The existence of amenities and kettle to make tea are important for the *Japanese and Chinese.*

Lack amenities (hair conditioning, toothbrush). (Japanese Review made in June 2019, Hotel nº37 – Independent Hotel Chain).

No kettle (Japanese Review made in December 2018, Lisboa Plaza and December 2019, Hotel nº40 – Independent Hotel Chain).

"No 4 starts hotel. No slippers, toothbrush, toothpaste, tea bag." (Chinese Review made in December 2018 - Hotel nº31 - Hotel Chain- included in the Top 10).

"Need to request kettle." (Chinese Review made in August 2018, Hotel nº18 - Hotel Chain- included in the Top 10).

“No bathtub, it cannot be a 4 stars hotel” (Review made in February 2019, Hotel nº 40 – Independent Hotel Chain).

c. Breakfast

“Good food in spite of, lack of vegetables” (Japanese Review made in September 2018, Hotel nº8 - Hotel Chain- included in the Top 10 of hotels).

“Breakfast with lack of vegetables” (Japanese Review made in February 2019, Hotel nº14 - Hotel Chain- included in the Top 10).

“The breakfast variety is poor. There were no vegetables.” (Japanese Review made in August 2021, Hotel nº10 - Hotel Chain- included in the Top 10).

“Monotonous Breakfast” (Japanese Review made in May 2019, Hotel nº33 – Hotel Chain).

“Breakfast is the same every day” (Japanese Review made in August 2022, Hotel nº5 - Hotel Chain- included in the Top 10).

“Monotonous breakfast for 8 days. (Chinese Review made in July 2019, Hotel nº20 - Hotel Chain- included in the Top 10).

d. Staff

“Although my room wasn’t ready when I checked in, I felt the staff were attentive enough to not make me feel uncomfortable”. (Japanese Positive Review made in August 2021, Hotel nº 40 – Independent Chain Hotel). The staff are kind (Japanese Positive Review, January 2022, Hotel nº 9 - Hotel Chain- included in the Top 10).

“Staff slow response” (Japanese Negative Review, September 2019) Hotel nº15 - Hotel Chain- included in the Top 10).

“Rude Service” (Japanese Negative Review, September 2018, Hotel nº29 - Hotel Chain- included in the Top 10).

“Enthusiastic breakfast staff.” (Chinese Positive Review, August 2018, Hotel nº21 - Hotel Chain- included in the Top 10).

For the same hotel,

“Front office staff attitude (too serious, no smiling faces).” (Chinese Negative Review, December 2018, Hotel nº21 - Hotel Chain- included in the Top 10).

e. Price

“Price” (Chinese Negative Review, April 2019, Holiday Inn Lisbon and February 2020, Hotel nº34 – Independent Hotel Chain).

Did not understand the invoice (Japanese Negative Review, October 2019, Hotel nº42 - Independent Hotel Chain).

4.1.4 Dimensions Summary

While the Chinese consider location, room features, facilities, price, or value for money more important than the Japanese do, the Japanese are more concerned about service items, and they are much more specific than Chinese. When Japanese refers to Room, it relates to clear processes, the efficiency of room equipment and its ease of use, whether there is a bathtub, hot water, and bathroom privacy more than pure room´s features. The

Japanese focus more on service than the Chinese do and, in line with what Liu et al. (2017) found, the Chinese focus is on features.

In summary, Japanese focus more in processes and people and Chinese more in Features and Price as Fig. 1.

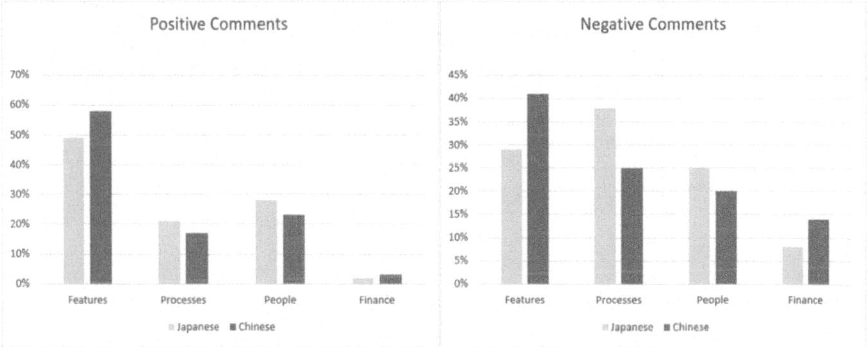

Fig. 1. Japanese and Chinese Hotels Comments Compared Adapted Balanced Scorecard Dimensions

The summary of differences and similarities concerning hotels attributes between Japanese and Chinese customers is in Table 5. Below.

Several improvements in processes and procedures could be made to delight the Japanese customers, for example, by introducing vegetables and Asian food at breakfast; providing clearer explanations of how to use the room equipment and verifying whether they are working properly before check-in; and clearly explained invoicing procedures and pricing. Staff could be trained to develop a better understanding of Asian culture and could learn, for instance, the importance of not only attributing a room with a bathtub to the Japanese, but one with bathroom privacy. Staff could be taught to be more aware of the elderly Asian's need for a kettle in which to boil water for tea, and the importance the Japanese attach to a hot bath. Another aspect of service that is important to the Japanese is the timely resolution of problems and clear, quick feedback on any issues they may have raised; it is important to bear in mind that the Japanese regard making excuses for any lack of service as impolite. Price, however, is not a big issue for the Japanese and they usually do not bargain for discounts, but it must be stressed that price and invoicing processes should be crystal clear. Concerning Chinese customers to be received by the staff with a smiling face demonstrating they enjoy what they are doing, it is important. Breakfast with Chinese food and room with soundproof and big dimension are also appreciated.

This research is in line with some other research such as that of Tsang and Ap (2007), which reports that Asians prefer the basic and practical aspects of service, giving high ratings for variables such as, "staff understanding your problems and needs", "dependable service" and responding effectively". Japanese customers focus more on service than Chinese customers do. In accordance with the findings of Liu et al. (2017) and Magnini et al. (2011), the focus of the Chinese is on room features and facilities. Peace-of-mind also plays a large role in customer satisfaction and has a significant

Table 5. Luxury Hotel Attributes differences and similarities between Japanese and Chinese customers

	Japanese	Common	Chinese
Location (near by)	Shops, Restaurants	Transports	Tourists Points
Facilities	Historical Design Appreciated	New and renovated	Historical Design - Outdated
Room	Bathroom with privacy Existence of Bathtub Peaceful Hot water flow Equipment easy to be used	Kettle Amenities Quiet and Soundproof Cleaning Equipment working well	The Dimension is important
Breakfast	Vegetables Peaceful	Diverse not monotonous Early Availability	Chinese food
Staff	Good information and advises about restaurants, transports…Excuse if a mistake occurs Proactive, solving problems	Kind, Friendly and Helpful	Enthusiastic and smiling
Price	Clear procedures about invoicing and pricing		The price itself

impact on loyalty and word-of- mouth recommendation, which means that providing hotel customers with a pleasurable, trouble-free experience not only satisfies them but also generates positive-word-mouth for the organization (Ko 2020; Lee 2019).

While the Chinese consider location, room features, facilities, price, or value for money more important than the Japanese do, the Japanese are more concerned about service items, clear processes, the efficiency of room equipment and its ease of use, whether there is a bathtub, hot water, and bathroom privacy. Comments on booking.com about "rude staff behavior" and the importance of taking the needs of older people into consideration such as, provision of a kettle to boil water in the hotel room are in line with Asian culture and the core virtue of Confucianism - filial piety - which leads to respect for the elderly. Young people in Asia are raised to respect the elderly. Scholars stress politeness as a critical dimension when Westerner's deal with the Japanese (Reisinger & Turner, 1998; Ziff-Levine, 1990), and have pointed out two major Japanese cultural attributes: one is a behavioral characteristic referred to as "polite inexplicitness," and the other is trust. To promise a service and not fulfil this promise is a big trust issue for the Japanese.

In comparison with the Chinese, the Japanese give more importance to a helpful employee attitude (apology for lacks or failure), cleanliness, the existence of bathtub, clear instructions on how to work equipment in the room.

The Chinese are more concerned with décor. They like modernity, with dated decor being considered negative and, in some cases, even in historic areas and renovated.

The Chinese also like smiling and enthusiastic employees (as this means, they are doing their job with happiness and joy) and give more importance to price/value for money. This is in line with the findings of Mohsin and Lockyer (2010), where the most important hotel attributes are the first contact with the hotel staff (front-office), helpful, friendly staff, room furnishings and value for money. Hospitality complaints mainly appear in the following categories: "rooms", "arrogant and/or clueless staff" and "failure to respond" (Ekiz et al., 2012). Room (such as room size, sleep quality), cleanliness, location (such as close to attractions and accessibility to public transport) and value (such as room price) are reasons for customer dissatisfaction with luxury 4- and 5-star hotels (Zhou et al., 2014, Li et al., 2020, Robbins et al., 2021). Hotel frontline employees play a key role in customer satisfaction and prompt customers to write positive online reviews. Regarding the two major categories of consumer complaints: room improvement and hotel staff attitudes, (Berezina, 2015), the Japanese focus more on service and the Chinese focus more on room features (Liu et al., 2017). The gap between Luxury Hotels Attributes versus Customer Expectations/Perceptions Gap can be summarized as Fig. 2. Below.

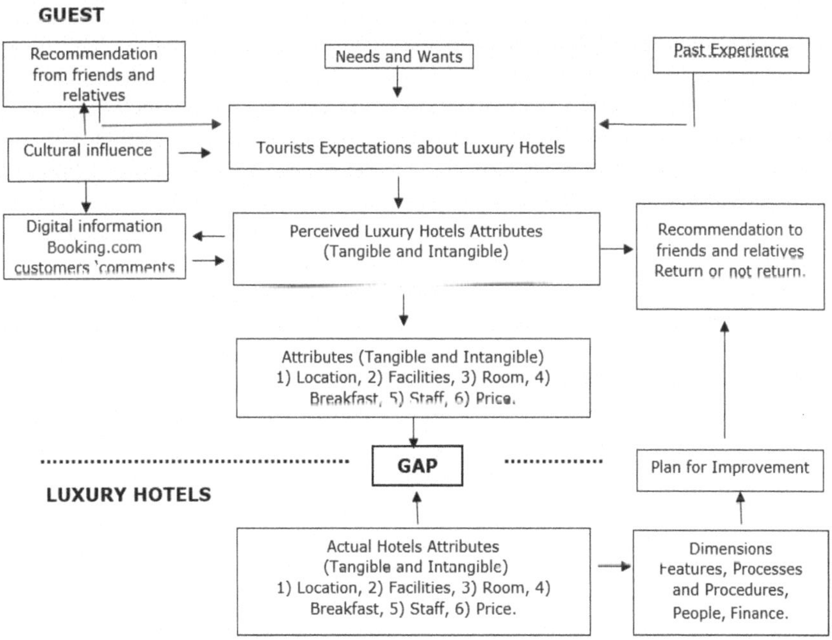

Fig. 2. Digital Information Analysis Luxury Hotels Attributes Versus Customer Expectations/Perceptions Gap

5 Conclusions

The research concludes that it is crucial for tourism and hospitality managers and employees to be aware of the influence that cultural differences have on tourists' perceptions and behavior. Hence the importance of designing cross-cultural training programs to minimize and avoid dissatisfaction arising among tourists. While these programs should focus on the specific preferences of major cultural groups, human resources managers and trainers will also need to identify which attributes of service and relational service are universal in nature in order to develop a better understanding of different guests' needs and expectations. (Tsang and Ap, 2007). Personal characteristics need to be considered when staging experiences for potential guests. Additionally, frontline employees should be trained and guided on how personal characteristics influence guests' expectations and behavior in the consumer experience process. Human interactions are a crucial part of their luxury hotel experiences. Four themes emerged as important factors affecting their hotel stay experience when human interactions with employees are considered, attitude, professional behavior, proactive service, and appearance (Wall et al., 2011).

5.1 Theoretical Implications

The six main Hotel Attributes (Tangible and Intangible): 1. Location; 2. Facilities; 3. Room (Bedroom, Bed and Bathroom); 4. Breakfast; 5. Staff / Service; 6. Price are aligned with similar attributes used in other studies of online review analysis such as that of Liu et al. (2017), Berezina et al. (2015), Zhang et al. (2015) and Magnini (2011). This study adds theoretical value, as these six attributes can be grouped as Features (attributes hard to change or improve in a short time such as, location and some facilities characteristics), Processes and Procedures, People and Finance. Concluding that Japanese are more concerned with hotels attributes linked to Processes and Procedures and People and Chinese are more concerned with attributes linked to features and finance.

5.2 Practical Implications

In Portugal, the Asian Market has hitherto been seen as predominantly Chinese, but what pleases Chinese customers differs from what pleases Japanese customers.

Using the digital tools as online travel agencies such as: booking.com, hotels can understand the customers´ perceptions and expectations, adapting their strategy and developing their human resources to improve service and customer satisfaction.

References

Ahani, A., Nilashi, M., Ibrahim, O., Sanzogni, L., Weaven, S.: Market segmentation and travel choice prediction in Spa hotels through TripAdvisor's online reviews. Int. J. Hosp. Manag. **80**, 52–77 (2019)

Assiouras, I., Skourtis, G., Koniordos, M., Giannopoulos, A.A.: Segmenting east asian tourists to greece by travel motivation. Asia Pac. J. Tourism Res. **20**(12), 1389–1410 (2015). https://doi.org/10.1080/10941665.2014.982140

Berezina, K., Bilgihan, A., Cobanoglu, C. and Okumus, F.: Understanding satisfied and dissatisfied hotel customers: text mining of online hotel reviews. J. Hosp. Market. Manage. **25**(1), 1–24 (2016). https://doi.org/10.1080/19368623.2015.983631

Bharwani, S., Jauhari, V.: An exploratory study of competencies required to co-create memorable customer experiences in the hospitality industry. Int. J. Contemp. Hosp. Manag. **25**(6), 823–843 (2013). https://doi.org/10.1108/IJCHM-05-2012-0065

Bitner, M.J.: Servicescapes: the impact of physical surroundings on customers and employees. J. Mark. **56**(2), 57–71 (1992)

Buttle, F., Bok, B.: Hotel marketing strategy and the theory of reasoned action. Int. J. Contemp. Hosp. Manag. **8**(3), 5 (1996). https://doi.org/10.1108/09596119610115943

Chand, M.: The impact of HRM practices on service quality, customer satisfaction and performance in the Indian hotel industry. Int. J. Hum. Resour. Manag. **21**(4), 551–566 (2010). https://doi.org/10.1080/09585191003612059

Charmaz, K.: Constructing grounded theory: A practical guide through qualitative analysis. Sage Publications, London (2006)

Choi, T.Y., Chu, R.: Levels of satisfaction among Asian and western travellers. Int. J. Qual. Reliabil. Manage. **17**(2), 116–132 (2000). https://doi.org/10.1108/02656710010304537

Crick, A.P., Spencer, A.: Hospitality quality: new directions and new challenges. Int. J. Contemp. Hosp. Manag. **23**(4), 463–478 (2011). https://doi.org/10.1108/09596111111129986

Cronin, J.J., Taylor, S.A.: Measuring quality: a re-examination and extension. J. Marketing **56**(3), 55–68 (1992). https://doi.org/10.1177/002224299205600304

Crotts, J.C., Erdmann, R.: Does national culture influence consumers' evaluation of travel services? a test of hofstede's model of cross-cultural differences. Manag. Serv. Qual. Int. J. **10**(6), 410–419 (2000). https://doi.org/10.1108/09604520010351167

Deloitte. Atlas da Hotelaria (2019)

Dhar, R.L.: Service quality and the training of employees: the mediating role of organizational commitment. Tour. Manage. **46**, 419–430 (2015). https://doi.org/10.1016/j.tourman.2014.08.001

Ekiz, E.H., Khoo-Lattimore, C., Memarzadeh, F.: Air the dirty laundry or hide the facts: The case of complaints about lodging services on travel-blogging sites. Tourism Manag. Perspect. **4**, 41–47 (2012)

El-Adly, M.I.: Modelling the relationship between hotel perceived value, customer satisfaction, and customer loyalty. J. Retail. Consumer Serv. **50**, 322–332 (2019). https://doi.org/10.1016/j.jretconser.2018.07.007

Fam, K S, Cheng, B.L., Cham, T-H, Yi, M.T.C., Ting, H.: The role of cultural differences in customer retention: evidence from the high-contact service industry. J. Hospit. Tourism Res. XX(X), 1–32 (2021). https://doi.org/10.1177/10963480211014944

González, M.E.A., Comesaña, L.R., Brea, J.A.F.: Assessing tourist behavioral intentions through perceived service quality and customer satisfaction. J. Bus. Res. **60**(2), 153–160 (2007). https://doi.org/10.1016/j.jbusres.2006.10.014

Gunasekar, S., Sudhakar, S.: How user-generated judgments of hotel attributes indicate guest satisfaction. J. Glob. Scholars Market. Sci. **29**(2), 180–195 (2019). https://doi.org/10.1080/21639159.2019.1577155

Heskett, J.L., Jones, T.O., Loveman, G.W., Sasser, W.E., Jr., Schlesinger, L.A.: The service profit chain: How leading companies link profit and growth to loyalty, satisfaction, and value. Free Press (1997)

Hu, N., Zhang, T., Gao, B., Bose, I.: What do hotel customers complain about? text analysis using structural topic model. Tour. Manage. **72**, 417–426 (2019). https://doi.org/10.1016/j.tourman.2019.01.002

Hu, Y., Liu, Y., Zhang, L., Zhu, D.: Exploring hotel service quality based on customer reviews via text mining. Sustainability **11**(6), 1516 (2019)

Johnson, G.D., Grier, S.A.: Understanding the influence of cross-cultural consumer-to-consumer interaction on consumer service satisfaction. J. Bus. Res. **66**(3), 306–313 (2013). https://doi.org/10.1016/j.jbusres.2011.08.010

Kim, H., Vogt, C.A., Knutson, B.J.: Relationships among customer satisfaction, delight, and loyalty in the hospitality industry. J. Hospital. Tourism Res. **39**(2), 170–197 (2014)

Kim, S., Elliot, S., Law, R., Chon, K.: Differences in expectations and perceptions between hospitality providers and international customers: the case of Korean and Japanese group tourists in thai hotels. Int. J. Hosp. Tour. Adm. **15**(2), 121–149 (2014). https://doi.org/10.1080/15256480.2014.901058

Ko, C.: Exploring hotel customer service experience. Int. J. Organ. Innov. **13**(2), 270–278 (2020)

Lee, B.Y., Park, S.Y.: The role of customer delight and customer equity for loyalty in upscale hotels. J. Hospital. Tourism Manage. **39**(April), 175–184 (2019) https://doi.org/10.1016/j.jhtm.2019.04.003

Kotler, P., Bowen, J.T., Makens, J.C.: Marketing for hospitality and tourism (2nd ed.). Prentice Hall (1998)

Lee, K., Zhao, J.: Japanese travelers' service preferences in U.S. hotels. J. Travel Tourism Market. **14**(2), 67–85 (2003). https://doi.org/10.1300/J073v14n02_04

Li, H., Liu, Y., Tan, C.W., Hu, F.: Comprehending customer satisfaction with hotels: data analysis of consumer-generated reviews. Int. J. Contemp. Hosp. Manag. **32**(5), 1713–1735 (2020). https://doi.org/10.1108/IJCHM-06-2019-0581

Li, Y., Xu, L., Tang, L., Wang, S., Li, L.: Big data and hotels: sentiment analysis of hotel customer online reviews. Int. J. Hosp. Manag. **89**, 102593 (2020)

Liu, B., Zhang, L., Guo, X., Hu, X.: A sentiment-enhanced hybrid recommender system for hotel recommendation. Inf. Manage. **54**(2), 276–288 (2017)

Liu, Y., Teichert, T., Rossi, M., Li, H., Hu, F.: Big data for big insights: Investigating language-specific drivers of hotel satisfaction with 412,784 user-generated reviews. Tour. Manage. **59**, 554–563 (2017). https://doi.org/10.1016/j.tourman.2016.08.012

Lu, C., Berchoux, C., Marek, M.W., Chen, B.: Service quality and customer satisfaction: qualitative research implications for luxury hotels. Int. J. Cultu. Tourism Hosp. Res. **9**(2), 168–182 (2015). https://doi.org/10.1108/IJCTHR-10-2014-0087

Lu, D., Zhu, W.: Study on the difficulties and countermeasures of hilton hotel group's cross-cultural management in China. Adv. Soc. Sci. Educ. Hum. Res. **315** (2019)

Magnini, V.P., Crotts, J.C., Zehrer, A.: Understanding customer delight: an application of travel blog analysis. J. Travel Res. **50**(5), 535–545 (2011). https://doi.org/10.1177/0047287510379162

Mohsin, A., Lockyer, T.: Customer perceptions of service quality in luxury hotels in New Delhi, India: an exploratory study. Int. J. Contemp. Hosp. Manag. **22**(2), 160–173 (2010). https://doi.org/10.1108/09596111011018160

Oliveira, A.S., Renda, A.I., Correia, M.B.: Online reviews: a pathway to improve hotel management. Dos Algarves: A Multidisciplinary E-J. **36**, 108–126 (2020). https://doi.org/10.18089/damej.2020.36.7

Oliveira, B., Alhinho, M., Correia, A., Silva, F.: Understanding luxury hotel customer experiences via online reviews: a text-mining approach. Tour. Hosp. Res. **20**(4), 440–455 (2020)

Oppermann, M.: Destination threshold potential and the law of repeat visitation. J. Travel Res. **37**(2), 131–137 (1998)

Ottenbacher, M.C.: Innovation management in the hospitality industry: different strategies for achieving success. J. Hosp. Tourism Res. **31**(4), 431–454 (2007). https://doi.org/10.1177/1096348007302352

Özdemir, C., Yolal, M.: Cross-cultural tourist behavior: an examination of tourists' behavior in guided tours. Tour. Hosp. Res. **17**(3), 314–324 (2017). https://doi.org/10.1177/1467358415589658

Phillips, P., Barnes, S., Zigan, K., Schegg, R.: Understanding the impact of online reviews on hotel performance: an empirical analysis. J. Travel Res. **56**(2), 235–249 (2017)

Pizam, A., Sussmann, S.: Does nationality affect tourist behavior? Ann. Tour. Res. **22**(4), 901–917 (1995). https://doi.org/10.1016/0160-7383(95)00023-5

Prasad, B.D.: Content analysis a method in social science research. Research methods for Social Work, New Delhi: Rawat, pp.173–193 (2008)

Prasad, K., Mahajan, V., Sai, L.P.: Customer satisfaction and loyalty in hotel industry: a review of literature. Int. J. Manage. Res. Rev. **4**(7), 682–695 (2014)

Prasad, K., Wirtz, P.W., Yu, L.: Measuring hotel guest satisfaction by using an online quality management system. J. Hosp. Market. Manag. **23**(4), 445–463 (2014). https://doi.org/10.1080/19368623.2013.805313

Prentice, K.: 'Service quality perceptions and customer loyalty in casinos. Int. J. Contemp. Hosp. Manag. **25**(1), 49–64 (2013)

Reisinger, Y., Turner, L.: Asian and western cultural differences. J. Int. Hosp. Leisure Tourism Manage. **1**(3), 21–35 (1998). https://doi.org/10.1300/j268v01n03_03

Robbins, R., Grandner, M., Knowlden, A., Severt, K.: Examining key hotel attributes for guest sleep and overall satisfaction. Tour. Hosp. Res. **21**(2), 144–155 (2021). https://doi.org/10.1177/1467358420961544

Schmid, Y., Pscherer, F.: Digital transformation affecting human resource activities: a mixed-methods approach. Hum. Inter. Emerg. Technol. Future Syst. 543–549 (2021)

Schmid, A., Pscherer, S.: Digital transformation in human resource management: a bibliometric and content analysis. J. Bus. Res. **131**, 300–316 (2021)

Sipe, L.J., Testa, M.R.: From Satisfied to memorable: an empirical study of service and experience dimensions on guest outcomes in the hospitality industry. J. Hosp. Market. Manag. **27**(2), 178–195 (2018). https://doi.org/10.1080/19368623.2017.1306820

Soper, T.: Hilton worldwide launches new website for Japanese travelers. Japan Today (2014)

Torres, E., Kline, S.: From customer satisfaction o customer delight: creating a new standard of service for the hotel industry. Int. J. Contemp. Hosp. Manag. **25**(5), 642–659 (2013)

Tsang, N.K.F., Ap, J.: Tourists' perceptions of relational quality service attributes: a cross-cultural study. J. Travel Res. **45**(3), 355–363 (2007). https://doi.org/10.1177/0047287506295911

Tsaur, S.H., Lin, Y.C.: Promoting service quality in tourist hotels: the role of HRM practices and service behavior. Tour. Manage. **25**(4), 471–481 (2004). https://doi.org/10.1016/S0261-5177(03)00117-1

Tsaur, S.H., Lin, C.T., Wu, C.S.: Cultural differences of service quality and behavioral intention in tourist hotels. J. Hosp. Leis. Mark. **13**(1), 41–63 (2005). https://doi.org/10.1300/J150v13n01_04

Walls, A., Okumus, F., Wang, Y., Kwun, D.J.W.: Understanding the consumer experience: an exploratory study of luxury hotels. J. Hosp. Market. Manag. **20**(2), 166–197 (2011). https://doi.org/10.1080/19368623.2011.536074

Wang, D., Kirilenko, A.P.: Topic modeling of online hotel reviews: a comparison of latent Dirichlet allocation and BERT-based approaches. Tour. Manage. **87**, 104390 (2021)

Wang, L., Kirilenko, A.P.: Do tourists from different countries interpret travel experience with the same feeling? sentiment analysis of tripadvisor reviews. In: Information and Communication Technologies in Tourism 2021, pp. 294–301. Springer, Cham (2021)

Weiermair, K.: Tourists' perceptions towards and satisfaction with service quality in the cross-cultural service encounter: implications for hospitality and tourism management. Manag. Serv. Qual. Int. J. **10**(6), 397–409 (2000). https://doi.org/10.1108/09604520010351220

Wikhamn, W.: Innovation, sustainable HRM and customer satisfaction. Int. J. Hosp. Manage. **76**(March 2018), 102–110 (2019). https://doi.org/10.1016/j.ijhm.2018.04.009

Wikhman, A.: Sustainable human resource management and customer satisfaction in the hotel industry [Master's thesis, University of Gothenburg] (2019)

Wu J., Yang T., Zhou, Z., Zhao N.: Consumers' affective needs matter: open innovation through mining luxury hotels' online reviews. Int. J. Hosp. Manag. **114** (2023). https://doi.org/10.1016/j.ijhm.2023.103556

Wu, J., Xu, Z., Zhang, S., Xie, K.L.: Identifying customer-driven innovation opportunities from online reviews: a sentiment–affect–need analysis framework. J. Bus. Res. **158**, 113632 (2023)

Xu, X.: Examining the relevance of online customer textual reviews on hotels' product and service attributes. J. Hosp. Tourism Res. **43**(1), 141–163 (2019). https://doi.org/10.1177/1096348018764573

Xu, X.: Examining the role of emotion in online hotel reviews. Int. J. Hosp. Manag. **77**, 438–447 (2019)

Xu, X., Wang, X., Li, Y., Haghighi, M.: Business intelligence in online customer textual reviews: understanding consumer perceptions and influential factors. Int. J. Inf. Manage. **37**(6), 673–683 (2017). https://doi.org/10.1016/j.ijinfomgt.2017.06.004

Yee, R.W.Y., Yeung, A.C.L., Cheng, T.C.E.: An empirical study of employee loyalty, service quality and firm performance in the service industry. Int. J. Prod. Econ. **124**(1), 109–120 (2009). https://doi.org/10.1016/j.ijpe.2009.10.015

Yee, R.W.Y., Yeung, A.C.L., Cheng, T.C.E.: The impact of employee satisfaction on quality and profitability in high-contact service industries. J. Oper. Manag. **26**(5), 651–668 (2009)

Zeithaml, V.A., Bitner, M.J., Gremler, D.D.: Services marketing: Integrating customer focus across the firm. McGraw Hill, New York, NY (2006)

Zeithaml, V.A., Bitner, M.J., Gremler, D.D.: Services marketing: Integrating customer focus across the firm 4th ed. McGraw-Hill (2006)

Zemke, D., Pullman, M.: Assessing the value of good design in hotels. Build. Res. Inform. **36**(6), 543–556 (2008). https://doi.org/10.1080/09613210802380993

Zhang, J., Chen, Z.: Exploring human resource management digital transformation in the digital age. J. Knowl. Econ. (2023). https://doi.org/10.1007/s13132-023-01214-y

Zhang, W., Chen, L.: The evolving impact of digital transformation on consumer behavior and HRM: a literature review and future research agenda. J. Bus. Res. **160**, 113930 (2023)

Zhang, Z., Li, H., Law, R.: Differences and similarities in perceptions of hotel experience: the role of national cultures. J. Travel Tour. Mark. **32**(1), S2–S14 (2015). https://doi.org/10.1080/10548408.2014.959153

Zhou, L., Ye, S., Pearce, P.L., Wu, M.Y.: Refreshing hotel satisfaction studies by reconfiguring customer review data. Int. J. Hosp. Manag. **38**, 1 (2014). https://doi.org/10.1016/j.ijhm.2013.12.004

Ziff-Levine, W.: The cultural logic gap: a Japanese tourism research experience. Tourism Manage. **11**(2), 105–110 (1990)

In-Person and Remote Employees and Information Security Policy Compliance

Joyce Y. Mui, Barbara A. W. Eversole(✉), and Cindy L. Crowder

Indiana State University, Terre Haute, IN 47809, USA
Barbara.Eversole@indstate.edu

Abstract. Many workers have started working remotely, with a significant increase during and after the COVID-19 pandemic. At the same time, increasing costs of data breaches and number of security incidents continue to be a concern for organizations seeking to protect their organization, systems, and data. Using the theoretical frameworks of the Theory of Planned Behavior (TPB) and the Social Bonds Theory (SBT), we sought to understand employee compliance with information security policies (ISP). Using phenomenology, we interviewed both in-person and remote workers in a variety of industries. We uncovered their experiences with their organization's implementation and enforcement of ISP, organizational culture and leadership attitudes shaping ISP compliance, and clarity and training of ISP for employees. Top factors that influenced employee compliance of ISPs of both in-person and remote workers included the automation of policies, hectic/busy times, efficiency, availability, training, and enforcement of the ISPs. Overall, participants reported positive relationships within their organization, regardless of whether they were in-person or remote; however, nearly all participants also noted that building relationships was easier to do in-person than remote, even as technology has bridged some of the gap between in-person and remote working relationships. We offer implications for practice and research.

Keywords: Information Security Policies · Security Compliance · Remote Workers

1 Introduction

Security and compliance are essential for consideration and implementation in organizations, regardless of industry, though of heightened importance for organizations in regulated industries. Failure to manage threats and risks to organizational assets and comply with information security policies (ISP) increases risks to the organization, including potential fines, bad press, and loss of customer trust (S. Ali et al., 2021). Data breaches may be the result of security and compliance failures and gaps, causing increases in organizational spending after data breach events to repair damage and upgrade information technology as a part of corrective actions (Choi et al., 2020).

Information security is how security is implemented to prevent risks or threats from becoming intrusions, incidents, or issues in an organization (R. Ali et al., 2021). Compliance is the paperwork and adherence to regulations, laws, and organizational policies

M. J. Sousa (Ed.): UFHRD 2024, ISEM 52, pp. 91–112, 2025.
https://doi.org/10.1007/978-3-031-92548-1_6

and procedures (Kim and Kim, 2017). Employee compliance to ISP has become a focus to reduce risk and increase security of organizational assets as employees make decisions daily on compliance (Nasirpouri et al., 2021). SANS Institute (2023) provided examples of ISP, including Acceptable Encryption Policy or Router and Switch Security Policy that were more technical policies for technical employees, as well as ISP that non-technical employees may be aware of, such as a Password Protection Policy or Email Policy.

$7.35 million is the average total cost of a data breach in the U.S. (Ponemon Institute, 2017). Symantec (2014) reported that from October 2013 to September 2014 there were a total of 255 data breaches and 656 million identities exposed within the 157 countries that they monitor. Identity Theft Resource Center (ITRC) (2023) reported that in 2022, 87-92% of all U.S. data breaches were the result of cyberattacks, with phishing and related exploits being the cause, followed by ransomware. Edwards et al. (2016) used a public data set from the Privacy Rights Clearinghouse (PRC), a non-profit organization for data privacy, to predict that data breaches may cumulatively cost up to $179 billion over the next three years. Employee behavior and actions impact the effectiveness of security and compliance in organizations, such that improving employee compliance to ISP increases security in organizations (Renaud et al., 2021). The phone and internet company Verizon (2022) reported that 82% of data breaches involved human elements, such as stolen credentials, errors, phishing, or misuse.

As a result of the pandemic, remote work increased by 33% amongst U.S. organizations, with the information industry having over 50% of employees working remotely full-time (BLS, 2022). One study conducted in 2020 showed that because of the pandemic, at least half of the employees sampled worked at home at least some of the time in July 2020, and one third worked at home all the time (Eurofound, 2020). In the U.S. during the pandemic, 71% of employees worked remotely (Pew Research Center, 2020), and after the pandemic in February 2022, 59% of employees, who can work remotely, were working remotely (Pew Research Center, 2022).

With the increased trend of remote work, organizations were also reporting that remote work has a strong correlation with data breaches. Employee technostress, or technology stress on individuals, increased because of COVID-19 as so many employees began working remotely (Singh et al., 2022). Meanwhile, increased stress has been found to lower ISP compliance (Trang and Nastjuk, 2021).

2 Research Questions

Our purpose was to describe and explain how the social bonds that leadership and management build with employees, and vice versa, influence employee attitudes and resulting behavior, specifically for ISP compliance. Hirschi's Social Bond Theory (SBT), also called Social Control Theory, claims that individuals who have social bonds or attachments to other persons within a community or society are less likely to have deviant behavior (Hirschi, 2017). The Theory of Planned Behavior (TPB) suggests that attitudes, norms, and perceived behavioral control contribute to individual intention and behavior (Ajzen, 1991). Thus, we examined attitudes, intention, and factors contributing to employee compliance behavior. Applying SBT allowed insights into whether remote employees possess or lack social bonds to others in their organization that may lead to less compliant or deviant behavior, e.g., non-compliance with ISP. Further, the relationship between in-person and remote employees and managers may influence attitudes and intention and be a factor in resulting employee compliance behavior. Then through application of the TPB, along with SBT, insights were gained into the lived experiences of in-person and remote workers, including ISP compliance and behavior in the workplace. The following research questions guided the inquiry:

RQ1. What are in-person employee experiences with ISP compliance?

RQ2. What are remote employee experiences with ISP compliance?

RQ3. What factors influence in-person employees to comply with ISP compliance?

RQ4. What factors influence remote employees to comply with ISP compliance?

RQ5. What impact does a relationship between a manager and employee influence ISP compliance?

3 Literature Review

Many studies have covered ISP compliance and the factors that contribute to increasing ISP compliance. ISP compliance is complicated as the compliance is influenced by factors that align to the TPB intention and resulting behavior. Some factors that influence ISP compliance include management/leadership (Hu et al., 2012; R. Ali et al., 2021), culture (R. Ali et al., 2021; Alshaikh, 2020; Gwebu et al., 2020), motivations, and employee behaviors (R. Ali et al., 2021). Some factors that increase ISP compliance may include norms (Bauer and Bernroider, 2017), attitudes, beliefs, individual capability (Carmi and Bouhnik, 2020), and employee social bonds (Ali et al., 2020; Dong et al., 2021). Conversely, ISP non-compliance may include factors such as technostress (Hwang et al., 2022; Nasirpouri et al., 2022) and deterrence mechanisms (Chen et al., 2020; Raddatz et al., 2020).

Remote work has been studied a bit less than ISP compliance, though a flurry of recent studies has been conducted since the start of the COVID-19 pandemic. Pandemic remote work studies include social isolation (Galanti et al., 2021) and stress (Chudzicka-Czupała et al., 2023; Toscano et al., 2022). SBT was selected as part of our theoretical framework to further build on and explore remote employees' and managers' relationships (Müller et al., 2022; Toscano et al., 2022). Best practices for remote work include having an organizational culture that supports security (Sidor-Rządkowska, 2022).

The Theory of Planned Behavior and Social Bond Theory were the theoretical frameworks for our study. TPB takes the perspective of explaining or predicting employee behavior via intention towards a behavior (Ajzen, 1991) that represents the intention of in-person and remote employees to comply with ISP. SBT uses the basis of attachments affection, commitment to conformity, and involvement in community to form social bonds to explain human behavior in terms of delinquency (Hirschi, 2017). Together, TPB is used to explain behavior and SBT to explain deviant behavior of employees.

Since the COVID-19 pandemic, more studies have been conducted because of the increase of remote workers. However, a gap in the literature exists with regards to ISP compliance and remote employees. More research is needed to evaluate in-person and remote employees and their ISP compliance. Studies, such as Ayyagari (2012), have shown that employee ISP compliance or employee behavior within their organizations is a cause of increasing data breaches. Thus, our study is needed to address this research gap.

4 Research Methodology

In this section we will describe our research design, study population, instruments and data collection, how we analyzed our data, and how we ensured validity and reliability.

4.1 Research Design

We utilized a phenomenological design method to study employee attitudes, intention, and factors towards ISP compliance. The phenomena were the experiences of those working in-person and remote as they navigated ISP compliance.

4.2 Study Population

Our population was voluntary participants who were able to spend approximately one hour on a video call with one of us. The pre-screening requirements were that participants affirmed that they were employed by an organization (not self-employed), work either 100% remote or in-person (or close to 100%) and were aware that there were ISP that their organization required them to follow. The selection of the sample population was a convenience sample (Creswell and Creswell, 2020), based on the selection criteria and the convenience and availability of the participant for the interview. The study sample population was recruited from one of the author's personal and professional networks within the U.S., though the participants were not known to us personally. We requested volunteers verbally, via email, and via a social media contact, through direct messages in LinkedIn.

4.3 Instruments and Data Collection

We recorded and transcribed interview conversations for clarity, though the primary data collection method was notes from the interview. An interview guide was used, containing approximately ten questions to prompt the participant, though the semi-structured interview also allowed for discovery and a free flow of conversation and information. Notes were typed live and after the interview directly into a word processor. We primarily used video interviews to reduce boundaries of location and increase the participant pool. The interviews extracted rich data from the participants, leaving room for the participants to add their experiences.

Data collected was kept confidential to us. Data was stored locally on one of our devices and backed up on Google Drive. We used the principle of least privilege to secure access to the participant's data within Google Drive, as well as using a default deny all external connections firewall on their device. Field notes and memos provided additional context to the research process, along with the data collected from participants.

4.4 Data Analysis

Interview transcripts were coded and analyzed with ATLAS.ti, qualitative research software. Although an automated transcription tool was used, transcripts were validated manually. Transcript data was imported into ATLAS.ti and a first level analysis using the word cloud feature was used to gather popular and relevant concepts within the interviews. We used ATLAS.ti manually to code inductively and deductively and was documented in a codebook which served as the code system. Finally, analysis and visualizations were completed, as appropriate, to establish relationships in the data, find patterns, and suggest further areas of research.

The coding began with deductive codes, based on the interview guide, which started the initial codebook. With coding line-by-line, inductive codes emerged which further categorized qualitative data and increased rigor of data analysis (Oliveira, 2022). The first level of coding used the data collected from the interviews to analyze the text within the data to identify relevant data excerpts through "speech acts" (Larsen and Adu, 2021). Labels or codes were applied to pieces of data, which included phrases or sentences, and may be "descriptive, theoretical, emotional, reflective" (Bhattacharya, 2017, p. 150). This initial line-by-line coding was both deductive—codes were created based on interview guide questions—and inductive or in vivo coding, where codes emerge from the participant's exact words (Rogers, 2018). We uploaded transcripts into ATLAS.ti to manually check results of the automated ATLAS.ti tool for initial coding. The purpose of the first cycle coding was to find phrases and concepts that were common, match demographic data to parsed excerpt data, and create categories of concepts (Larsen and Adu, 2021).

Our next level of coding was focused on creating categories and themes of concepts that were used to answer the study's research questions. The second cycle coding identified emerging themes based on commonalities and relationships (Deterding and Waters, 2021), and used focused coding in the second cycle to search for the frequent or significant codes (Rogers, 2018). "Themes, or categories, are the classification of more discrete concepts," whereas expressions were symbolic of a theme (Ryan and Bernard, 2003, p. 87). Then, the emerging themes were transformed into meaningful themes (Larsen and Adu, 2021). Themes were identified using various techniques including repetitions, metaphors and analogies, transitions, similarities and differences, linguistic connectors, or missing data (Ryan and Bernard, 2003).

4.5 Validity and Reliability

We achieved validity by maintaining and ensuring accuracy of documents and recordings of the interviews and inviting participants to comment on transcripts and research findings. One of us audited records from our research process. We maintained a clear interview process, including documentation on any issues that arose during our research to provide adequate information that would allow replication. We also completed an audit trail.

5 Findings

We conducted 20 semi-structured interviews, with twelve mostly remote participants and eight mostly in-person participants. Our findings include participant demographics, and themes are presented by research question.

5.1 Demographics

Figure 1 illustrates the demographics of our participants. Table 1 includes gender.

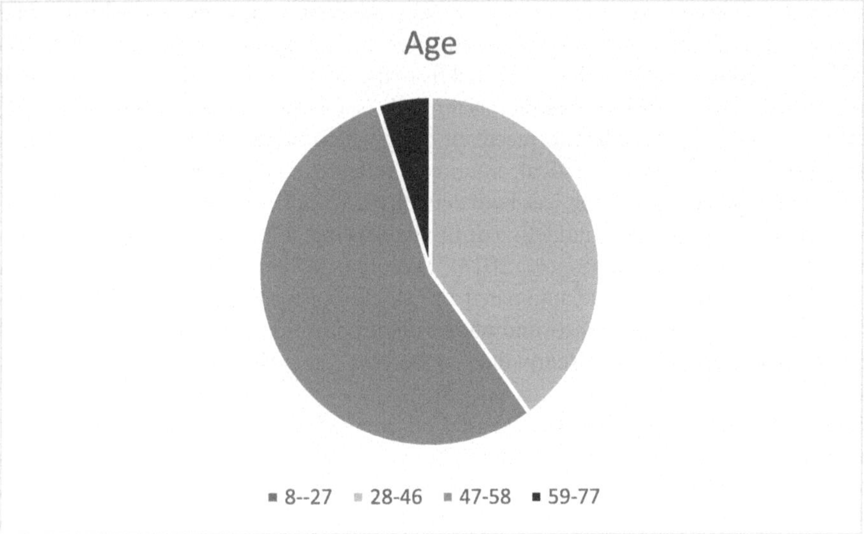

Fig. 1. Participant demographics generation and age

Table 1. Participant demographics gender

Generation	Male	Percentage Male	Female	Percentage Female	Total
Millennials	7	87.5%	1	12.5%	8
Generation X	9	82%	2	18%	11
Baby Boomers	1	100%	0	0%	1
Total	17	85%	3	15%	20

Note. One of the Generation X female participants worked in-person. The other two female participants, one Millennial and one Generation X, worked remotely

Figure 2 illustrates the location of our participants and Fig. 3 their titles.

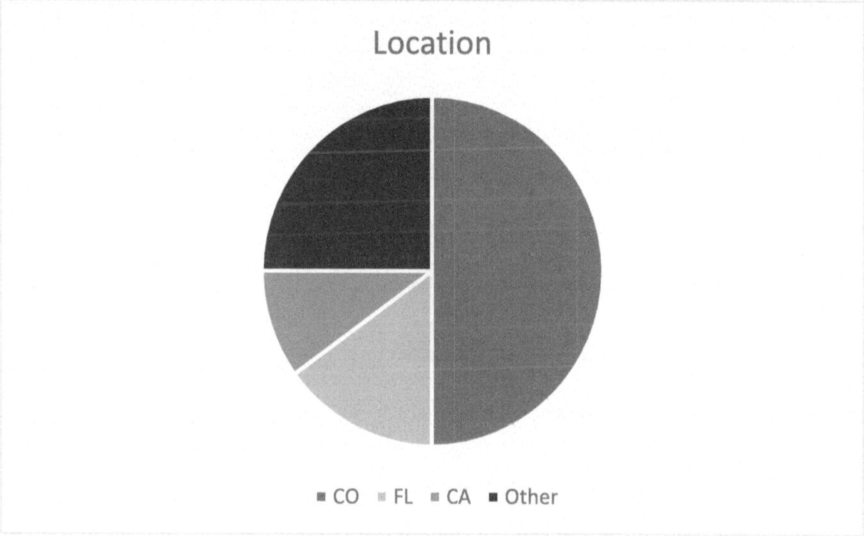

Fig. 2. Location of participants

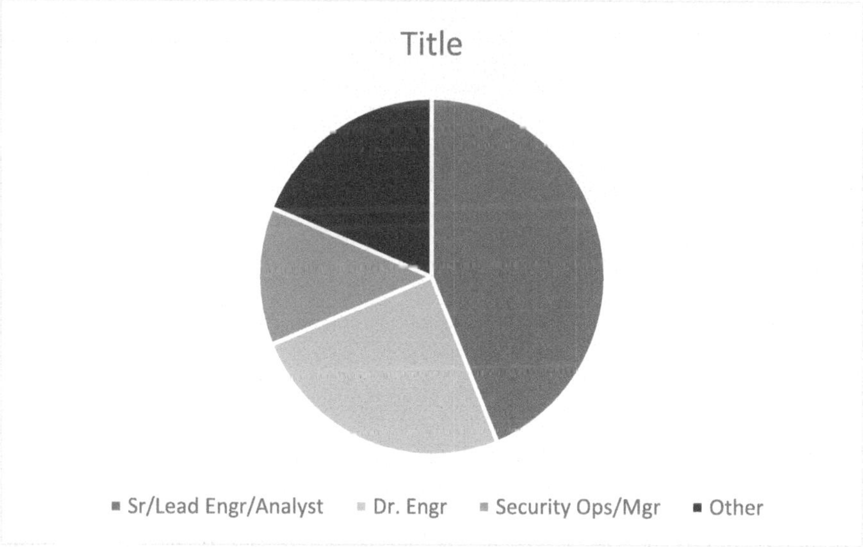

Fig. 3. Participant job titles. The lead systems analyst participant acted in an engineering-similar role, thus was grouped with the senior/lead engineers.

Figure 4 illustrates the industries in which our participants worked, while Fig. 5 notes their self-ratings of ISP familiarity.

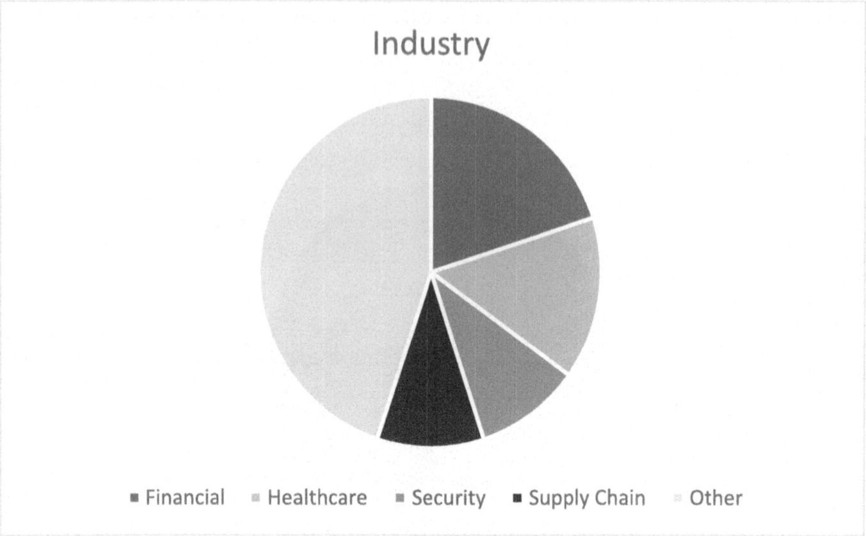

Fig. 4. Industry in which our participants worked.

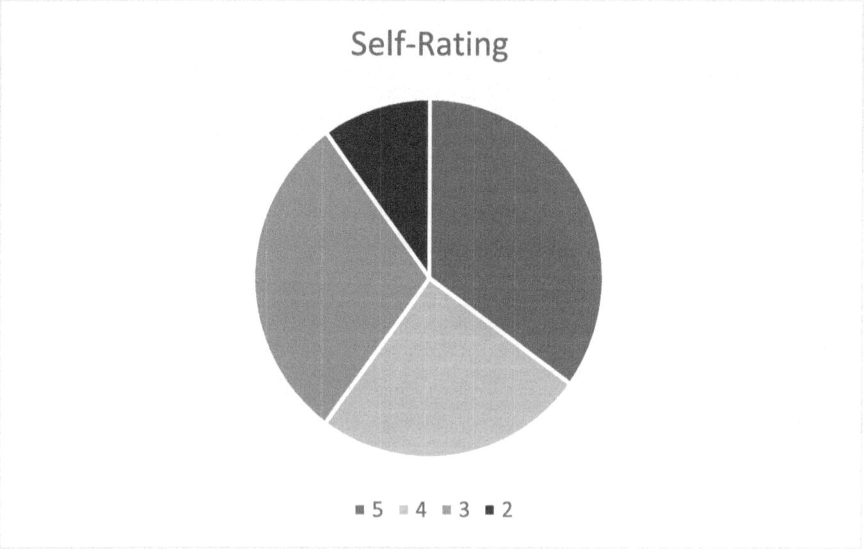

Fig. 5. Participant self-ratings. The average ISP familiarity self-rating was 3.85 out of 5.

5.2 Research Questions

Findings are presented by research question, with the themes that emerged from the data.

RQ1. What are in-person employee experiences with ISP compliance? The remote lived experiences with ISP compliance contains the themes of ISP intention, organization ISP stance, result of non-compliance to ISP, and types of ISPs.

ISP Intention. The intention of ISPs in organizations is the purpose of why the documents exist.

88% of participants noted that the intention of ISPs is to protect data/customers/organizations and systems. Irvin remarked that ISPs "protect the company from cyber-attacks, breaches, ransomware" and "protect patients, identifying information, and patient health information." Isaac said that ISP were to ensure "data security' and that "person's information is protected." Iver said that ISP were to protect data and reduce risk of "being compromised," while Isadora contributed that ISP intention was to protect confidentiality and privacy.

Organizational ISP Stance. ISP stance was determined by the participant's answer to the interview question of if ISP compliance was mandatory or nice to have in their organization.

63% commented it was nice to have. Isaac commented, "smaller ma and pa type organizations don't always do that [mandatory ISPs]." Ike noted that they have "never had anything be enforced" leading them to believe that ISP compliance is a nice to have. Ilya commented:

> *Ilya: "Both my companies [that I've worked for] were very, very lax on this whole ISP thing."*

50% said it was mandatory. Iver provided a caveat on data type, "mandatory compliance depending on the classification level." Irvin said that ISP compliance was "mandatory, but it's because it's enforced."

Result of Non-Compliance to ISPs. In-person study participants noted their organization's response or potential response to non-compliance to ISPs.

88% noted a security incident, which is any incident, including an actual data breach, which breaks ISPs. Ilya described an actual security incident and data breach:

> *Ilya: "So, we were actually breached, and they held our information up for ransom, for Bitcoin. So, it was some guy overseas, I think, in like Saudi Arabia or something, and we were able to track the IP back and say, hey, you guys have our information. So, we [the organization] had to negotiate a deal, but they [the blackmailer] said our company had to pay… so yeah, our company had to pay."*

50% of participants noted results of non-compliance to ISPs may result in termination, possibly depending on the severity or intent of the incident. Isaac said a potential result might be "employment loss," and Irvin agreed, providing an example:

> *Irvin: "One person was terminated when they refused to do their information security training."*

38% mentioned additional coaching or training. Ivan mentioned "coaching or training," while Iver agreed that their organization provides a counseling session in the event of non-compliance. Irvin said:

Irvin: "We will basically sit them down and get them remedial training."

Types of ISPs. Participants didn't necessarily know the names of policies and many simply provided examples of a policy that they knew about in their organization that impacted them. Figure 6 illustrates the examples.

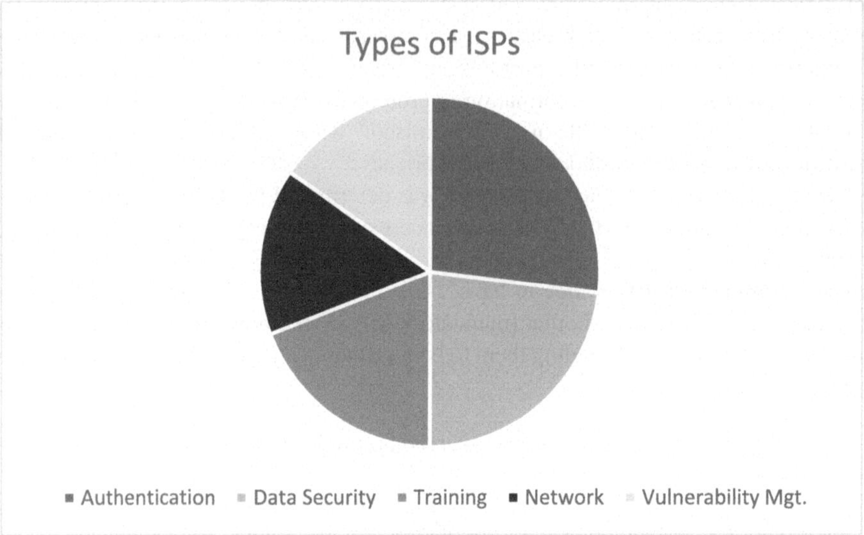

Fig. 6. Types of ISPs mentioned by in-person study participants.

The overwhelming majority of the participants cited passwords and Multi-Factor Authentication (MFA) as their example, which included using a vendor application or phone to obtain a code. Ira spoke about passwords and having a "different password for each application."

Data security policies including encryption, handling, and classification, topped the number of non-repeat and unique mentions by participants. Notably, methods of data security varied from participants, which included encryption of data in transit, via email, mobile devices and on servers, as well as data masking, access, classification, scrubbing identifiable information, and data loss prevention tools. Isadora contributed that the security training included "what information is okay to share. And what information isn't okay to share."

Methods of training included classes and videos.

The comments about the network include physical access restrictions to the network, remote access, VPN, network encryption, and firewalls blocking access to external websites. Irvin stated:

Irvin: "If you are not in America, you cannot access any of our information, no matter what."

Stopping or reacting to phishing attempts were at the top of the examples of vulnerability management. Other tools and methods included endpoint protection, antivirus, software management tools, or patches.

RQ2: What are remote employee experiences with ISP compliance? As with RQ1, the remote lived experiences with ISP compliance contains the themes of ISP intention, organization ISP stance, result of non-compliance to ISP, and types of ISPs.

ISP Intention. The intention of ISPs in organizations is the purpose of why the documents exist.

67% noted it was to meet regulations. Rupert commented, "NIST 800–53 is just a good general place to start for us. We have HIPAA as well that we have to comply with. We have state-based regulations that we have to deal with. We have international compliance and regulatory requirements."

50% said it was to protect data/customers/organization & systems. Remington said "[ISPs] are protective to the organization. It's to protect proprietary data. It's to protect personal information." Rupert agreed that ISPs intend to prevent "IP (intellectual property) leaking."

Organizational ISP Stance. ISP stance was determined by the participant's answer to the interview question of if ISP compliance was mandatory or nice to have in their organization.

50% of the participants noted ISPs were mandatory. Raven noted that "ISPs are applicable to every employee in our organization." Rhett also provided a caveat to mandatory ISP compliance, depending on the leader:

> *Rhett: "It all depends on the commander, how bad do they want to drop the hammer?".*

Result of Non-Compliance to ISPs. Participants noted their organization's response or potential response to non compliance to ISPs.

75% noted a security incident. Rusty described a data breach at their organization, caused by a customer, and said in the aftermath of the breach, the organization and employees learned a lot and "really tightened up."

75% said termination. Roman shared that, "People have been fired for tripping DLP (data loss prevention)" in their organization, and Rex said if the incident was at a:

> *Rex: "...criminal level where you're going in clicking on these things and then using that information to open up credit cards or something that would be instant termination."*

25% noted additional coaching or training. Rupert said his organization seeks to:

> *Rupert: "...educate and improve as opposed to, to punish... and turn a situation to something that's a learning opportunity."*

Rex's organization has employees review training in the event of non-compliance.

Types of ISPs. Many participants didn't know the names of policies and instead provided examples of the policy as they understood it. Figure 7 illustrates the examples.

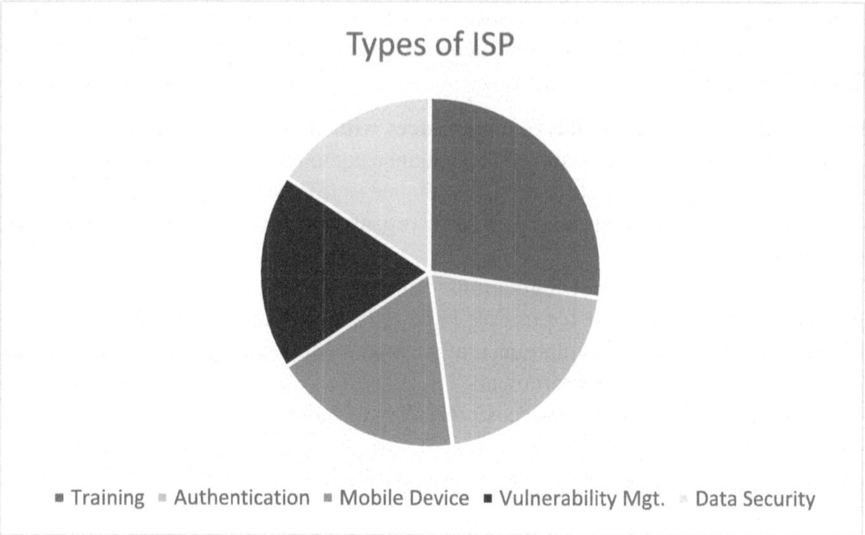

Fig. 7. Types of ISP

Most participants reported annual training requirements from their organization, though one participant in a security organization said the training was bi-weekly. Methods of training included classes, videos, gamification, and quizzes. Some of the participants said that security training was included in the onboarding with their organizations.

A few participants mentioned session limits or timeouts with authentication. Rhett mentioned using a Common Access Card (CAC):

Rhett: "There's no access to anything without a CAC card. There are no passwords anymore. if you lose your CAC card, you have to call IT, and they might give you a temporary user and password to log in that way."

Remington mentioned that in their organization if you "sit idle for too long, it logs you out." Rachel described their organization's authentication as:

Rachel: "…passwordless. So, everything is facial recognition. Fingerprints."

Policy examples included use of USB or thumb drives, how or what is possible to download, restrictions of administrator access, network restrictions, and hardening on a mobile device. Rachel stated:

Rachel: "You can't download software that's going to put our company at risk of having some type of data breach."

Management examples include stopping or reacting to phishing attempts. Tools and methods that participants mentioned that assist with vulnerability management include endpoint protection, antivirus, software management tools, or patches.

Methods of data security varied, including encryption of data in transit, via email, mobile devices and on servers, as well as data masking, access, classification, scrubbing identifiable information, and data loss prevention tools. Rex said:

> *Rex: "…any kind of banking, information, credit, or information, any of that cannot be sent electronically."*

Rachel contributed that their organization takes data security seriously and "masks even our own employees."

RQ3: What factors influence in-person employees to comply with ISP compliance? In-person participants contributed to narratives about their lived experiences, in their current and past organizations, regarding decisions and actions influencing ISP compliance.

Availability (100%). Ivan said, "the fact that an ISP exists is obviously helpful," though acknowledged sometimes a "policy just wasn't written." Irvin commented that "all of our employees have access to our ISP documents." Some participants noted that they had full access and availability to their organization's documented ISPs. Isaac said, "certain layers of the ISP that are totally hidden," and Ivan agreed:

> *Ivan: "Our entire ISP is not something we disclose, and it should not be disclosed to every associate."*

Proper Training (88%). Proper training is where employees understand the why behind the training and how to manage different scenarios. To support proper security training, Ivan and Isadora commented:

> *Ivan: that you need, "…to understand a policy in order to truly buy into the policy."*
>
> *Isadora: "Training is really not equitable across the board. You know some people are being taught and some aren't depending on the employee type."*

Clarity (75%). Ira said, "a little bit difficult to kind of track down the right person quickly" to ask questions about ISPs or their availability. Ivan suggested:

Ivan: Iif there is a scenario that is gray and somebody truly doesn't understand the aspect of the policy, that should be brought to the IT Security Group, or whoever is in charge of owning that policy."

In Irvin's organization, they explained that the ISPs "had not been customized to our specific circumstances," making it more difficult to comply.

Efficiency (75%). Ira and Ivan said:

> *Ira: "It usually comes down to time. If it's going to cost me time to follow a particular policy… I focus more on trying to meet the goal of the project or try to get the task done."*
>
> *Ivan: "There are very few aspects of an ISP that makes your day-to-day job easier… There's a human nature element that may make people not want to follow it."*

*Hectic/Busy (75%).*For timelines, Irvin said:

Irvin: "If you are overloaded, you're going to focus on the things that are most important and to a lot of people the ISP is not the most important."

Ike provided a real-world example of how easy it is to get caught up:

> *Ike: In those times where you're more distracted, more exhausted, it'd be easier to [say], "Oh, shoot! Let me check this really quick while I'm thinking about it. I'm online. Let me do this." [But] you might be breaking the rules...*

Integrity (63%). Ivan said:

> *Ivan: "You know what they say about integrity, integrity is best when nobody's looking. Do the right thing."*

Ike described themselves as a "rule follower" who tries to "stick to the policy."

Automated (50%). Ira described that in their organization, technically automated ISPs force compliance:

> *Ira: "If you don't follow the ISP, you can't log on, or the system just won't work."*
>
> *Ivan: If a company has a lot of money and can automate things, then certainly tasks to comply with an ISP might be easier, more easily accomplished."*

Participants also suggested ways to increase ISP compliance. Availability (100%), Proper Training (88%) and Automation (50%) were the most popular suggestions and already covered under factors in compliance.

RQ4: What factors influence remote employees to comply with ISP compliance?

Remote participants contributed to narratives about their lived experiences, in their current and past organizations, regarding decisions and actions influencing ISP compliance.

Availability (83%). Ray and Roman said:

> *Ray: "If you can't find it [an ISP], it's easy to ask the relevant folks."*
>
> *Roman: "I have not looked up [ISP] policies in the last 5 years. I'm probably more detached."*

Regarding ISP availability, Rhys commented:

> *Rhys: If I'm not aware of the ISP, I'm less likely to follow it" and "employees generally want to do the right thing for the company, and therefore you just have to make it easy for them to find the information security policies."*

Efficiency (83%). Reid explained:

> *Reid: "Security is often seen as like a barrier to protect productivity. It's usually just not wanting to be slowed down and wanting things to be more convenient."*

On the topic of being too strict with ISPs, Ray and Raven said:

Ray: "Most security choices you make are just that trade off of like convenience versus security."

Raven: "The easier the ISP is to follow, the more likely it is to be followed."

Clarity (75%). Reid, Rachel and Raven said:

Reid: "If it were clear to employees that they were outright forbidden from using their own device [in an ISP], they probably wouldn't."

Rachel: "If they could make it easier for us to understand the differences between the regulations and the countries, for the different information security policies that would be helpful."

Raven: that if ISPs aren't clear then employees will say, "The policy wasn't clear, and so I didn't know what to do in this situation."

Automated (58%). As Rufus said:

Rufus: "I would put security policy into that bin of things that people don't get excited about and don't want to focus the brain power on and those are exactly the things that you should automate, because they are the things where mistakes get made."

Rupert recommended that organizations "automate the control or make it a hard control as opposed to a soft control."

Enforcement (58%). Raven claimed that enforcement depends on the manager, where some managers care, and some don't care to enforce ISPs. Rhett remarked:

Rhett: "You still have to monitor your people because your people are what's going to slip in ISP is going to be. The people get that all ***these systems are only as good as people that follow them [ISPs] [emphasis added].****"*

Hectic/Busy (58%). Rusty and Raven expressed:

Rusty: "It's very simple to get kind of caught up when things get kind of hectic for junior guys. They may just cave into it for someone like me [senior engineer]. I can't do it."

Raven: "Once they [employees] start getting busy, start to ignore ISPs, and then we have to have a conversation with them that consists of like you can't send data like that over email like you just can't."

Proper Training (58%). Reid noted, "I've had gigs in the past where management was essentially saying, like, look. Just do the ISP training say, yes, a yes, and it'll be out of your hair" making the case for the support of proper security training, rather than just checking it off the list of things to do. Similarly, Rachel added:

Rachel: "***if you don't have the right people that understand the right policies like coming into review, I think that's where the mistakes happen*** [emphasis added]."

Leadership (33%). Raven explained:

Raven: We have some leaders who are definitively, this is gonna sound bad, but definitively on our side, the side of the security team basically saying, like, you guys have to actually like, look at these and follow these. And it matters. We have some other leaders at the same level that are a lot less concerned with security and a lot more concerned with profits and sales, and the sort of money-making side of the business, and to them ISPs are just kind of a hassle that gets in the way of what they want to do because they want to move quickly.

Organizational Culture (33%). Raven explained:

Raven: "Once a culture becomes a culture, it's very difficult to sway it one way or the other, especially rapidly," but that their organization is "still working on getting other people to agree that they [ISPs] are useful or helpful."

Increase Organizational Compliance. Availability, automation, enforcement, and proper training were the most often mentioned suggestions to increase ISP compliance from remote participants. Rachel's organization attempted to make compliance more fun with "gamification." Rupert provided an example at their organization:

Rupert: "We do keep track of phishing rates by some business units, and every now and then we'll have them [business units] ask us... to have a competition between our business units to see who has the lowest rates."

RQ5: What impact does a relationship between a manager and employee influence ISP compliance? The theme of organizational relationships directly applies to this RQ. All participants typically reported good relationships with their managers and co-workers, even if they were not best of friends. One participant was ambivalent towards leadership saying that they were not around or involved to have a relationship.

Co-Workers. Participants overall were friendly or friends with co-workers. Ira suggested that there was a "small company vibe within our department." Ivan was more diplomatic and said, "I think you're influenced by everyone around you, good at or indifferent. You know you just have to choose what influences to follow" for ISP compliance. Remington said, "I work remote, but some of the people that I've worked with [in-person] over the years have become good friends," but in their current remote organization "it's that physical geographical distance that you can't really get over" to become good friends with co-workers.

Good Vibes. Reid said they feel "super connected" to their co-workers and manager because of constant communication through Slack or phone calls. Isaac spreads good vibes in the office to "make everyone's day better that meets me than what they started with."

Manager/Leadership. Rusty commented that the support from their management and leadership team drives them to work harder and double and triple check that everything is done correctly. He added:

Rusty: "I'm enthusiastic every single day when I come in, and it's because the guys that I work with, and my manager and my manager before him. "I value them [referring to their manager and leadership] and want them to look good."

Remington said their manager is:

Remington: "someone who's genuinely just feels like they're trying to eliminate obstacles and enable the resources and freedoms to get the tasks done, you need to which I really appreciate."

6 Discussion

In this section we discuss the relevance of our findings to the literature by research question along with implications for practice and research.

6.1 Relevance of Findings to Literature

We relate our findings to literature by research question.

RQ1: What are in-person employee experiences with ISP compliance? The in-person experiences of participants varied. Some of the in-person participants were intimately involved with ISPs, although about half of the in-person participants were only aware of a couple of ISPs. Bauer and Bernroider (2017) suggested that personal norms and attitudes towards security and compliance, along with social norms are factors in compliance. The results of our study align with this concept, as each study participant had a different experience, story, and attitude towards ISP compliance, where each participant brought unique perspectives to the study. Many participants shared their experiences and perspectives, some of which included how colleagues, managers, leadership, or the overall organization's culture influenced them.

RQ2: What are remote employee experiences with ISP compliance? The remote experiences of study participants also varied. Previously, our literature review found a gap in literature for remote employees, specifically, around ISP compliance. Our study's lived experiences begin to contribute knowledge about remote employees' experiences with ISP compliance.

RQ3: What factors influence in-person employees to comply with ISP compliance? Suggestions for increasing ISP compliance also overlap with factors that influence ISP compliance, while a few factors and suggestions also overlap with principles of human resources development (HRD). Six factors were also suggestions to increase ISP compliance: ISP automation, availability, enforcement, leadership, organizational culture, and proper training. HRD emphasizes learning, performance, and change where mechanisms to achieve these goals include leadership, organizational culture, training, and rewards (Gilley and Maycunich, 2000). Alshaikh (2020) found organizational culture drove employee behavior and compliance in organizations.

Comparing our study's ISP compliance factors with previous literature from R. Ali et al. (2021), there are many similarities. Similar to previous research studies such Hofeditz et al. (2017), the study participants revealed that ISP compliance and possible workarounds depend on if the ISP enables work or is disruptive. In-person participants noted that ISPs should minimize the effort required for employees to follow ISPs to maximize compliance. The in-person study participants noted proper training with ISPs and security as necessary to ensure employees understand why the ISPs were the way they were and how they contributed to security of the organization systems and data. Our finding for in-person experience agrees with the previous literature review finding where He and Zhang (2019) shared that minimizing security fatigue was important for compliance. As described by Jalali et al. (2020), employee noncompliant behavior increases with a higher workload. Kim and Han (2019) found that employees were more complaint when they understood the benefits and costs of noncompliance.

RQ4: What factors influence remote employees to comply with ISP compliance? Specifically for remote employees, we found that a literature gap exists for remote employees and ISP compliance. Our results and data begin a conversation to further extend how remote employees may increase ISP compliance.

RQ5: What impact does a relationship between a manager and employee influence ISP compliance? Most study participants agree that their managers influence their decisions including ISP compliance; thus, our results from study participants agreed with past studies such as Hu et al. (2012) who found that management participation influences employees directly or indirectly via the TPB. Moreover, SBT postulates that behavior is influenced by individual relationships and that negative or delinquent behavior is due to weak social bonds (Mears and Stafford, 2022). Since study participants reported overall positive relationships with co-workers and managers, the direct results of delinquency resulting in noncompliance via SBT were not seen as we expected, though our results agree that the strong social bonds appear to influence employees with compliance. Some participants mentioned that the strength of their relationship with their manager increased their due diligence in decision-making and ISP compliance. Our results agreed with the findings from Toscano et al. (2022) that strong trust and quality of an employee and manager relationship predicts higher job performance.

6.2 Implications for Practice

Encouraging complaint behavior should be tackled from multiple fronts.

Information Security Awareness and Training. Information security awareness and training intending to ensure knowledge is shared, though not all training provides enough information. Study participants said that training should be improved from obligatory training into thoughtful and more challenging training so that employees can understand the why behind the ISPs. Other suggestions included employee collaboration in developing ISPs (Nicolas-Rocca et al., 2014), real-world examples and feedback (Bauer et al., 2017), relating security back to employee personal lives (He and Zhang, 2019), and making security training situational (Jaeger and Eckhardt, 2021).

Carrot not the Stick. Organizations should consider rewards programs to entice employees to learn about and comply with ISPs as well as punitive measures. Study participants suggested rewards to increase ISP compliance which aligns with the findings from Liang et al. (2013) that high promotion focused employees were motivated by rewards. On the other hand, perceived sanction celerity or perception of enforcement of ISP compliance strongly influenced employee compliance (Raddatz et al., 2020). Study participants commented that security incidents should be handled on a case-by-case basis, depending on the severity, where many cases will simply result in remedial training or coaching.

Leadership and Organizational Culture. Organizational culture that is supported by leadership and values ISP compliance, influences employee ISP compliance (Hu et al., 2012). One participant pointed out that leadership and organizational culture are linked and when leadership models ISP compliance, then the organizational culture will follow.

6.3 Recommendations for Future Research

Future research could seek to quantify differences between in-person, remote, and hybrid populations to differentiate possible training or management tools that each population may respond to increase ISP compliance. Research could compare populations with different lengths of remote work and their effectiveness with remote work and ISP compliance levels and attitudes. Research may also investigate the correlation between level of ISP compliance and size of organization or publicly versus privately owned organizations. Research on the impact on Gen Z may be studied and compared, for remote, hybrid, and in-person populations to gather data on their experiences. Questions surrounding Gen Z and the value of in-person communication and mentorship may be answered over time.

Social media and ISP compliance may be another research area to discover how decisions are made when employees post on social media. A future investigation may include a phishing test or experiment with the different generations in the workforce or a study to catalog the number and types of security incidents aligned with their generation. Finally, a future study may include what rewards were most effective and measuring any increase in the results of employee ISP compliance.

References

1. Ajzen, I.: The theory of planned behavior. Organ. Behav. Hum. Decis. Process. **50**(2), 179–211 (1991)
2. Alshaikh, M.: Developing cybersecurity culture to influence employee behavior: a practice perspective. Comput. Secur. **98**, 102003 (2020)
3. Ayyagari, R.: An exploratory analysis of data breaches from 2005–2011: trends and insights. J. Inf. Privacy Secur. **8**(2), 33–56 (2012)
4. Bauer, S., Bernroider, E.: From information security awareness to reasoned compliant action: analyzing information security policy compliance in a large banking organization. Data Base Adv. Inf. Syst. **48**(3), 44–68 (2017)
5. Bauer, S., Bernroider, E.W.N., Chudzikowski, K.: Prevention is better than cure! Designing information security awareness programs to overcome users' non-compliance with information security policies in banks. Comput. Secur. **68**, 145–159 (2017)
6. Bhattacharya, K.: Fundamentals of qualitative research: a practical guide. Routledge/Taylor and Francis Group, New York, NY (2017)
7. Carmi, G., Bouhnik, D.: The effect of rational based beliefs and awareness on employee compliance with information security procedures: a case study of a financial corporation in Israel. Interdisc. J. Inf. Knowl. Manage., **15**, 109–125 (2020)

8. Choi, S.J., Johnson, M.E., Lee, J.: An event study of data breaches and hospital IT spending. Health Policy Technol. **9**(3), 372–378 (2020)
9. Chudzicka-Czupała, A., Żywiołek-Szeja, M., Paliga, M., Grabowski, D., Krauze, N.: Remote and on-site work stress severity during the COVID-19 pandemic: comparison and selected conditions. Int. J. Occup. Med. Environ. Health **36**(1), 96–111 (2023)
10. Creswell, J. W., Creswell, J. D.: Research design: qualitative, quantitative, and mixed methods approaches (5th ed.). Sage (2020)
11. Deterding, N.M., Waters, M.C.: Flexible coding of in-depth interviews: a twenty-first-century approach. Sociol. Methods Res. **50**(2), 708–739 (2021)
12. Dolezel, D., McLeod, A.: Managing security risk: modeling the root causes of data breaches. Health Prog. **38**(4), 322–330 (2019)
13. Dong, K., Ali, R.F., Dominic, P.D.D., Ali, S.E.A.: The effect of organizational information security climate on information security policy compliance: the mediating effect of social bonding towards healthcare nurses. Sustainability (Basel, Switzerland) **13**(5), 2800 (2021)
14. Edwards, B., Hofmeyr, S., Forrest, S.: Hype and heavy tails: a closer look at data breaches. J. Cybersecur. (Oxford) **2**(1), 3–14 (2016)
15. Eurofound.: Living, working and COVID-19. Publications Office of the European Union, Luxembourg (2020)
16. Galanti, T., Guidetti, G., Mazzei, E., Zappalà, S., Toscano, F.: Work from home during the COVID-19 outbreak: the impact on employees' remote work productivity, engagement and stress. J. Occup. Environ. Med. **63**(7), e426–e432 (2021)
17. Gilley, J.W., Maycunich, A.: Organizational learning performance and change: an introduction to strategic human resource development. Perseus Publishing, Cambridge, MA (2000)
18. Gwebu, K.L., Wang, J., Hu, M.Y.: Information security policy noncompliance: an integrative social influence model. Inf. Syst. J. **30**(2), 220–269 (2020)
19. Hannah, D.R., Robertson, K.: Why and how do employees break and bend confidential information protection rules? J. Manage. Stud. **52**(3), 381–413 (2015)
20. He, W., Zhang, Z.: Enterprise cybersecurity training and awareness programs: Recommendations for success. J. Organ. Comput. Electron. Commer. **29**(4), 249–257 (2019)
21. Hirschi, T.: Causes of delinquency (1st ed.). Routledge (2017)
22. Hofeditz, M., Nienaber, A.M., Dysvik, A., Schewe, G.: "Want to" versus "have to": intrinsic and extrinsic motivators as predictors of compliance behavior intention. Hum. Resour. Manage. **56**(1), 25–49 (2017)
23. Hu, Q., Dinev, T., Hart, P., Cooke, D.: Managing employee compliance with information security policies: the critical role of top management and organizational culture. Decis. Sci. **43**(4), 615–660 (2012)
24. Hwang, I., Kim, S., Rebman, C.: Impact of regulatory focus on security technostress and organizational outcomes: the moderating effect of security technostress inhibitors. Inf. Technol. People **35**(7), 2043–2074 (2022)
25. Identity Theft Resource Center (ITRC): 2022 data breach report. Idthefcenter.org. https://www.idtheftcenter.org/wp-content/uploads/2023/01/ITRC_2022-Data-Breach-Report_Final-1.pdf (January, 2023)
26. Jaeger, L., Eckhardt, A.: Eyes wide open: the role of situational information security awareness for security-related behaviour. Inf. Syst. J. (Oxford, England) **31**(3), 429–472 (2021)
27. Jalali, M. S., Bruckes, M., Westmattelmann, D., Schewe, G.: Why employees (still) click on phishing links: investigation in hospitals. J. Med. Internet Res., **22**(1) (2020)
28. Kim, H.L., Han, J.: Do employees in a "good" company comply better with information security policy? A corporate social responsibility perspective. Inf. Technol. People **32**(4), 858–875 (2019)
29. Kim, S.S., Kim, Y.J.: The effect of compliance knowledge and compliance support systems on information security compliance behavior. J. Knowl. Manag. **21**(4), 986–1010 (2017)

30. Larsen, H. G., Adu, P.: The theoretical framework in phenomenological research: development and application. Taylor and Francis (2021)
31. Liang, H., Xue, Y., Wu, L.: Ensuring employees' IT compliance: carrot or stick? Inf. Syst. Res. **24**(2), 279–294 (2013)
32. Mears, D. P., Stafford, M. C.: A reconceptualization of social bond theory to predict change sequences in offending. Crime and Delinquency, 1112872210880 (2022)
33. Meisner, M.: Financial consequences of cyber-attacks leading to data breaches in healthcare sector. Copernican J. Finan. Account. **6**(3), 63–73 (2018)
34. Müller, T., Schuberth, F., Bergsiek, M., Henseler, J.: How can the transition from office to telework be managed? The impact of tasks and workplace suitability on collaboration and work performance. Front. Psychol. **13**, 987530 (2022)
35. Nasirpouri Shadbad, F., Biros, D.: Understanding employee information security policy compliance from role theory perspective. J. Comput. Inf. Syst., 1–10 (2021)
36. Nasirpouri Shadbad, F., Biros, D.: Technostress and its influence on employee information security policy compliance. Inf. Technol. People, **35**(1), 119–141 (2022)
37. Oliveira, G.: Developing a codebook for qualitative data analysis: insights from a study on learning transfer between university and the workplace. Int. J. Res. Method Educ., 1–13 (2002)
38. Pew Research Center.: How the coronavirus outbreak has – and hasn't – changed the way Americans work. Pew Research Center. https://www.pewresearch.org/social-trends/2020/12/09/how-the-coronavirus-outbreak-has-and-hasnt-changed-the-way-americans-work/ (December 9, 2020)
39. Pew Research Center.: COVID-19 pandemic continues to reshape work in America. Pew Research Center. https://www.pewresearch.org/social-trends/2022/02/16/covid-19-pandemic-continues-to-reshape-work-in-america/ (February 16, 2022)
40. Ponemon Institute.: 2017 cost of data breach study: United States. Ponemon Institute. https://www.ponemon.org/userfiles/filemanager/qrylc104ssftu5sxcz32/ (June, 2017)
41. Privacy Rights Clearinghouse (PRC).: What is a data breach? Privacyrights.org. https://privacyrights.org/resources/whats-data-breach (July 19, 2019)
42. Raddatz, N.I., Marett, K., Trinkle, B.; S: The impact of awareness of being monitored on computer usage policy compliance: an agency view. J. Inf. Syst. **34**(1), 135–149 (2020)
43. Renaud, K., Flowerday, S., Dupuis, M.: Moving from employee compliance to employee success in the cyber security domain. Comput. Fraud Secur. **2021**(4), 16–19 (2021)
44. Rogers, R.: Coding and writing analytic memos on qualitative data: a review of Johnny Saldaña's the coding manual for qualitative researchers. Qual. Rep. **23**(4), 889–892 (2018)
45. Ryan, G.W., Bernard, H.R.: Techniques to identify themes. Field Methods **15**(1), 85–109 (2003)
46. SANS Institute.: Security policy templates. SANS Institute (2023)
47. Sidor-Rządkowska, M.: Human - the weakest or the strongest link? The role of organisational culture in ensuring security of remote work. J. Mod. Sci. **49**(2), 608–620 (2022)
48. Singh, P., Bala, H., Dey, B.L., Filieri, R.: Enforced remote working: the impact of digital platform-induced stress and remote working experience on technology exhaustion and subjective wellbeing. J. Bus. Res. **151**, 269–286 (2022)
49. Symantec.: Symantec intelligence report. Broadcom. https://docs.broadcom.com/doc/intelligence-report-sept-14-en (September, 2014)
50. Toscano, F., Bigliardi, E., Polevaya, M.V., Kamneva, E.V., Zappalà, S.: Working remotely during the COVID-19 pandemic: work-related psychosocial factors, work satisfaction, and job performance among Russian employees. Psychol. Russia: State of the Art **15**(1), 3–19 (2022)

51. Trang, S., Nastjuk, I: Examining the role of stress and information security policy design in information security compliance behaviour: an experimental study of in-task behaviour. Comput. Secur., **104**, 102222 (2021)
52. Verizon. Data breach investigations report. Verizon. https://www.verizon.com/business/resources/T7dd/reports/dbir/2022-data-breach-investigations-report-dbir.pdf (2022)

A History of Incivility, Workplace Bullying, Cyberbullying and Hate Crimes: A Toxic Continuum of Behavior

Barbara A. W. Eversole(✉)

Indiana State University, Terre Haute, IN 47809, USA
baweversole@gmail.com

Abstract. Incivility means rude, discourteous, impolite behaviour (Alola and Alola, 2018). Reio, Jr. Once stated, "The direct and indirect costs of incivility to a society or a culture can be substantial" (2011, p. 54). The cumulative effect of incivility when it happens at work (workplace incivility) affects individual, team, and organizational performance (Estes and Wang, 2008). Incivility at work is part of a toxic continuum leading to workplace violence (Estes and Wang, 2008; Ghosh, Dierkes and Falletta, 2011; Kormanik, 2011). One need only look at recent headlines to see the shocking violence that has gripped our world. We seem to have a particular predilection for violence as a way to express frustration. Hate crimes can be considered as a form of workplace violence (Jacobs and Scott, 2011). HRD scholars began studying incivility around 2008–2011, but its roots go back further than that in the management literature. Moreover, incivility and workplace bullying have evolved to cyberbullying in the workplace (Escartin et al., 2021; Piotrowski, 2012). By investigating the history of the study of toxic behavior, this paper presents new insights into the violence of today's headlines, as well as implications for HRD research and practice today.

Keywords: Incivility · workplace bullying · workplace violence

1 Introduction

This paper introduces the topics of workplace incivility, bullying, and cyberbullying and defines the terms and constructs for these topics. It reviews the history of incivility and workplace violence in the four decades from 1980–2020. Argyres et al. (2020) and Koselleck (2004) suggest that historical research can inform future expectations and lead to the development of strategies and organizational strengths.

The paper then reviews the most recent toxic continuum of incivility to workplace bullying and cyberbullying to workplace violence and hate crimes in the last few years. The paper also discusses a theoretical framework for incivility. Finally, the paper discusses implications for research and practice. Implications for research include antecedents and mediating variables associated with emotions, personality, intersectionality and uncivil behavior and the association between incivility and hate crimes. Implications for practice include training & development programs, organization development cultural change efforts, and performance management to monitor behavior and arrange consequences.

M. J. Sousa (Ed.): UFHRD 2024, ISEM 52, pp. 113–128, 2025.
https://doi.org/10.1007/978-3-031-92548-1_7

2 The History of Incivilities and Crime

In 1981, Skogan and Maxfield showed that incivilities such as broken windows, drinking in public, disorderly youth, and prostitution would increase the amount of fear that a community felt about crime. However, when people are afraid of crime, they are also less civil; desirous of revenge and punishment, they create a culture of incivility. Wilson and Kelling (1982) also theorized that minor "incivilities" can cause fear called the broken window theory. There was a distinction between social and physical incivilities, with social incivilities including public drunkenness or drug use, noisiness, or verbal harassment of women. Physical incivilities included littering, graffiti, vacant or unkempt houses or lots, or abandoned cars. These incivilities cause fear both because they represent a disregard of the public order and a lack of ability of local officials to be able to control the problem.

Hausknecht (1983) noted, "crime by definition is uncivil behavior, (p. 76)." Greenberg et al. (1985) went further and stated that these incivilities would not only increase fear but would also increase the amount of crime itself, because potential criminals would recognize that there was deterioration in the community so that members would not care about what goes on and would not "be motivated to confront strangers, intervene in a crime, or call the police (p. 82)." Other scholars have similarly noted this relationship (Sherman, 1986; Skogan, 1986). In 1986 Taylor and Hale wrote a chapter on "Testing an alternative model of the fear of crime".

Rohe (1988) noted that the perception of the presence of gangs and drug users (social incivilities) had an impact on residents' fear of crime, as did physical incivilities such as poor maintenance of the grounds, but not to the same extent. Sampson and Cohen's (1988) study suggested support for the theory that controlling incivilities through 'proactive policing' might also control crimes like robbery. The presence of commercial property and social class can also add to incivilities. Parker (1988) found that age, race, and marital status all had significant effects on the fear of crime. In 1992, LaGrange et al. found that physical and social incivilities were related to risk perceptions; while incivilities in general were related to the fear of crime.

In current times, the Broken Window theory has led cities to criminalize uncivil behaviors such as begging, drinking in the streets, sleeping rough and loitering as disorderly conduct and anti-social behavior, although these behaviors may not be that harmful (Newburn & Jones, 2007; Pospech, 2020). In 1998, Carter defined what it meant to be civil: "It is morally proper to treat our fellow citizens with respect…Our duty to follow those standards does not depend on whether we happen to agree with or even like each other" (p. 15). It is a way to be respectful in society with fellow citizens.

3 Incivility and Workplace Bullying

Anti-social behaviors that cause interpersonal mistreatment at work, such as bullying, violence, harassment, and aggression, started being studied by researchers at the turn of the century (Cortina et al., 2001). They studied direct forms of aggression leading to active intention to harm, such as in cases of workplace violence caused by unhappy employees (Griffin, O'Leary, and Collins, 1998; Leather, et al., 1999; VandenBos and

Bulatao, 1996). In 1996, Hornstein described frequent, toxic behaviors of supervisors behaving badly; name-calling, throwing and smashing things, yelling obscenities, lying, being rude, spitting, ridiculing and even assaulting subordinates physically. Psychological harm to others was also studied at that time (Baron and Neuman, 1996; Folger and Baron, 1996). Then in 1999, Andersson and Pearson introduced the construct of workplace incivility, which they defined, in contrast to direct intention to harm, as "low-intensity deviant behavior with ambiguous intent to harm the target, in violation of workplace norms for mutual respect. Uncivil behaviors are characteristically rude and discourteous, displaying lack of regard for others" (p. 457). Another way to describe workplace incivility is "low intensity negative behaviors that violate workplace norms of respect (Welbourne, Miranda, and Gangadharan, 2020, p. 335)." This breach of norms and respect becomes rudeness and unfairness (Meier and Gross, 2015).

3.1 Incivility

Almost everyone has experienced incivility (Alola and Alola, 2018). In fact, Porath and Pearson (2013) stated they have polled employees over 14 years and 98% reported that they have experienced incivility at work. 50% said they experienced it weekly in 2011; up from 25% in 1998. By 2016, it had risen to 62% (Porath, 2022). In 2022, Porath conducted a new global survey of more than 2000 frontline employees and those who observed them in over 25 industries. She found that 76% of them experienced incivility at least once a month; 78% observed incivility at work at least once a month with 70% observing it 2–3 times a month.

Frontline employees may be experiencing incivility at a greater rate than general employees. Research by the Society of Human Resource Management (Gonzales, 2024) found in a survey of more than 1000 US employees that 66% say that they either experienced incivility themselves or witnessed it within the past month. 57% either experienced it or witnessed it within the past week.

Workplace incivility is bad for both the organization and its employees (Porath and Pearson, 2013), affecting their job satisfaction negatively (Cortina et al., 2001). Examples include side talk, ignoring calls, messages, and greetings, and giving credit to others for their work (Pearson, Andersson, and Wegner, 2001), talking with someone unprofessionally, ignoring someone's ideas, or excluding a colleague from an important meeting (Welbourne, Miranda, and Gangadharan, 2020). Other uncivil behaviors include talking behind someone's back, giving someone the silent treatment, overruling decisions without any rationale, making demeaning and condescending remarks, disrupting meetings, giving someone negative eye contact or dirty looks, publicly reprimanding an employee, insulting someone, or sexually harassing others (Johnson and Indvik, 2001). Other behaviors include treating someone like a child, undermining someone's credibility in public, sending a mean or nasty note, or berating someone for something they didn't do (Pearson et al., 2000). It's important to note that even though there is an ambiguous intent to harm, with the instigator often stating that the target is over-sensitive or misinterpreting, that doesn't mean that no harm occurs (Estes and Wang, 2008).

Incivility is only stressful after several occurrences along with a negative outcome (Andersson and Pearson, 1999). Alola and Alola (2018) studied the effect of resilience on workplace incivility and intention to turnover. They found that resilience mediated the relationship, and also found that workplace incivility had a detrimental effect on employees.

Studies have shown that when employees experience incivility, their job satisfaction, physical and psychological well-being, and affective commitment are all decreased, while their intention to turnover is increased (Cortina et al., 2001; Lim and Cortina, 2005; Tremmel and Sonnentag, 2018) as well as increased exhaustion (Rhee, Hur, and Kim, 2017; van Jaarsveld, Walker, and Skarlicki, 2010). Victims of incivility are more likely to engage in incivility themselves (Gallus, et al., 2014) and counterproductive work behavior (Welbourne and Sariol, 2017). Incivility affects well-being by causing stress from negative emotions (Bunk and Magley, 2013; Lim, Cortina and Magley, 2008). This stressful experience is both over the long term due to repeated incidents and also during the day the incivility occurred (Lim et al., 2018).

Incivility can also occur between co-workers and affect bystanders. Sinclair (2021) did a study investigating the reactions of bystanders when one co-worker was condescending to another. Women were more likely to support and defend the co-worker, especially if gender discrimination was involved.

Classroom Incivilities. Incivility occurs in higher education institutions (HEIs) and it has a negative impact on retention and job satisfaction of faculty as well as the commitment of students to the HEI (Hirschy and Braxton, 2004). Boice (1996) did a study of incivilities occurring in the classroom (CI). Over five years, he observed five main classroom incivilities in a large research university. The first was the mannerisms of the teachers being distancing and aloof. The second was teachers using fast-paced lectures that discouraged students from participating. The third was the indifference and noise of the students in class. The fourth was the practice of students coming in early and leaving late to class. The fifth was the behaviors of the students themselves: sarcasm, rude gestures, etc. High CI meant students did not pay attention or take notes; teachers were not enthusiastic, warm, approachable, organized, or clear. Interestingly, he found that the amount of warmth and approachability displayed by teachers in the first few days was a key initiator of CI. Irwin et al. (2021) found that the most common uncivil behaviors by students in the lecture format were non-participation, absenteeism, and ignoring.

Customer Incivilities. Another type of incivility is customer incivility (Van Jaarsveld, Walker, and Skarlicki, 2010). This type of incivility occurs when a customer service representative interacts with a customer in such a way that violates social norms, such as making disparaging remarks. When customer service representatives do not engage in this type of incivility, they are more likely to earn sales and repeat purchases (Jaramillo and Grisaffe, 2009). I experienced customer incivility recently when traveling in Germany while writing this book. We arrived at our hotel late due to train connections running late. According to Expedia, the hotel had 24/7 reception, but no one was there. Luckily, the owner arrived about an hour later, and chastised us for arriving late as reception closed at 10 pm (we had arrived at 10:30). We had an argument about the reception hours. I was tired, he was frustrated. Neither of us had empathy for the other; however, I felt that as the customer, he should have been more service-oriented toward me as I had a reasonable expectation of service given what Expedia had told me. In the end, he calmed down and was a more gracious, civil host.

Customer incivility also occurs the other way around when customers are uncivil toward customer service employees (Beck and Voorhees, 2018). Porath (2022) found in her global survey of over 2000 people in 25 industries, that 73% reported that it's not unusual for customers to be uncivil. Moreover, 78% believe that customer incivility is worse than five years ago.

Beck and Voorhees (2018) studied bystander reactions of customers who witnessed other customers being uncivil to service employees, and found that they were unlikely to intervene, even feeling some satisfaction if they felt the employee deserved the uncivil behavior. If the abusive behavior is repetitive and online, it rises to the level of customer cyberbullying (D'Cruz and Noronha, 2014).

Examples of Incivility. In 1988, The Wall Street Journal carried the headline, "Judgment Day: Texaco and Icahn Hurl Last Gibes and Prepare to Meet their Holders: Friday Vote Ends Proxy War of High Financial Drama and Immense Incivility" (Sullivan, 1988). Carl Icahn and Texaco management hurled "slogans, slurs, and solicitors" at each other (p. 1) over Icahn's takeover bid. "Texaco management, Mr. Icahn says, 'has taken a company with great assets and run it into the ground.' Mr. Icahn's financial analysis, says Texaco, 'is sketchy, sloppy, and *wrong*'" (p.1).

Uncivil behavior has even reached the judicial branch. In 1991, a survey of the 7th US judicial circuit found that 41% of the judges and lawyers responded that bad manners prevailed in the courts (Hansen, 1991). Judges thought lawyers were uncooperative and rude, while lawyers thought that judges were impolite and arrogant. In 2021, a judge was suspended after she was found to exhibit uncontrollable incivility, which included behaviors such as pushing employees, yelling at other judges and employees, and mishandling paperwork over a period of 8 years (Knezevich, 2021). And in the 2020 election debate between candidates Trump and Biden, the incivility that occurred was described as surprising even the most jaded viewers, with Trump being compared to "an adolescent bully in a lunchroom cafeteria" (Zarracina & Petras, 2020, p. 1). They talked over each other and the moderator for almost 25% of the debate. Indeed, in an opinion piece written in 2019, Darling wrote for USA Today that "It is not our beliefs themselves that seem to be fueling incivility but a failure to apply what we say we believe," (p. 1) when discussing that "to be uncivil is to be un-Christian. (p. 2)."

3.2 Incivility Theory

Schilpzand et al. (2016) proposed three different theoretical frameworks that could be used to develop a unifying theoretical framework for incivility theory. The first theory is the social interactionist view, such as from Andersson and Pearson (1999), and similar to trait activation theory (Tett and Guterman, 2000; Tett and Burnett, 2003). The social interactionist theory applies to the interactions of targets, observers, and instigators of incivility based on personality traits and situational cues relevant to them.

The second theory Schilpzand et al. (2016) identified is the transactional model of stress (Folkman and Lazarus, 1985; Lazarus and Folkman, 1984). This has evolved into the appraisal theory of incivility. Appraisal theory postulates that we need to understand the targets' appraisal of a stressor in order to understand how the stressor will affect the target (Bunk and Magley, 2013; Cortina and Magley, 2009). This theory is increasingly being used to understand antisocial behavior, aggression and incivility (Cortina and Magley, 2009; Douglas, et al., 2008; Porath, MacInnis, and Folkes, 2010; Sinclair, Martin and Croll, 2002).

The third theory Schilpzand et al. (2016) identified to explain incivility is affective events theory (AET, Weiss and Cropanzano, 1996), which focuses on the emotional reactions which then lead to specific behaviors, such as health and turnover intentions (Lim et al., 2008). AET can be used to explain behavior of the target, observer, and instigator; thus, Tracy and Robins' (2006) appraisal model of self-conscious emotions was used by Miranda, Welbourne and Sariol (2020) in their study measuring shame and guilt in observers of workplace incivility. Eversole and Graham (2012) suggested Goffman's stigma theory to explain the instigator's affect in uncivil interactions.

3.3 Workplace Bullying

Workplace bullying goes further and actually creates a toxic environment that impedes the learning and development of employees (Reio Jr. And Ghosh, 2009). Farley, Coyne and D'Cruz (2018) noted that researchers began to study workplace bullying in the late 1990's, after being ignored for most of the decade (Rayner and Cooper, 1997).

Violence, verbal aggression, and bullying are different from incivility because there is a clear intention to hurt the target (Hershcovis, 2011). Bullying occurs when an employee is subjected to negative acts by supervisors, co-workers, or subordinates repeatedly (Einarsen, 2000). Supervisors are usually the main perpetrators of workplace bullying (Hoel, 2013).

Workplace bullying has been studied as a construct with many facets and forms of behavior (Escartin et al., 2010). It has also been studied as a way to discriminate (Fevre et al., 2012; Fevre et al., 2013) and as a gendered construct (Salin and Hoel, 2013). The interactionist framework (Samnani and Singh, 2016) theorizes that bullying has impacts beyond the individual affected to work groups and the organizational system itself. Valentine, Giacalone, and Fleischman (2021) found that workplace bullying indeed leads to mistrust of others and a political view of jobs, leading to decreased work group effectiveness and ultimately decreased organizational effectiveness. Bullying is distinguished from other forms of aggression by the persistence and repetition of the behavior, as the perpetrator systematically acts negatively toward the target over time (Nielsen et al., 2015).

3.4 Cyberbullying

Farley, Coyne and D'Cruz (2018) noted that cyberbullying is repeated, aggressive and power-oriented, and therefore the same as workplace bullying. Nevertheless, because it is done on the web, it is experienced differently. In some companies, an employee can be terminated for posting something discriminatory, abusive, or prejudiced online.

Workplace cyberbullying can be defined as behavior that is repetitive and persistent, where there is a power imbalance, and where there is intent to harm and that takes place via technology at work (Farley, Coyne and D'Cruz, 2018). Farley et al. (2016) defines workplace cyberbullying as.

> a situation where over time, an individual is repeatedly subjected to perceived negative acts conducted through technology (e.g., phone, email, web sites, social media) which are related to their work context. In this situation the target of workplace cyberbullying has difficulty defending him or herself against these actions (p. 299).

Piotrowski (2012) offers this definition.

> Cyberbullying is largely viewed as inappropriate, unwanted social exchange behaviors initiated by a perpetrator via online or wireless communication technology and devices (p. 45).

Cyberbullying differs from regular bullying in that it is indirect rather than face-to-face, the target and perpetrator are often not in the same location, the bullying acts remain visible for a long time, anytime, anywhere, and by anyone, there are many different bystander roles (Farley, Coyne and D'Cruz, 2018) and viruses can be spread (Forssell, 2016). However, it involves many of the same behaviors such as name calling, threats and teasing (Farley, 2015, as cited in Farley, Coyne and D'Cruz, 2018). In a study of a large multi-national corporation, managers are more likely to engage in email bullying toward direct reports, including work-related insults, intimidation, and personal threats (Baruch, 2005).

Research is still being conducted into to whether cyberbullying is simply workplace bullying carried out by technology at work, or whether it is something quite different. Farley, Coyne and D'Cruz (2018) concluded that behaviors in both types of bullying are the same, the only difference being whether they occur online or off.

Research shows that cyberbullying is less prevalent than workplace bullying (Farley, Coyne and D'Cruz, 2018), yet it correlates more strongly with job dissatisfaction (Coyne et al., 2016). Cyberbullying may be more risky in the workplace than traditional bullying, as it leaves a footprint, but this does not seem to stop the perpetrators (Farley, Coyne and D'Cruz, 2018). Borstoff et al. (2007) however, claims that cyberbullying is the preferred mode of harassment of employees.

Both bullying and cyberbullying can be prevented using Hershcovis, Reich and Niven's (2015) framework for intervention at the primary, secondary and tertiary levels. Prevention is the goal at the primary level, providing targets with resources is the goal of the secondary level, and reducing impact is the goal of the tertiary level.

4 Workplace Violence and Hate Crimes

"Going postal" is part of the lexicon of workplace violence since the shooting in the workplace in Oklahoma in 1986. This was followed by Timothy McVeigh's terrorist attack against a federal office building in 1995. In 1996, Heinemann noted the connection between uncivil speech and hate crimes on college campuses. He related disrespect, dehumanization, name calling and threats to "hate incidents" such as racism, sexism, and homophobia, as well as religious intolerance.

Moffat (2001) argued that incivility in our society is actually an indicator of the decay of society. Going further, he noted "Rudeness Kills!" (p. 66), offering examples of road rage, airline confrontations, and cell phone disagreements. He quotes columnist William Raspberry: "Am I suggesting that ordinary incivility is partly to blame for the deaths of school children? In a word, yes. I'm saying that adult irascibility--from political intemperance to road rage--can poison our social and civic atmosphere," (as cited in Moffat, 2001, p. 67). In 2000, Pearson et al. (2000) cited a national survey where 90% of those polled believed that incivility was a contributing factor in violence.

Lutgen-Sandvik (2003) noted that incivility may lead to workplace violence and aggression. 25% of incivility instigators have threatened a colleague (Pearson et al., 2000). Another study (Martin and Lavan, 2010) investigated 45 cases in the US that went to court and found that 20% of the cases that involved bullying also involved physical violence. Workplace violence differs from incivility since it is "purposive and goal-directed," (Inness, Barling and Turner, 2005, p. 732). Campus shootings at Virginia Tech in 2007 and other universities have kept workplace violence in the news.

According to the Bureau of Labor Statistics, there were 392 workplace homicides in 2020. There were also 37,060 nonfatal injuries in the workplace resulting from an intentional injury by another person. Nearly one in ten (8.5 per cent or 277 million) persons in employment has experienced physical violence and harassment at work in their working life. Men were more likely than women to report experiencing physical violence and harassment.

Beghini (2020) offered the following statistics from a global survey on workplace violence:

- Nearly one in ten (8.5 per cent or 277 million) employees has experienced physical violence and harassment at work
- More than 60% of victims of violence and harassment at work said it has happened to them more than once, and for most of them, the last incident took place less than last five years ago.

According to the US Department of Justice, a crime becomes a hate crime when "offenders choose a victim because of some characteristic—for example, race, ethnicity, or religion—and provide evidence that hate prompted them to commit the crime" (2005, p.1). Jacobs and Scott (2011) noted that hate crimes are a form of workplace violence; although actual estimates of the prevalence of hate crimes are difficult to obtain, they are pervasive and disruptive to organizations. It seems imperative that incivility be dealt with before it escalates to bullying, workplace violence, and hate crimes.

5 Implications for Research and Practice

I would now like to offer some implications for research and practice for the reduction of incivility and workplace bullying in organizations.

5.1 Implications for Research

Research should be conducted into what situations lead to workplace bullying (Hershcovis et al., 2015). The authors noted that poor leadership, perceived injustice, role ambiguity and high demands are all factors that may lead to workplace bullying. Mediating variables such as emotional regulation may affect how employees react to workplace bullying (Niven, Sprigg, and Armitage, 2013; Niven et al., 2013). If employees do not take the bullying personally, the negative effects of the bullying are buffered, thereby reducing the severity of the consequences of the bullying on the employees' well-being. In 1999, Pearson and Gibson presented a study entitled "Workplace Incivility: The Target's Eye View" at the Academy of Management ('Incivility…', 1999) where they noted that at least 25% of workers were somewhat angry at work. Gibson gave an interview where he stated, " 'an undercurrent of anger and resentment' could potentially lead to bursts of rage typified by some workplace violence incidents" (p. 8).

Miner et al. (2018) also suggested more avenues for theoretical expansion of the construct of incivility: 1) Assessing the intent of the perpetrator to harm; 2) whether there are any positive effects of incivility at the individual, work group, or organizational level; variations in the role, understanding, and agreement of norms in the organization; an exploration of the different types of responses to incivility; and an exploration of incivility as an organizational phenomenon beyond the instigator-target relationship.

5.2 Implications for Practice

Leiter et al. (2012) found support for the notion that interventions can decrease incivility and increase civility in organizations. HRD interventions can include Training and Development, Organization Development and Cultural Change initiatives, and Performance Management initiatives.

Training and Development. Training and Development can be an important tool to increase civility in organizations, particularly to improve empathy skills (Porath and Pearson, 2013). Empathy is a skill that can be learned (Gerdes, et al., 2011). Skill development in stress management, mindfulness, difficult conversations, negotiations, active listening, giving and receiving feedback and dealing with difficult employees will help employees deal with incivility in their organization (Porath and Pearson, 2013).

Role-playing is another technique that can increase empathy skills Porath and Pearson (2013) gave an example of doctors in a role-playing exercise. In a study of 1500 doctors who participated in role-playing how to have better conversations with their patients, their empathy improved significantly. This not only improved the experience of the patients, but also reduced the emotional exhaustion and burnout of the participating doctors, and this effect continued for another three months.

Your employees can also be trained in de-escalation of uncivil conversations, including protocols of what to do when they happen (Porath and Pearson, 2013). Employees should also be trained in how to manage conflict, interpersonal skills, and expectations around incivility (Estes and Wang, 2008).

Estes and Wang (2008) noted that the importance of leadership development by HRD practitioners cannot be overstated when it comes to increasing civility in organizations:

> Because leadership development is a central mission of HRD, HRD practitioners must assume full responsibility for developing organizational leaders who are highly conscious of the impact of incivility, have zero tolerance of uncivil behavior

> at work, and are committed to creating an incivility-free work environment. As HRD practitioners, we must constantly remind, coach, and educate organizational leaders to make sure that they provide opportunities within the organization to build the positive psychological capital of their own and their employees through successful practice and performance (pg. 235).

Gedro and Wang (2013) described a successful, three-part training session in workplace incivility they used in an academic setting. The first part used PowerPoint presentations to explain incivility terms. The second part presented a "continuum of civility" they adapted from Schwehm (2011), including workplace incivility statements that they could rate from 1–10. The third part included case studies that participants considered in small groups, ending with the trainers offering ideas for creating civil cultures.

Another training program that has been successful in preventing workplace bullying is the CREW (Civility, Respect, Engagement in the Workforce) program in the US Department of Veteran Affairs (Hershcovis, et al., 2015). It utilizes role plays and discussions to increase employees' awareness of their impact at work. Two studies have shown that the program has improved workplace civility (Osatuke, et al., 2009), as well as improving burnout, organizational commitment, trust, and job satisfaction (Leiter, et al., 2012).

Organization Development Cultural Change Initiatives. As noted, it is not healthy to ignore incivility in your organization. As Pearson and Porath (2005) wrote.

> When incivility thrives, targets suffer and organizations lose. When incivility cascades within and beyond organizational boundaries, it can malign organizational interactions, tarnish the company's reputation, and create spillover effects that diminish customer satisfaction and bottom-line objectives. To ignore civility invites norms that erode cooperation through unbridled individual self-interest and organizational deterioration (pg. 16).

One way to change the culture of an organization is to change who you hire. Porath and Pearson (2013) suggest using behavioral interviewing techniques to find candidates who are equipped to deal with incivility, including people who are self-aware, interested in learning, and motivated toward self-improvement.

Having signs around the organization spelling out the company's policies around civility, such as standards of respect, can help create a culture of civility (Porath and Pearson, 2013). Companies can also enact code-of-conduct agreements, which notes unacceptable behavior such as making offensive statements, swearing, or yelling (Porath and Pearson, 2013). Incivility expectation can be included in the organization's mission statement (Estes and Wang, 2013).

Coaching is also a way to create a more civil workplace (Porath and Pearson, 2013). Norms and expectations for how employees should interact with each other should be set, and consequences for what should happen when these norms and expectations are breached. An example might be a template of what to say, such as '*Either you stop [the problematic behavior], or [the result of the behavior]*'.

Porath and Pearson (2013) also notes the importance of coaching employees to have empathy for the person being uncivil and think about why they may be behaving that way.

The uncivil person maybe be dealing with challenging stressful situations or frustrations that are causing their behavior. Having empathy for them may make it easier to react in a civil way to them instead of being uncivil back to them.

Organizations should also coach their leaders to model self-care to reduce incivility in organizations (Porath and Pearson, 2013) and to model respectful behavior in the organization (Estes and Wang, 2008). When leaders model self-care and work sustainably, employees at their organizations are more focused and engaged, and much more likely to be retained. Having a satisfying life outside of work also helped employees deal with incivility at work, as well as doing better at work and better health (Porath and Pearson, 2013). Porath and Pearson (2013) noted that incivility is contagious, but so is civility—so an organization can take steps to make sure they have a civil culture.

Adopting a zero-tolerance policy for bullying can have good results for job satisfaction. Meloni and Austin (2011) reported that employee satisfaction had significantly improved on a survey in a hospital after they adopted a zero-tolerance bullying program.

Hershcovis et al. (2015) noted that role ambiguity, high workload, poor leadership, and injustice can lead to workplace bullying and aggression; thus, creating workplace policies and systems that can counteract these situational factors should lead to less bullying cultures. These might include policies that ensure just and fair treatment for all employees, along with clarity of roles and reasonable workloads. Support for leaders along with leadership training would help prevent poor leadership.

Other tertiary interventions at the organizational level include mediation by a third party (Saam, 2010) to prevent escalation or using counseling or debriefing for the targets or even the perpetrator (Herschcovis, et al., 2015).

Performance Management to Monitor Behavior or Arrange Consequences. Using the performance management system to reward and recognize civility is also a way to increase civility in your organization (Porath and Pearson, 2013). Recognition increases well-being and retention and decreases burnout. Performance reviews should not just focus on results but should also look at how those results are being achieved, which includes handling incivility. This means using the performance management system to identify issues employees may have with incivility by using performance criteria to arrange consequences as well (Estes and Wang, 2008).

References

Alola, U.V., Alola, A.A.: Can resilience help?: coping with job stressor. Academic J. Econ. Stud. **4**(1), 141–152 (2018)

Andersen, L.M., Pearson, C.M.: Tit for tat? the spiraling effect of incivility in the workplace. Acad. Manag. Rev. **24**, 452–471 (1999)

Argyres, N.S., De Massis, A., Foss, N.J., Frattini, F., Jones, G., Silverman, B.S.: History-informed strategy research: the promise of history and historical research methods in advancing strategy scholarship. Strateg. Manag. J. **41**(3), 343–368 (2020)

Beck, J.M., Vorhees, C.M.: Witnesses of customer incivility: sympathy and schadenfreude for poorly-treated service employees. Winter AMA Proceedings. K-21 (2018)

Baron, R.A., Neuman, J.H.: Workplace violence and workplace aggression: evidence on their relative frequency and potential causes. Aggress. Behav. Offic. J. Int. Soc. Res. Aggress. **22**(3), 161–173 (1996)

Baruch, Y.: Bullying on the net: adverse behavior on e-mail and its impact. Inf. Manage. **42**(2), 361–371 (2005)
Beghini, V., Cattaneo, U., Dugan, A., Pozzan, E.: Experiences of violence and harassment at work: a global first survey. international labor organization. https://webapps.ilo.org/wcmsp5/groups/public/---dgreports/---dcomm/documents/publication/wcms_863095.pdf Accessed 30 Apr 2024
Boice, B.: Classroom incivilities. Res. High. Educ. **37**(4), 453–486 (1996)
Borstorff, P., Graham, G., Marker, M.: E-Harassment: employee perceptions of e-technology as a source of harassment. J. Appl. Manag. Entrep. **12**(2), 44–60 (2007)
Bunk, J.A., Magley, V.J.: The role of appraisals and emotions in understanding experiences of workplace incivility. J. Occup. Health Psychol. **18**(1), 87–105 (2013)
Bureau of Labor Statistics, U.S. Department of Labor: Workplace violence: homicides and non-fatal intentional injuries by another person in 2020. Econ. Daily (2020). https://www.bls.gov/opub/ted/2022/workplace-violence-homicides-and-nonfatal-intentional-injuries-by-another-person-in-2020.htm Accessed 29 Apr 2024
Carter, S.L.: Civility: Manners, morals and the etiquette of democracy. Basic Books, New York (2008)
Cortina, L.M., Magley, V.J.: Patterns and profiles of response to incivility in the workplace. J. Occup. Health Psychol. **14**, 272–288 (2009)
Cortina, L.M., Magley, V.J., Williams, J.H., Langhout, R.D.: Incivility in the workplace: Incidence and impact. J. Occup. Health Psychol. **6**(1), 64–80 (2001)
Coyne, I., Farley, S., Axtell, C., Sprigg, C., Best, L., Kwok, O.: Understanding the relationship between experiencing workplace cyberbullying, employee mental strain and job satisfaction: a dysempowerment approach. Int. J. Hum. Res. Manage. **28**(7), 945–972 (2016)
Darling, D.: Don't blame incivility on religion: christian principles are an antidote to nastiness. USA Today, October 4. https://usatoday.com/story/opinion/2019/10/04/christians-return-to-core-convictions-make-america-more-civil-column/3845002002/ Accessed 24 Apr 2024 (2019)
Douglas, S.C., Kiewitz, C., Martinko, M.J., Harvey, P., Kim, Y., Chun, J.U.: Cognitions, emotions, and evaluations: an elaboration likelihood model for workplace aggression. Acad. Manag. Rev. **33**, 425–452 (2008)
Einarsen, S.: Harassment and bullying at work: a review of the scandinavian approach. Aggress. Violent. Beh. **5**(4), 379–401 (2000)
Escartín, J., Rodríguez-Carballeira, Á., Gómez-Benito, J., Zapf, D.: Development and validation of the workplace bullying scale EAPA-T. Int. J. Clinic. Health Psychol. **10**, 519–539 (2010)
Escartín, J., Vranjes, I., Baillien, E., Notelaers, G.: Workplace bullying and cyberbullying scales: an overview. In: D'Cruz, P., Noronha, E., Notelaers, G., Rayner, C., (eds). Concepts, Approaches and Methods, pp. 325–368. Springer Link, Singapore (2021)
Estes, B., Wang, J.: Workplace incivility: impacts on individual and organizational performance. Hum. Resour. Dev. Rev. **7**(2), 218–240 (2008)
Eversole, B.A.W., Graham, C.M.: Workplace incivility: antecedents of instigator affect and potential implications. In: J. Wang (ed.) Proceedings from the 2012 Academy of Human Resource Development Conference. AHRD, St. Paul, MN. (2012)
Farley, S., Coyne, I., Axtell, C., Sprigg, C.: Design, development and validation of a workplace cyberbullying measure, the WCM. Work Stress. **30**(4), 293–317 (2016)
Farley, S., Coyne, I., D'Cruz, P.: Cyberbullying at work: Understanding the influence of technology. In: D'Cruz, P., Noronha, E., Notelaers, G., Rayner, C. (eds.) Concepts, Approaches and Methods, pp. 233–263. Springer, Singapore. (2018)
Fevre, R., Lewis, D., Robinson, A., Jones, T.: Trouble at work. Bloomsbury, London (2012)
Fevre, R., Robinson, A., Lewis, D., Jones, T.: The ill-treatment of employees with disabilities in British workplaces. Work Employ Soc. **27**(2), 288–307 (2013)

Folger, R., Baron, R.A.: Violence and hostility at work: A model of reactions to perceived injustice. In: VandenBos, G.R., Bulatao, E.Q. (eds.) Violence on the job: Identifying risks and developing solutions, pp. 51–86. American Psychological Association, Washington, D.C. (1996)

Folkman, S., Lazarus, R.S.: If it changes it must be a process: study of emotion and coping during three stages of a college examination. J. Pers. Soc. Psychol. **48**, 150–170 (1985)

Forssell, R.: Exploring cyberbullying and face-to-face bullying in working life–Prevalence, targets and expressions. Comput. Hum. Behav. **58**, 454–460 (2016)

Gallus, J.A., Bunk, J.A., Matthews, R.A., Barnes-Farrell, J.L., Magley, V.J.: An eye for an eye? exploring the relationship between workplace incivility experiences and perpetration. J. Occup. Health Psychol. **19**(2), 143–154 (2014)

Gedro, J., Wang, J.: Creating civil and respectful organizations through the scholar-practitioner bridge. Adv. Dev. Hum. Resour. **15**(3), 284–295 (2013)

Gerdes, K.E., Segal, E.A., Jackson, K.F., Mullins, J.L.: Teaching empathy: a framework rooted in social cognitive neuroscience and social justice. J. Soc. Work. Educ. **47**(1), 109–131 (2011)

Ghosh, R., Dierkes, S., Falletta, S.: Incivility spiral in mentoring relationships: reconceptualizing negative mentoring as deviant workplace behavior. Adv. Dev. Hum. Resour. **13**(1), 22–40 (2011)

Goffman, E.: Stigma: Notes on the management of spoiled identity. Simon & Shuster, New York (1986)

Gonzales, M.: Workplace incivility is more common than you think. https://www.shrm.org/topics-tools/news/inclusion-equity-diversity/workplace-incivility-shrm-research-2024#:~:text=The%20survey%20of%20over%201%2C000,work%20within%20the%20past%20week Accessed 24 Apr 2024. (2024)

Greenberg, S.W.: Informal citizen action and crime prevention at the neighborhood level: Synthesis and assessment of the research. US Government Printing Office, Washington DC (1985)

Griffin, R.W., O'Leary-Kelly, A., Collins, J.M.: Dysfunctional behavior in organizations: Violence and deviant behavior. JAI Press, Stamford, CT (1998)

Hansen, M.: Incivility a problem, survey says. ABA J. **77**(7), 22 (1991)

Hausknecht, M.: Crime and the culture of incivility. Dissent **30**, 75–78 (1983)

Heinemann, R.L.: Addressing campus-wide communication incivility in the basic course. a case study. In: 82nd SCA Annual Conference. (1996)

Hershcovis, M.S.: "Incivility, social undermining, bullying… oh my!": a call to reconcile constructs within workplace aggression research. J. Organ. Behav. **32**(3), 499–519

Hershcovis, M.S., Reich, T.C., Niven, K.: Workplace bullying: causes, consequences, and intervention strategies. Society for Industrial and Organizational Psychology, London, UK, SIOP White Paper Series (2015)

Hirschy, A.S., Braxton, J.M.: Effects of student classroom incivilities on students. New Dir. Teach. Learn. **99**, 67–76 (2004)

Hoel, H.: Workplace bullying in the United Kingdom. In: The Japan Institute for Labour Policy and Training (ed.), Workplace Bullying and Harassment JILPT Report 12, 61–76. (2013)

Hornstein, H.A.: Brutal bosses and their prey. Riverhead, New York (1996)

Incivility: Gateway to violence. Security **37**(2), 8 (2000)

Inness, M., Barling, J., Turner, N.: Understanding supervisor targeted aggression: a within-person between-jobs design. J. Appl. Psychol. **90**, 731–739 (2005)

Jacobs, J.L., Scott, C.L.: Hate crimes as one aspect of workplace violence: recommendations for HRD. Adv. Dev. Hum. Resour. **13**(1), 85–98 (2011)

Jaramillo, F., Grisaffe, D.B.: Does customer orientation impact objective sales performance? Insights from a longitudinal model in direct selling. J. Person. Sell. Sales Manage. **29**, 67–78 (2009)

Knezevich, A.: 'Uncontrollable incivility: Baltimore judge suspended without pay by Maryland Court of Appeals.' The Baltimore Sun. (July 1, 2021)

Kelling, G.L., Wilson, J.Q.: Broken windows. Atl. Mon. **249**(3), 29–38 (1982)

Kormanik, M.B.: Workplace violence: assessing organizational awareness and planning interventions. Adv. Dev. Hum. Resour. **13**(1), 114–127 (2011)

Koselleck, R.: Futures past: On the semantics of historical time. Columbia University Press, New York (2004)

LaGrange, R.L., Ferraro, K.F., Supancic, M.: Perceived risk and fear of crime: Role of social and physical incivilities. J. Res. Crime Delinquency **29**(3), 311–334 (1992)

Lazarus, R.S., Folkman, S.: Stress, appraisal, and coping. Springer, New York (1984)

Leather, P., Brady, C., Lawrence, C., Beale, D., Cox, T.: Work-related violence: Assessment and intervention. Routledge, London (1999)

Leiter, M.P., Day, A., Gilin-Oore, D., Laschinger, H.: Getting better and staying better: assessing civility, incivility, distress and job attitudes one year after a civility intervention. J. Occup. Health Psychol. **17**, 425–434 (2012)

Lim, S., Cortina, L.M.: Interpersonal mistreatment in the workplace: the interface and impact of general incivility and sexual harassment. J. Appl. Psychol. **90**, 483–496 (2005)

Lim, S., Cortina, L.M., Magley, V.J.: Personal and workgroup incivility: Impact on work and health outcomes. J. Appl. Psychol. **93**(1), 95–107 (2008)

Lim, S., Ilies, R., Koopman, J., Christoforou, P., Arvey, R.D.: Emotional mechanisms linking incivility at work to aggression and withdrawal at home: an experience-sampling study. J. Manag. **44**(7), 2888–2908 (2018)

Lutgen-Sandvik, P.: The communication cycle of employee emotional abuse. Manag. Commun. Q. **16**, 471–502 (2003)

Meier, L.L., Gross, S.: Episodes of incivility between subordinates and supervisors: examining the role of self-control and time with an interaction-record diary study. J. Organ. Behav. **36**(8), 1096–1113 (2015)

Meloni, M., Austin, M.: Implementation and outcomes of a zero tolerance of bullying and harassment program. Aust. Health Rev. **35**, 92–94 (2011)

Miner, K.N., Diaz, I., Wooderson, R.L., McDonald, J.N., Smittick, A.L., Lomeli, L.C.: A workplace incivility roadmap: Identifying theoretical speedbumps and alternative routes for future research. J. Occup. Health Psychol. **23**(3), 320–337 (2018)

Miranda, G.A., Welbourne, J.L., Sariol, A.M.: Feeling shame and guilt when observing workplace incivility: elicitors and behavioral responses. Hum. Resour. Dev. Q. **31**(4), 371–392 (2020)

Moffat, R.C.: Incivility as a barometer of societal decay. Florida Philosophic. Rev. **1**(1), 63–80 (2001)

Nielsen, M.B., Hoel, H., Zapf, D., Einarsen, S.: Exposure to aggression in the workplace. In: Clarke, S., Probst, T.M., Guldenmund, F., Passmore, J., (eds.), The Wiley Blackwell Handbook of the Psychology of Occupational Safety and Workplace Health, pp. 205–227. Wiley Blackwell, Chichester. (2015)

Niven, K., Sprigg, C.A., Armitage, C.J.: Does emotion regulation protect employees from the negative effects of workplace aggression? Eur. J. Work Organ. Psy. **22**, 88–106 (2013)

Niven, K., Sprigg, C.A., Armitage, C.J., Satchwell, A.: Ruminative thinking exacerbates the effects of workplace violence. J. Occup. Organ. Psychol. **86**, 67–84 (2013)

Osatuke, K., Moore, S., Ward, C., Dyrenforth, S., Belton, L.: Civility, respect, engagement in the workforce (CREW): nationwide organization development intervention at veterans health administration. J. Appl. Behav. Sci. **45**, 384–410 (2009)

Parker, K.D.: Black-white differences in perceptions of fear of crime. J. Soc. Psychol. **128**(4), 487–494 (1988)

Pearson, C.M., Andersson, L.M., Porath, C.L.: Assessing and attacking workplace incivility'. Organ. Dyn. **29**, 123–137 (2000)

Pearson, C.M., Andersson, L.M., Wegner, J.W.: When workers flout convention: a study of workplace incivility. Hum. Relations **54**, 1387–1419 (2000)

Piotrowski, C.: From workplace bullying to cyberbullying: the enigma of e-harassment in modern organizations. Organ. Dev. J. **30**(4), 44 (2012)

Porath, C.: Frontline work when everyone is angry. Harvard Business Review https://hbr.org/2022/11/frontline-work-when-everyone-is-angry Accessed 24 Apr 24 (2022)

Porath, C., Pearson, C.: The price of incivility. Harv. Bus. Rev. **91**(1–2), 114–121 (2013)

Porath, C.L., MacInnis, D.J., Folkes, V.S.: Witnessing incivility among employees: effects on consumer anger and negative inferences about companies. J. Consumer Res. **37**, 292–303 (2010)

Rayner, C., Cooper, C.: Workplace bullying: myth or reality—can we afford to ignore it? Leadersh. Org. Dev. J. **18**(4), 211–214 (1997)

Reio, T.G., Jr.: Supervisor and coworker incivility: testing the work frustration-aggression model. Adv. Dev. Hum. Resour. **13**(1), 54–868 (2011)

Reio, T.G., Jr., Ghosh, R.: Antecedents and outcomes of workplace incivility: Implications for research and practice. Hum. Resour. Dev. Q. **20**, 237–264 (2009)

Rhee, S.-Y., Hur, W.-M., Kim, M.: The relationship of coworker incivility to job performance and the moderating role of self-efficacy and compassion at work: the job demands-resources (JD-R) approach. J. Bus. Psychol. **32**, 711–726 (2017)

Rohe, W.M.: Fear of crime in public housing. Environ. Behav. **20**(6), 700–720 (1988)

Saam, N.J.: Interventions in workplace bullying: a multilevel approach. Eur. J. Work Organ. Psy. **19**(1), 51–75 (2010)

Salin, D., Hoel, H.: Workplace bullying as a gendered phenomenon. J. Manag. Psychol. **28**(3), 235–251 (2013)

Samnani, A., Singh, P.: Workplace bullying: considering the interaction between individual and work environment. J. Bus. Ethics **139**, 537–549 (2016)

Sampson, R.J., Cohen, J.: Deterrent effects of the police on crime: a replication and theoretical extension. Law Soc. Rev. **22**(1), 163–189 (1988)

Schilpzand, P., De Pater, I.E., Erez, A.: Workplace incivility: a review of the literature and agenda for future research. J. Organ. Behav. **37**, S57–S88 (2016)

Schwehm, K.: Civility in the workplace (2011). https://www.slideserve.com/gregoryemerson/civility-in-the-workplace-powerpoint-ppt-presentation#google_vignette Accessed 5 Mar 2024

Sherman, L.: Fear of crime and neighborhood change. In: Reiss, A.J., Tonry, M. (eds.) Crime and Justice, 8, Communities and Crime, pp. 343–386. University of Chicago, Chicago, IL. (1986)

Sinclair, S.: Bystander reactions to workplace incivility: the role of gender and discrimination claims. Eur. J. Psychol. **17**(1), 134–144 (2021)

Sinclair, R.R., Martin, J.E., Croll, L.W.: A threat-appraisal perspective on employees' fears about antisocial workplace behavior. J. Occup. Health Psychol. **7**, 37–56 (2002)

Skogan, W.: Fear of crime and neighborhood change. In: Reiss, A.J., Tonry, M. (eds.) Crime and Justice, 8, Communities and Crime, pp. 203–229. University of Chicago Press, Chicago, IL. (1986)

Skogan, W., Maxfield, M.: Coping with crime: Individual and neighborhood reactions. Sage, Beverly Hills, CA (1981)

Sullivan, A.: Judgment day: texaco and icahn hurl last gibes and prepare to meet their holders; friday vote ends proxy war of high financial drama and immense incivility; Slogans, slurs and solicitors. Wall Street J. **1**, 6. (1988)

Taylor, R.B., Hale, M.: Testing alternative models of the fear of crime. J. Crim. Law Criminol. **77**(1), 151–189 (1986)

Tett, R.P., Gutterman, H.A.: Situation trait relevance, trait expression, and cross-situational consistency: testing a principle of trait activation. J. Res. Pers. **34**, 397–423 (2000)

Tett, R.P., Burnett, D.D.: A personality trait-based interactionist model of job performance. J. Appl. Psychol. **88**, 500–517 (2003)

Tracy, J.L., Robins, R.W.: Appraisal antecedents of shame and guilt: support for a theoretical model. Pers. Soc. Psychol. Bull. **32**(10), 1339–1351 (2006)

Tremmel, S., Sonnentag, S.: A sorrow halved? a daily diary study on talking about experienced workplace incivility and next morning negative affect. J. Occup. Health Psychol. **23**, 568–583 (2018)

Valentine, S.R., Giacalone, R.A., Fleischman, G.: Workplace bullying, socially aversive attitudes, reduced work group effectiveness, and organizational frustration. Hum. Resour. Dev. Q. **32**, 131–153 (2021)

VandenBos, G.R., Bulatao, E.Q. (eds.): Violence on the job: Identifying risks and developing solutions. American Psychological Association, Washington, D.C. (1996)

Van Jaarsveld, D.D., Walker, D.D., Skarlicki, D.P.: (2010) 'The role of job demands and emotional exhaustion in the relationship between customer and employee incivility.' J. Manag. **36**, 1486–1504 (2010)

Weiss, H.M., Cropanzano, R.: Affective events theory. Res. Organ. Behav. **18**(1), 1–74 (1996)

Welbourne, J.L., Miranda, G., Gangadharan, A.: Effects of employee personality on the relationships between experienced incivility, emotional exhaustion, and perpetrated incivility. Int. J. Stress. Manag. **27**(4), 335–345 (2020)

Welbourne, J.L., Sariol, A.M.: When does incivility lead to counterproductive work behavior? roles of job involvement, task interdependence, and gender. J. Occup. Health Psychol. **22**, 194–206 (2017)

Wilson, J.Q., Kelling, G.L.: Broken windows. Atlantic Monthly 249 (1982). https://www.theatlantic.com/past/politics/crime/windows.htm Accessed 22 Apr 2024

Zarracina, J., Petras, P.: Trump, biden debate was exchange of insults and incivility. See for yourself. USA Today (2020). https://usatoday.com/in-depth/news/2020/10/01/2020-debate-trump-biden-presidential-debate-campaign/5879363002/ Accessed 22 Apr 2024

Author, F.: Article title. J. **2**(5), 99–110 (2016)

Author, F., Author, S.: Title of a proceedings paper. In: Editor, F., Editor, S. (eds.) CONFERENCE 2016, LNCS, vol. 9999, pp. 1–13. Springer, Heidelberg (2016)

Author, F., Author, S., Author, T.: Book title, 2nd edn. Publisher, Location (1999)

Author, F.: Contribution title. In: 9th International Proceedings on Proceedings, pp. 1–2. Publisher, Location (2010)

LNCS Homepage, http://www.springer.com/lncs Accessed 21 Nov 2016

Impact of Generative AI on Supply Chain and Human Resources

Otacilio Moreira[1](✉), Bruno Santos Cezario[2], André Luis Azevedo Guedes[2,3], Aurélio Lamare Soares Murta[4], and Maria Carolina Rodrigues[1,5]

[1] Universidade Federal Fluminense (UFF), Niterói, RJ 24210-240, Brazil
otaciliomoreira@id.uff.br
[2] Programa Pós-Graduação Em Desenvolvimento Local, Centro Universitário Augusto Motta, Rio de Janeiro, RJ 21041-010, Brazil
brunoscezario@souunisuam.com.br
[3] Programa Pós-Graduação Em Engenharia Civil, UFF, Niterói, RJ 24210-240, Brazil
[4] Diretor Do CASI, Universidade Federal Fluminense, Niterói, RJ 24210-240, Brazil
aureliomurta@id.uff.br
[5] CinTurs - Universidade Algarve - Campus Gambelas, Faro, Portugal

Abstract. Generative Artificial Intelligence (Gen-AI) is an emerging technology that could revolutionize supply chain management (SCM). However, there are concerns about the benefits, challenges, and implications of Gen-AI for businesses and workers. Research on the impact of Gen-AI on SCM is important in helping companies make informed decisions about adopting this technology. Therefore, the objective of this article was to understand the aspects of Gen-AI's influence on the management of SCM processes and people, exploring the benefits, challenges, and emerging trends. The methodology used for this study was based on field research and, as for the objectives, qualitative exploratory. Data were collected from 160 professionals working in SCM processes in Rio de Janeiro through an online questionnaire. The main results show that 25% of respondents already use Gen-AI in their processes and 17% have projects evaluating its use. The most significant benefit pointed out was increased efficiency. Technology can also improve the overall performance of SCM, but risks and ethical considerations need to be considered. Organizational learning is an accelerating factor of results in the implementation of Gen-AI in SCM. This study offers insights for academics and practitioners working in supply chains. The technology has further demonstrated a potential to improve the efficiency and performance of supply chain management (SCM), but some limitations need to be overcome for its better adoption.

Keywords: Generative Artificial Intelligence · Supply Chain Management · Organizational Learning

1 Introduction

The concept of supply chain management supporting a new business model, according to Moreira and Silvestre (2015) [1], integrates the activities of planning, control, and execution of the flow of goods, services and information – to increase efficiency, reducing

M. J. Sousa (Ed.): UFHRD 2024, ISEM 52, pp. 129–143, 2025.
https://doi.org/10.1007/978-3-031-92548-1_8

costs and better meeting customer demands – from the origin (through demand forecasting processes, sourcing, procurement, production, delivery logistics, customer service; each playing a crucial role in transforming training data into generative outcomes) to the point of consumption.

These supply chains are referred to as traditional, for Dürr (2018) [2] and Rausch-Phan et al. (2022) [3], due to its conventional models that have long been the cornerstone of logistics and distribution systems. Compared to those of the Industrial Revolution 4.0. For this, they consider them informal, and inefficient. These authors emphasize the need for another more effective and efficient model; through artificial intelligence, making them a fully integrated and transparent network for all parties involved.

Through our experience of more than thirty-five years in the industrial segment, the creative process begins through the process sheets (a simplification of the human-machine diagram that establishes the step-by-step to be followed for production or manufacturing in the work centers), which will serve as the basis for generative AI learning. These operations are then converted into digital data and organized into training datasets through automation and human decision-making.

In the context of generative Artificial Intelligence (AI), Martins Junior et al. (2023) [4], highlight the advancement of Large-Scale Language Models, also known as Large Language Models (LLMs), for example, OpenAI's GPT-3. According to the authors, these models, fed and trained with large volumes of text, demonstrate an impressive ability to understand and generate human language fluidly and accurately, creating original content, and elevating human-machine interactions to unprecedented levels of sophistication.

The results obtained through the survey respondents, the supply chains of companies in Rio de Janeiro, are using caution when using generative AI, especially when it comes to large language models (LLMs) such as ChatGPT, as there are associated risks, as well as the benefits of integrating these technologies into digital transformation plans.

In their research, Zhang et al. (2017 [5] and Li et al. (2023) [6], point out that risks commonly identified by managers include: responses contrary to company policies or brand image; lack of knowledge about current organizational knowledge; difficulties in differentiating fabricated responses from creative outputs; audit trails and unclear citation sources; possible threat for exposure of trade secrets; and the cost of using proprietary LLMs; as well as, unrealistic expectations of LLMs as a comprehensive solution for enterprise content without human involvement.

Also, according to Li et al. (2023 [6], in the training phase, the LLM model uses books, articles, websites, and code repositories to learn statistical patterns, word relationships, and contextual information from various sources. In the inference phase, they are used for a variety of tasks, such as chatbots, translation, writing assistance, coding, planning, and composing poems and stories.

From this new reality for Organizations, the question that is intended to be answered is – what is the current state of Generative AI in supply chain management and its impacts on human resources in industries in Rio de Janeiro? The main objective is to investigate the aspects of Gen-AI's influence on the management of processes and people in the Supply Chain of Rio de Janeiro, pointing out the benefits, challenges, and emerging trends, treating the supply chain as a collection of interconnected processes that translate training data into generation. The division of AI-generated in these processes reveals all the places where companies and users make decisions that have direct consequences on customer service (deadlines, quality, costs).

This article can assist academics, practitioners, and policymakers by offering an understanding of the implications and benefits of Generative AI in the Supply Chain for adopters and non-adopters alike. In addition to the immediate operational impacts, the indispensable role of organizational learning processes in navigating the complexities of successful adoption of Gen-AI in the Operations and Supply Chain Management domain is highlighted.

The sequence of the article is organized as follows. In Sect. 2, we provide the theoretical framework for supply chain optimization and current LLM technology. In Sect. 3, we describe the methodology employed. Section 4 evaluates the questionnaire answered by the managers of the companies in Rio de Janeiro. In Sect. 5, we describe our findings. In Sect. 6 the conclusions and future work.

2 Theoretical Framework

In this section, we provide a brief background on decision-making in supply chain operations. We use a supply chain example, which will be useful in explaining our approach in response to the problem question. Next, we describe the current capabilities and limitations of LLMs.

2.1 The Supply Chain's Decision-Making Process

We started by reinforcing the degree of complexity of supply chains, whose integration, according to Moreira and Silvestre (2015 [1], includes the activities of planning, controlling, and executing the flow of goods, services, and information from the origin (through the processes of demand forecasting, sourcing, procurement, production, delivery logistics, customer service; each playing a crucial role in transforming training data into generative outcomes) – to the point of consumption. This complexity requires decision-making on multiple timescales, from determining the number and location of factories to selecting suppliers, managing inventory levels, and optimizing transportation routes for intermediate and finished goods.

AI, according to Alto (2023) [7], encompasses the creation of systems capable of performing tasks with human-like intelligence and interacting with the ecosystem. To better understand the relationship between AI, ML, DL, and generative AI, Alto considers AI as the foundation, while ML, DL, and generative AI represent increasingly specialized and focused areas.

Considered by Ozdemir (2023)[8] as a practice to create specific instructions or questions to elicit the desired answers from a generative language model, Large Scale Language Models (LLM) are a subarea of AI used in the field of Prompt Engineering development. The author states that all LLMs on the market have been pre-trained on a large corpus of text data and specific tasks related to language modelling. A period in which the LLM developed learning and understanding of general language and the relationships between words in different tasks.

The use of artificial intelligence (AI) according to Doanh et al. (2023) [9], in the supply chain manufacturing process, offers a complex approach to improve operational results. The authors state that manufacturers can achieve considerable improvements in efficiency and safety by applying advanced quality control techniques, streamlining processes, and maximizing the use of resources. Generative AI provides a proactive approach to defect identification and maintenance, resolving issues quickly through data analysis, anomaly detection, and real-time insights. Additionally, by allowing businesses to match supply to demand, AI's market analysis capabilities improve inventory control and marketing strategies.

AI has been having a revolutionary effect on production in several areas, including market analysis, skills analysis, defect diagnosis, and maintenance forecasting (Javaid et al., 2022) [10]. Artificial Intelligence (AI), according to the authors, provides the necessary skills for manufacturing, which helps in the strategic planning of labour; Its market analysis helps in accurately matching production to customer demand, and by optimizing revenues and resource efficiency, this integrated approach provides companies with a competitive advantage in the world market.

The creation of a generative AI model therefore involves selecting a technical architecture, assembling training datasets, and employing a training algorithm to code features from the data, according to Lee et al. (2023) [11]. This process, according to the authors, both scientific and artistic, requires significant resources and results in a model they refer to as a "base model" or "pre-trained" and can be adjusted to improve performance or adaptation to specific domains.

2.2 Generative AI in the Supply Chain

As companies grow in size, the role of supply chain management becomes increasingly important as, rather than just assisting in the flow of resources and information, the current emphasis is strategic on optimizing supply against demand Yandrapalli (2023) [12]. According to the author, as AI-based solutions become more widely available, companies are achieving unprecedented levels of performance in supply chain management. He cites that when compared to non-adopters, successful AI deployments have demonstrated significant improvements in service levels, a 35% reduction in cost levels, and a 15% reduction in logistics costs. Yandrapalli concludes that incorporating AI into supply chain processes is a technological advancement and a strategic requirement that provides companies with the ability to navigate inefficient and competitive supply chain environments.

The integration of big data technologies in supply chain management research is evidenced through the publications found in this domain; however, according to Meriton et al. (2020) [13], the practical adoption of these technologies has not corresponded to optimistic projections. Their study identified emerging trends and assessed the potential added value of generative AI in enhancing SCM practices.

Meriton's findings highlight that despite some imperfections in the existing literature, the study underscores the importance of recognizing the value-added potential of AI and that it will increasingly play a significant role in distinguishing between successful and unsuccessful companies in the future of supply chain management.

Academic articles, according to Michel-Villarreal et al. (2023) [14], have increasingly highlighted both the potential obstacles and advantages associated with generative AI, concerns often revolve around issues of correctness, reliability, and plagiarism rates. Accuracy and reliability challenges result from the model's reliance on biased data for training, potentially outdated knowledge, and the generation of false or inaccurate information, including the provision of false references.

The emerging literature on generative AI is already responsible for far-reaching changes in various industries while bringing both opportunities and challenges (Fosso Wamba et al., 2023) [15]. However, according to the authors, the technology has several disadvantages and limitations – from a financial point of view, it cannot yet effectively cope with data synthesis and privacy issues – despite its potential to support research in this field.

Through AI systems, according to Dijmărescu (2023) [16], companies can analyze large amounts of data, identify patterns, and generate insights, but they cannot replace strategic thinking and human interpersonal skills. For the author, the main areas of application of the supply chain in the manufacturing industry – predictive maintenance, quality control, optimization of supply chain processes, mass customization, autonomous operations, and sustainability – have to be developed in parallel. She cites the following benefits:

- Improving supply chain resilience,
- Real-time visibility,
- Prevention of potential disruptions and.
- Optimization of purchasing processes for the industrial sector.

Because it cannot replace human experience and decision-making in every situation Dijmărescu points it as a limitation. She suggests a solution – to build a robust supply chain, based on a holistic approach that combines artificial intelligence with human expertise and risk management strategies.

With a particular focus on text-based emotion detection, Acheampong et al. (2021) [17] in their research, offer a thorough overview of the models that currently set the standard for Natural Language Processing (NLP) applications; including Generative Pretraining (GPT) models and Bidirectional Encoder Representations of Transformers (BERT), which is currently the most researched model for text-based emotion detection. The authors suggest the use of robust semi-supervised learning architectures, highlighting the effectiveness of this method and recommending the GPT model for this use, highlighting its cutting-edge performance in similar tasks and lexical robustness.

2.3 Generative AI Training

Generative AI training is a multi-step process that begins with collecting data from creative tasks that serve as the foundation for AI learning (e.g., the process sheets mentioned earlier). According to Martins & Martins (2023) [4], they are converted into digital data files, and then these individual data are combined into training datasets, which are structured by a combination of automation and human decision-making.

The authors detail the walkthrough, mentioning that at first, the builder selects a technical architecture, then assembles training datasets, and uses a training algorithm to encode the data features into the model. The resulting model is called a "base model" or "pre-trained model" and can be tuned to improve performance or adapt to specific problem domains.

For illustrative purposes only, it is up to the authors to demonstrate that in the process of training a model for AI, it was didactically divided into four stages:

1. Data collection: This can be done from a variety of sources, including human-generated data and machine-generated data.
2. Data structuring: This involves normalizing and pre-processing the training data so that the model can understand it.
3. Model training: This involves using the training data to teach the model how to generate similar data. This is done through machine learning (ML) algorithms, such as linear regression or artificial neural networks.
4. Model evaluation and tuning: performance improvement after initial training. This can be done through techniques such as "cross-validation" and "regularization," which help prevent the model from becoming too complex and causing overfitting of the training data. In short, it can be said that the machine adjusts so much to the data that it becomes unable to generalize the learned patterns to new data.

As you can see, the process of training an AI model is a complex task and requires a deep knowledge of the areas related to computing, data science, and machine learning (ML). The steps described above are merely illustrative and serve to provide a macro-overview that can be adapted by technicians, according to the specific tasks where each model will be used in the practice of organizations.

For the co-authors of this article, Azevedo Guedes and Cezario (2024) [18] it is still necessary to "overcome the challenges and limitations in the process of training Artificial Intelligence (AI) models", as iterative training must apply techniques that allow a "controlled loop" until the model meets the objectives and/or performance requirements, which are usually preliminarily defined in the AI roadmap, this is also true for Generative AI (Gen-AI). Adjustments are not uncommon during this process of continuous improvement.

Some authors who have published works in AI in the last twenty years point out that in addition to the technical issues and the various types of algorithms used, there should be concerns about the social impacts and ethical issues of AI models (Ludermir, 2021 [19]; Sichman, 2021 [20]).

Nowadays, the impacts are present in several fields of knowledge and in several areas, which go far beyond the supply chain. For the SCM area, it is possible to highlight that investments in training and education are essential for employees at different hierarchical levels, especially those linked to decision-making, to understand how to apply Gen-AI in their daily work. Generative Artificial Intelligence (AI) has diverse applications in different industry sectors.

To illustrate some of the examples, it is possible to cite the report "The Generative AI Dossier" produced by the DELOITTE AI INSTITUTE in 2023, which points to at least six types of industries where generative AI is being used to: automate tasks, propose product design, perform demand forecasting, assist in production planning, improve customer service, help generate creative content, facilitate the screening of various data, develop new formulas for medicines, assist in the diagnosis of diseases, personalization of treatments, among many other applications.

Gen-AI's use cases are diverse and serve a multitude of industry types, but it is not the scope of this study to delve into the topic.

The main objective is to investigate the aspects of Gen-AI's influence on the management of processes and people in Supply Chains in Rio de Janeiro, exploring the benefits, challenges and emerging trends.

3 Method

The methodology of field research was used in terms of the procedure and qualitative exploratory research in terms of objectives, using a structured questionnaire as the main data collection tool. To this end, the questionnaires were analyzed, whose electronic access, each participant answered individually and voluntarily, through Google forms, thus avoiding any bias that could interfere with the results. Supply chain professionals and MBA students from a Federal University in Rio de Janeiro were invited from a diverse group of 240 participants between December 22, 2023, and January 10, 2024. The responses provided insights into their experiences, challenges, and potential opportunities related to integrating generative AI into supply chain processes; as well as the suggestion of some performance indicators to measure progress in the use of generative AI. The forms assessed respondents' experience and knowledge of generative AI. Upon receipt of the questionnaire responses, the results were analyzed with pie charts and frequency tables with proportion. The results were inferred, generating the discussions presented.

4 Results

This section presents the results of 160 questionnaires received, out of the 240 expected.

Participants' positions: transport administrator (7%); planning analyst (7%); expert buyer (7%); logistics coordinator (9%); Senior Strategic Analyst (7%); logistics analyst (10%); senior S&OP analyst (7%); supply analyst (11%); head of purchasing (7%); full buyer (4%); offshore buyer (4%); Supply Chain Managers (12%); maintenance manager (5%) Fig. 1.

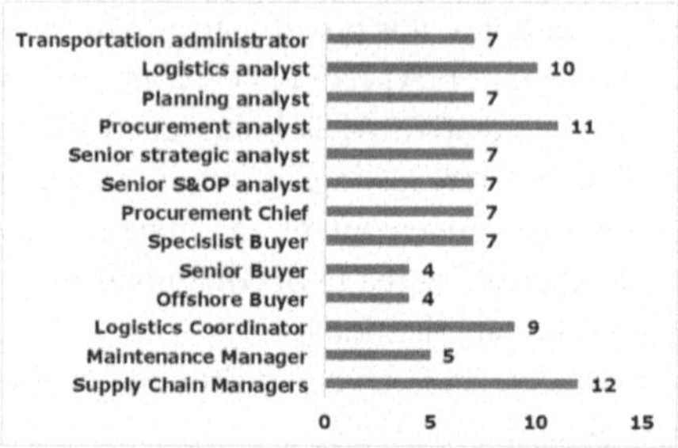

Fig. 1. Percentage of participants by position in companies. Source- the author

In response to what is your level of experience in terms of years of working in supply chain management? 57% reported 2 to 10 years and 47% 10 to 20 years.

The main areas of activity of the informed participants were Insurance Sector, Hospital, Railway, Retail, Building Maintenance, Information Technology, Oil and Gas, Municipal Government, Agribusiness, Transportation, Energy.The questions were answered in the following areas:

I. Use of generative AI
   1. Does your organization already use generative AI systems in supply chain management?
   2. What is the potential impact of this technology on supply chain management?
II. Generative model training
   1. If applicable, what does the process of training generative AI models look like in your organization?
   2. What specific challenges or successful strategies would you like to share?
III. Improved operational efficiency
   1. How has generative AI contributed to operational efficiency in supply chain management?
IV. Performance Indicators
   1. What metrics or indicators to measure the impact of generative AI on supply chain management do you suggest?

Considering that the impacts of innovative improvements in the Supply Chain, from the application of management systems (MRP I; MRP II, ERP) and other ICT-based technologies took about 15 to 20 years to consolidate in the country, our expectation was low, as highlighted by Meriton – the practical adoption of these technologies has not corresponded to optimistic projections.

Results of Survey Respondents' Participation

I. **Use of generative AI under way**

Does your organization already use generative AI systems in supply chain management? (Fig. 2)

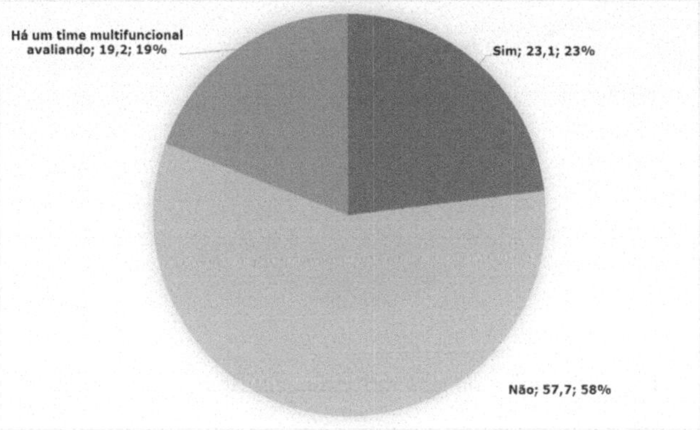

Fig. 2. Use of generative AI by participants. Source: the author

The graph shows that most companies still do not use generative AI in the Supply Chain, but that 23% use it and 19% are in the process of evaluating the project, through a cross-functional team. We can associate this response with the different improvement projects that, over the last 30 years, have sought to optimize Supply Chain processes (paperless factory, lean, Just in time, among others); making the analysis of where to use generative AI more selective and time-consuming. And, perhaps, with more investments in human resources and IT.

What is the potential impact of this technology on supply chain management? (Fig. 3)

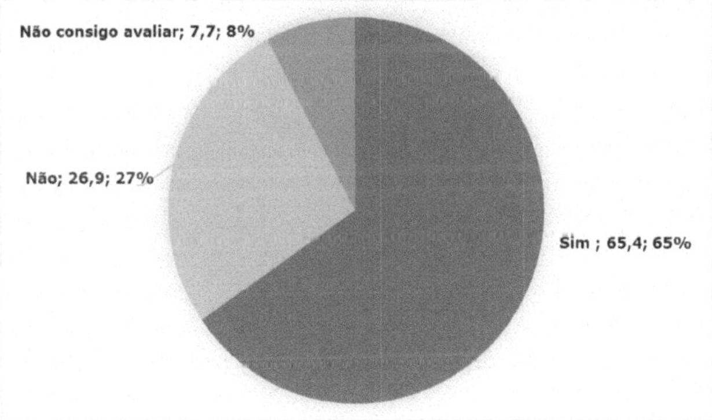

Fig. 3. Potential impact of generative AI on participants' Supply Chain. Source: the author

The answers to the second question demonstrate that many participating professionals are aware of the high impact of generative AI and are waiting for the directors to make company's investments decision so that they can get the projects underway.

II. **Generative model training**

If applicable, what does the process of training generative AI models look like in your organization? (Fig. 4)

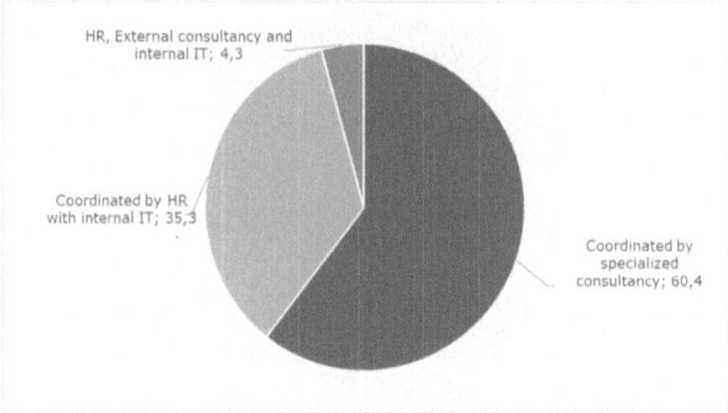

Fig. 4. Training on generative models. Source: the author

Regarding the training of professionals who will be involved in the projects (a fundamental investment for the formation of a critical mass of AI professionals), it is observed that many companies (65%) are using specialized consultancies, while approximately 36% are using internal IT personnel with HR coordination.

III. Improved operational efficiency.

How does generative AI contribute to operational efficiency in supply chain management? (Fig. 5)

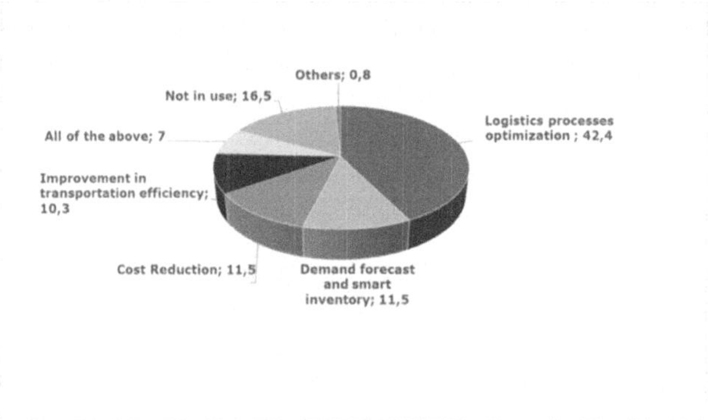

Fig. 5. Improvement in operational efficiency noted by participants. Source: the author

When evaluating the benefits of generative AI for organizations, 42% of respondents believe that investments will optimize logistics processes; This was followed by a better demand forecast and cost reduction (11.5%). Apparently, participants live amid complex logistics processes and that investments in generative AI can bring efficiency, effectiveness that result in better customer service.

IV. Performance Indicators

What metrics or indicators to measure the impact of generative AI on supply chain management do you suggest?

1. I can't Give an Opinion (57%)

2. Customer satisfaction. (30%).
3. Still in the process of implementation (15%).
4. Return on Investment (8%).

Regarding the suggestions for performance indicators and metrics that can assess whether the project is returning the expected benefits, the majority answered that they cannot give an opinion, in line with the answer to the first question in which the majority stated that they do not use generative AI. However, 30% say they are measuring the level of customer service, which aligns with concern about return on investment.

4.1 In-Depth Analysis

Based on the opinions expressed by the participants, it was decided to carry out an analysis that could clarify the beneficial impacts on Supply Chain processes and activities. In this way, the reader can assess the potential that Generative AI projects in the supply chain have to transform their various processes, providing significant improvements in terms of efficiency, flexibility, and resilience.

Demand Forecasting: The use of Gen-AI can increase the accuracy of demand forecasts. By analyzing large historical datasets, consumer behavior, and external factors, generative models can identify complex patterns and provide more accurate predictions; resulting in more efficient inventory management, reducing costs associated with excess or lack of products.
Routing optimization: Gen-AI can optimize resource allocation by considering variables such as real-time traffic conditions, storage capacity, and delivery times, generative algorithms can adjust routes in real-time, minimizing logistics costs and improving operational efficiency.
Inventory and Warehouse Management: Gen-AI can, by predicting seasonal demands, identify buying patterns and suggest replenishment strategies. In addition, it can optimize the storage layout, facilitating automated picking and packing and reducing cycle time.
Predictive Maintenance: Gen-AI can predict equipment failures and suggest maintenance interventions before problems occur; reducing the occurrence of unscheduled downtime, increasing asset life, and minimizing maintenance costs.
Supply Chain Collaboration: Gen-AI can facilitate collaboration between supply chain actors by analyzing shared data such as demand forecasts, production capacity, and inventory availability, suggesting dynamic adjustments and improving overall chain efficiency.
Production Customization: Gen- AI ability to generate models and patterns can be applied to production customization, optimizing production in smaller batches and customized products, quickly adapting to changing consumer preferences, without compromising efficiency.

5 Conclusions

This project aimed to understand the aspects of the influence of AI Gen. in the management of processes and people in Supply Chains in Rio de Janeiro, exploring the benefits, challenges and emerging trends. The qualitative nature of the study allowed for meaningful exploration, through the perspectives of the participants, shedding light on the current state and future implications of generative AI in the local supply chain landscape.

It highlighted the inherent complexity of supply chains, addressing their integration, through the crucial processes from the origin to the point of consumption. Decision-making in this context takes place on various time scales, from strategic to operational issues.

The introduction of Artificial Intelligence (AI) into the supply chain, especially through generative AI, is touted as a complex approach that can significantly improve operational outcomes (as pointed out by field survey respondents).

Generative AI, in particular, is recognized for its ability to generate entirely new content, such as images, natural language, music, and video. The importance of Large Scale Language Models (LLM) in generating specific instructions to elicit desired responses from generative language models is emphasized.

When applied in the supply chain, AI, according to studies by Doanh et al. (2023), offers substantial operational improvements, including advancements in efficiency, safety, quality control, and market analysis. Generative AI is highlighted as a proactive tool in identifying and maintaining defects, resolving problems quickly through real-time data analysis. In addition, its market analytical capabilities improve inventory control and marketing strategies.

By analyzing the literature on generative AI in the supply chain, the study highlights the significant role of AI in efficient supply chain management, resilience, and optimization. Results from Yandrapalli (2023) indicate improvements in service levels and reduced logistics costs with the successful deployment of AI. However, Meriton et al. (2020) warn that despite optimistic projections, the practical adoption of big data technologies, including generative AI, in supply chain management practices has fallen short of expectations.

Ethical and trustworthiness concerns associated with generative AI have been found, particularly in the educational context, where the use of ChatGPT can present challenges related to correctness, trustworthiness, and plagiarism rates. In addition, financial limitations, data synthesis, and privacy issues are mentioned as challenges to be overcome.

Training in generative AI has emerged as an important part of its use, describing it as a complex process that involves data collection and structuring, model training, and continuous evaluation. The role of generative AI emerges as a complement to human skills, highlighting the importance of combining artificial intelligence with human expertise to build robust supply chains.

The revolutionary influence of generative AI on the supply chain, offering significant benefits but recognizing the need to address ethical, technical, and operational challenges to maximize its potential in supply chain management practices in the specific context of Rio de Janeiro.

As benefits, generative AI has been noted to promise to increase productivity, reduce time spent on repetitive activities, launch new research initiatives, and facilitate the acquisition, synthesis, and summary of large amounts of data.

The results of the field survey indicate that most companies have yet to incorporate generative AI systems into their supply chain operations. This scenario can be attributed to the complexity of the decision-making process, which involves careful evaluation and possible investments in human resources and information technology. The presence of previous improvement projects over the past few decades may be contributing to a more selective approach to the implementation of generative AI.

Regarding the potential impact of technology, the participating professionals recognize its high impact but are waiting for strategic decisions from the directors to start projects. This delay may reflect a growing awareness of the potential benefits of generative AI in optimizing logistics processes, forecasting demand, and reducing costs.

On the other hand, risks commonly identified by managers include: responses contrary to company policies or brand image; lack of knowledge about current organizational knowledge; difficulties in differentiating fabricated responses from creative outputs; audit trails and unclear citation sources; possible threat for exposure of trade secrets; and the cost of using proprietary LLMs; as well as, unrealistic expectations of LLMs as a comprehensive solution for enterprise content without human involvement.

As for the training process, the survey highlights a diverse approach. Most companies are turning to specialized consultancies to train their professionals, indicating the need for specialized knowledge in this emerging field. In addition, a significant portion uses in-house IT personnel, coordinated by the human resources department. This highlights the importance of training a critical mass of AI professionals to ensure successful implementation.

When evaluating how generative AI contributes to operational efficiency in supply chain management, participants highlight the optimization of logistics processes as the most significant benefit. Better demand forecasting and cost reduction are also mentioned, indicating an expectation that generative AI can bring efficiency and effectiveness to processes, resulting in improved customer service.

When asked about metrics or indicators to measure the impact of generative AI, many participants are unable to give an opinion, aligning with the lack of widespread use of the technology. However, those who are measuring indicate a focus on the level of customer service, indicating a practical concern for return on investment and delivering perceived value to customers.

Future research is expected to take an interdisciplinary approach, addressing the technical, social, economic, and ethical aspects of artificial intelligence in the industrial sector. A long-term challenge will be to evaluate the effects of AI applications on different facets of human activity and life, both positively and negatively. It is suggested to expand the research to other states to get a more comprehensive view. Another suggestion for future work is the analysis of the role of AI Gen. in promoting sustainable practices in the supply chain.

References

1. Moreira, O. J.: Cadeia De Suprementos Um Novo Modelo De Gestão Empresarial. Scortecci Editora (2006)
2. Dürr, J.: Global value chains and traditional supply chains: evidence from Guatemala. Dev Policy Rev. **36** (2018). https://doi.org/10.1111/dpr.12348
3. Rausch-Phan, M.T., Siegfried, P.: Traditional Supply Chain Management. In: Sustainable supply chain management, pp. 7–15. Springer International Publishing, Cham (2022)
4. Martins, W., Martins, V. G.: Desafios e Inovações da IA Generativa: Large Language Model (LLM) e Engenharia de Prompt. In: Tecnologias do Amanhã, 1st ed. Letra e Forma Editora, pp 154–172 (2023)
5. Zhang, X., Yue, J.: Measurement model and its application of enterprise innovation capability based on matter element extension theory. Procedia Eng **174**, 275–280 (2017). https://doi.org/10.1016/j.proeng.2017.01.136
6. Li, B., Mellou, K., Zhang, B., et al.: Large Language Models for Supply Chain Optimization (2023)
7. Alto, V.: ULTIMATE GUIDE TO CHATGPT AND OPENAI harness the capabilities of openai's large language model... for productivity and innovation with gpt technolog. PACKT PUBLISHING LIMITED, S.l (2023)
8. Ozdemir, S.: Quick start guide to large language models: strategies and best practices for using ChatGPT and other LLMs. Addison-Wesley, Hoboken, New Jersey (2024)
9. Doanh, D.C., Dufek, Z., Ejdys, J., et al.: Generative AI in the manufacturing process: theoretical considerations. Eng. Manag. Prod. Serv. **15**, 76–89 (2023). https://doi.org/10.2478/emj-2023-0029
10. Javaid, M., Haleem, A., Singh, R.P., Suman, R.: Artificial intelligence applications for industry 4.0: a literature-based study. J. Ind. Integr. Manag. **07**, 83–111 (2022). https://doi.org/10.1142/S2424862221300040
11. Lee, K., Cooper, A.F., Grimmelmann, J.: Talkin' 'Bout AI generation: copyright and the generative AI supply chain. SSRN Electron. J. (2023) https://doi.org/10.2139/ssrn.4523551
12. Yandrapalli, V.: Revolutionizing supply chains using power of generative AI. Int. J. Res. Publ. Rev., **4**, 1556–1562 (2023). https://doi.org/10.55248/gengpi.4.1223.123417
13. Meriton, R., Bhandal, R., Graham, G., Brown, A.: An examination of the generative mechanisms of value in big data-enabled supply chain management research. Int. J. Prod. Res. **59**, 7283–7310 (2021). https://doi.org/10.1080/00207543.2020.1832273
14. Michel-Villarreal, R., Vilalta-Perdomo, E., Salinas-Navarro, D.E., et al.: Challenges and opportunities of generative AI for higher education as explained by ChatGPT. Educ. Sci. **13**, 856 (2023). https://doi.org/10.3390/educsci13090856
15. Fosso Wamba, S., Queiroz, M. M.: Chiappetta Jabbour, C. J., Shi, C. (Victor): Are both generative AI and ChatGPT game changers for 21st-Century operations and supply chain excellence? Int. J. Prod. Econ., **265**, 109015 (2023). https://doi.org/10.1016/j.ijpe.2023.109015
16. Bucharest University of Economic Studies, Bucharest, Romania, Dijmărescu, E.: AI Trends: Salient Aspects for the Manufacturing Sector and Its Global Supply Chain. In: New Trends in Sustainable Business and Consumption. Editura ASE, pp. 168–175 (2023)
17. Acheampong, F.A., Nunoo-Mensah, H., Chen, W.: Transformer models for text-based emotion detection: a review of BERT-based approaches. Artif. Intell. Rev. **54**, 5789–5829 (2021). https://doi.org/10.1007/s10462-021-09958-2
18. Souza, R.M., Cezario, B.S., Affonso, E.O.T., et al.: My human rights smart city: improving human rights transparency identification system. Sustainability **16**, 1274 (2024). https://doi.org/10.3390/su16031274

19. Ludermir, T.B.: Inteligência Artificial e Aprendizado de Máquina: estado atual e tendências. Estud Av **35**, 85–94 (2021). https://doi.org/10.1590/s0103-4014.2021.35101.007
20. Sichman, J.S.: Inteligência Artificial e sociedade: avanços e riscos. Estud. Av **35**, 37–50 (2021). https://doi.org/10.1590/s0103-4014.2021.35101.004

Web Bibliography

21. Deloitte AI Institute: The Generative AI Dossier. *Deloitte Consulting*, 12 Sept 2023. https://www2.deloitte.com/content/dam/Deloitte/th/Documents/deloitte-consulting/generative-AI-dossier.pdf. Accessed 6 Jan 2024

Organizational Culture and Turnover Intentions: The Role of Affective Commitment

Ana Lúcia Dias and Ana Palma-Moreira(✉)

Universidade Europeia, Quinta Do Bom Nome, Estr. da Correia 53, 1500-210 Lisboa, Portugal
ana.moreira@universidadeeuropeia.pt

Abstract. This study aimed to investigate the effect of organizational culture on turnover intentions and whether this relationship is mediated by affective commitment. For this purpose, the following hypotheses were formulated: (1) organizational culture (support culture, innovation culture, goal culture, rule culture) has a negative and significant effect on turnover intentions; (2) organizational culture (support culture, innovation culture, goal culture, rule culture) has a positive and significant effect on affective commitment; (3) affective commitment has a negative and significant effect on turnover intentions; (4) affective commitment mediates the relationship between organizational culture (support culture, innovation culture, goal culture, rule culture) and turnover intentions. A total of 345 individuals working in Portuguese organizations participated in this study. The results indicate that all dimensions of organizational culture have a negative and significant effect on turnover intentions and a positive and significant effect on affective commitment, with support culture having the strongest effect on both affective commitment and turnover intentions. Affective commitment has a negative and significant effect on turnover intentions. Affective commitment partially mediates the relationship between three dimensions of organizational culture (support culture, goal culture, and rule culture) and turnover intentions. The mediation effect of affective commitment in the relationship between innovation culture and turnover intentions is total. These results indicate that organizations should foster a strong organizational culture, especially support culture, so that their employees feel more affectively committed to the organization and reduce their turnover intentions.

Keywords: Organizational Culture · Turnover Intentions · Affective Commitment

1 Introduction

The success and sustainability of an organization depends on its human resources management. One of the major challenges of human resources management is related to high employee turnover. This study explores the interconnection between employee turnover, emphasizing the influence of organizational culture and employee affective commitment on turnover intentions.

M. J. Sousa (Ed.): UFHRD 2024, ISEM 52, pp. 144–155, 2025.
https://doi.org/10.1007/978-3-031-92548-1_9

Besides being an inherent reality of the contemporary corporate environment, employee turnover triggers a series of profound implications beyond the simple replacement of human resources. In addition, turnover intentions are an indicator of voluntary turnover, as they are an antecedent to voluntary departure from the organization. Understanding these motivations is vital to anticipating and mitigating the factors that drive departures, contributing to constructing preventive and proactive strategies. In this way, we will deepen our understanding of how employee turnover transcends the dynamics of talent coming and going, uncovering its ramifications for organizational culture, affective commitment and intentions to leave.

This study aims to investigate the relationship between organizational culture and turnover intentions and whether affective organizational commitment mediates this relationship.

2 Literature Review

2.1 Organizational Culture

Organizational culture is the set of attitudes, values, norms, intentions and beliefs shared by all the employees of an organization [1]. Employees' decisions in their professional environment are guided by the organization's culture where they work [2].

In this study, we will focus on the contrasting values model developed by [3], which considers the existence of four types of culture:

(1) In a supportive or clan culture, shared values and goals lead to improved teamwork. Participation in organizational programmes and shared commitments gives them power.
(2) Innovative culture or adhocracy is characterized as a temporary and dynamic culture in which employees require a high level of specialization. By accepting changes and challenges, employees adapt and support innovation in the organization.
(3) Bureaucratic or hierarchical culture stands out for its clear hierarchical structure, distinct division of powers, and impersonality. It is based on standardizing processes, following formal rules and policies, clearly dividing tasks, and a predominance of strictly formal employee relationships.
(4) Rational or market culture, which focuses externally, seeking to create competitive advantages and achieve goals. It is a results-oriented and competitive culture based on achieving objectives and is directed and orientated toward customers and their needs.

2.2 Turnover Intentions

Turnover intentions are an employee's intention to leave the organization where they work very shortly [4] and can influence the quality of their work [5].

Employee turnover has significant costs for organizations. In addition to the direct costs involved in recruiting, selecting and training new employees, there are indirect costs associated with new employees' inexperience and unfamiliarity with operating procedures, which can result in errors and delays and negatively impact the company's revenue and reputation [6]. By reducing turnover, organizations can implement strategies such as ongoing training programs, career development opportunities, encouraging retention, recognition and rewards policies, a positive and inclusive work environment, investing in the well-being and thus increasing employee satisfaction, engagement and loyalty to the company and thus reducing the costs associated with turnover [7].

Organizational Culture and Turnover Intentions. Organizational culture significantly influences employees' attitudes and behaviours [8]. When employees identify with the culture of the organization where they work, their turnover intentions decrease [4].

In this way, Organizational Culture plays a fundamental and significant role in employees' decision to stay with or leave the organization [5].

Hypothesis 1: Organizational culture (support, innovation, rules and objectives) negatively and significantly affects turnover intentions.

2.3 Affective Commitment

Affective commitment is related to the will to remain in the organization since it concerns the emotional connection to it [9]. The association between affective commitment and the workplace is positive [10]. It is associated with the emotional attachment and positive bond employees develop with the organization [9] and changes throughout the different career stages [11]. Age and tenure in the organization influence the formation and evolution of this commitment [12].

Employees with a solid affective commitment tend to have higher levels of job satisfaction, greater job involvement and a more positive attitude towards the organization [9].

Organizational Culture and Affective Commitment. The relationship between organizational culture and employees' affective commitment to the organization where they work is strong [13]. Organizational culture is one of the most important factors for employees to feel emotionally attached to the organization, developing a high emotional commitment to it [14]. Employees in an organization feel valued when the prevailing culture promotes trust, transparency and collaboration, leading them to contribute to its success [13]. It significantly influences employees' readiness for change and their affective commitment to organizational change [13]. It is a significant indicator of improved loyalty and retention [14]. This relationship can be interpreted based on social norm theory [15] and the premise of reciprocity [16] because employees will feel more emotionally attached to their work organization when they perceive a good organizational culture.

Hypothesis 2 - Organizational culture (support, innovation, rules and objectives) positively and significantly affects affective commitment.

Affective Commitment and Turnover Intentions. Affective commitment significantly influences employees' turnover intentions [17]. Employees have high turnover intentions when their affective commitment to the organization where they work is low [4]. This can be done through strategies that promote a positive working environment, which values and recognizes employees' work and offers opportunities for professional development [17].

The organization must ensure a favourable environment, reinforcing its relationship with employees so that their affective commitment increases, and turnover intentions decrease. It should also create professional development opportunities that allow employees to acquire new competencies and grow professionally within the organization [12].

Hypothesis 3 - Affective commitment negatively and significantly affects turnover intentions.

Organizational Culture, Affective Commitment and Turnover Intentions. High employee turnover can harm an organization that relies heavily on its employees to achieve competitiveness [5]. Organizations must adopt strategies that promote organizational commitment and employees' psychological capital to reduce their intention to leave [5]. Organizational culture and affective commitment significantly influence employees' turnover intentions [18]. Cultivating a positive culture and promoting employee commitment are essential strategies for reducing turnover and maintaining a productive and healthy work environment [19].

Each type of culture (supportive culture, goal culture, innovation culture and rules culture) can impact employees' affective commitment and satisfaction differently, affecting their intentions to leave and look for new opportunities in other organizations [8].

Hypothesis 4 - Affective commitment has a mediating effect on the relationship between organizational culture (support, innovation, rules and objectives) and turnover intentions.

Figure 1 presents the research model that summarizes the hypotheses formulated in this study.

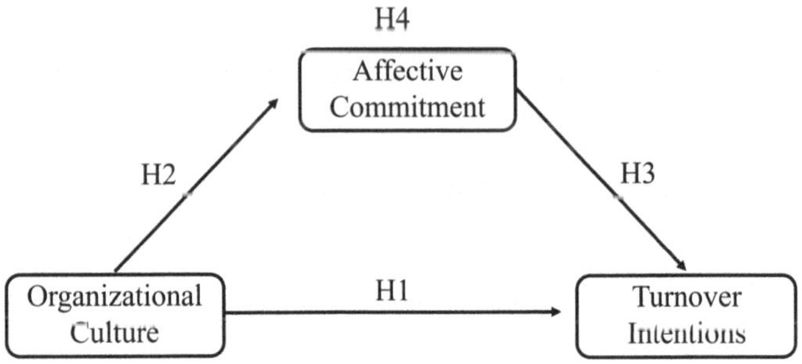

Fig.1. Research Model

3 Method

3.1 Data Collection Procedure

This study used a non-probabilistic, convenient, and intentional snowball sampling process [20]. The data was collected simultaneously, making it an exploratory and cross-sectional study.

The questionnaire was created on Google Forms, and the link was shared with the researchers' contacts via email, LinkedIn, WhatsApp, and Instagram.

Before starting to answer the questionnaire, the participants had to read the informed consent form, which guaranteed the confidentiality of the data. The participants also had to answer a question about whether they wanted to take part in the study.

The questionnaire included sociodemographic information to characterize the sample. In addition to these questions, participants were asked to answer three scales (organizational culture, affective commitment, and exit intentions). Data collection took place between January and March 2024.

3.2 Participants

A total of 349 participants took part in this study, 345 of whom were considered in the subsequent statistical analyses because they met the participation conditions. The participants who voluntarily participated in this study were between 19 and 70 ($M = 37.95$, $SD = 11.79$). As for gender, 57.1% of the participants were female, and 42.9% were male. Of these, 22.3% had a level of education equal to or less than the 12th grade, 47.5% had a bachelor's degree, and 30.1% had a master's degree or higher. Regarding seniority, 18% have been with the organization for less than a year, 56.2% between 1 and 10 years, 12.7% between 11 and 20 years and 13% more than 20 years. About employment, 13.9% have an uncertain term contract, 12.5% a fixed term contract, 65.8% an open-ended contract and 7.8% another type of contract.

3.3 Data Analysis Procedure

After collecting the data, it was imported into SPSS Statistics 29 software for further analysis. The first step was to assess the metric properties of the instruments used in this study. Confirmatory factor analyses (CFA) were carried out using AMOS Graphics 29 software to check the validity of the organizational culture, affective commitment, and exit intention instruments.

The reliability of the dimensions of all the instruments used in this study was assessed by calculating Cronbach's alpha coefficient, whose value should vary between 0 and 1, excluding negative values [21]. A value of 0.70 is considered the minimum acceptable in organizational studies [22]. As for the sensitivity study, various measures of central tendency, dispersion and shape were calculated for the different items of the scales used. This approach will allow normality to be analysed for all the items and scales in question. The items should have responses distributed at all points on the scale and not have the median close to the limits. In addition, the items' absolute values of asymmetry and kurtosis should be less than 2 and 7, respectively, as [23] recommended. Descriptive statistics were then analysed for the sample and the variables under study.

The formulated hypotheses were tested using simple and multiple linear regressions, considering a significance level of 0.05.

3.4 Instruments

To measure the perception of organizational culture, the organizational culture subscale of the FOCUS (first organizational culture unified search) instrument will be used, validated and adapted for the Portuguese population [24]. This instrument consists of 35 items anchored on a 6-point Likert scale. This questionnaire consists of four dimensions: culture of innovation, culture of support, culture of objectives and culture of rules.. A four-factor CFA was carried out to test the validity of this instrument. The fit indices obtained are adequate or very close to adequate ($\chi2/gl = 1.98$; GFI = .86; CFI = .94; TLI = .93; RMSEA = .053; SRMR = .088), which confirms the factor structure of this instrument. However, item 30 had to be removed as it had a low factor weight. The composite reliability values range from .87 (goals culture) to .94 (support culture), which indicates that all the dimensions have good composite reliability. Regarding convergent validity, Cronbach's alpha values vary between .87 (rules culture) and .94 (support culture). As for convergent validity, the AVE values vary between .47 (innovation culture) and .57 (support culture). As can be seen, only the innovation culture dimension has an AVE value below .50. However, as Cronbach's alpha value for this dimension is above .70, this dimension can be considered acceptable convergent validity [25].

To measure turnover intentions, we used the questionnaire developed by [26] and adapted for the Portuguese population by [27], consisting of 6 items, classified on a 5-point scale (from 1 "Not at all applicable" to 5 "Totally applicable"). A one-factor CFA was carried out. The fit indices obtained are adequate ($\chi2/gl = 2.65$; GFI = .99; CFI = .99; TLI = .98; RMSEA = .069; SRMR = .027), which confirms this instrument's factor structure. This instrument has a composite reliability of .91 and a good convergent value (AVE = .64). It has a Cronbach's alpha of 0.92, indicating good internal consistency.

The affective commitment was measured using the affective organizational commitment component of the questionnaire developed by [28]. It consists of 6 items classified on a 7-point Likert scale (from 1 "Strongly Disagree" to 7 "Strongly Agree"). A one-factor CFA was carried out. The fit indices obtained are adequate ($\chi2/gl = 3.03$; GFI = .99; CFI = .99; TLI = .98; RMSEA = .077; SRMR = .037), which confirms this instrument's factor structure. This instrument has a composite reliability of .90 and a good convergent value (AVE = .61). It has a Cronbach's alpha of 0.92, indicating good internal consistency.

4 Results

4.1 Descriptive Statistics of the Variables Under Study

To understand the answers obtained by the participants, the descriptive statistics of the variables under study were analysed.

The participants in this study have a high perception of a supportive culture, goals culture and rules culture, as their answers are significantly above the centre point of the scale (3.5). As for turnover intentions, the participants' answers are significantly below the scale's central point (3), indicating that the participants in this study have low intentions to leave the organization where they are working. Finally, affective commitment is significantly above the scale's central point, indicating they are highly committed to the organization (Table 1).

Table 1. Descriptive statistics of the variables under study

	t	df	p	Mean	SD
Support Culture	6.24***	344	<0.001	3.82	0.97
Innovation Culture	−0.62	344	0.268	3.47	0.89
Goals Culture	4.51***	344	<0.001	3.76	1.06
Rules Culture	12.90***	344	<0.001	4.08	0.83
Turnover Intentions	−9.19***	344	<0.001	2.47	1.06
Affective Commitment	7.11***	344	<0.001	4.59	1.52

Note. *** $p < 0.001$

4.2 Association Between the Variables in Study

The association between the variables under study was carried out using Pearson's correlations. Affective commitment is positively and significantly associated with support culture, innovation culture, Goals culture, and rules culture (Table 2). Turnover intentions are negatively associated with support culture, innovation culture, goals culture, rules culture, and affective commitment (Table 2).

Table 2. Association between the Variables in Study

	1.1	1.2	1.3	1.4	2	3
1.1 Support Culture	--					
1.2 Innovation Culture	.58***	--				
1.3 Goals Culture	.58***	.72***	--			
1.4 Rules Culture	.57***	.56***	.60***	--		
1 Affective Commitment	.58***	.39***	.43***	.31***	--	
2. Turnover Intentions	–.45***	–.31***	–.36***	–.33***	–.70***	--

Note. *** $p < 0.001$

4.3 Hypotheses

Hypotheses 1 and 2 were tested using multiple linear regressions using the stepwise method.

The results show that only support culture ($\beta = -0.37$, $p < 0.001$) and goal culture ($\beta = -0.15$, $p = 0.012$) negatively and significantly affect turnover intentions (Table 3). The model explains 21% of the variability of the dependent variable. The model is statistically significant ($F(2, 342) = 47.51$, $p < 0.001$) (Table 3). Hypothesis 1 was partially supported.

Table 3. Results of hypothesis 1

Independent Variable	Dependent Variable	F	p	R2a	β	p
Support Culture	Turnover Intentions	47.51***	<0.001	0.21	–.37***	<0.001
Goals Culture					–.15*	0.012

Note. * $p < .05$; *** $p < .001$

Table 4. Results of hypothesis 2

Independent Variable	Dependent Variable	F	p	R2a	β	p
Support Culture	Affective commitment	92.44***	<0.001	0.35	0.50***	<0.001
Goals Culture					0.15**	0.006

Note. ** $p < .01$; *** $p < .001$

The results show that only support culture (β = 0.50, p <0.001) and goal culture (β = –0.15, p = 0.006) positively and significantly affect affective commitment (Table 4). The model explains 35% of the variability of the dependent variable. The model is statistically significant (F (2, 342) = 92.44, p <0.001) (Table 4). Hypothesis 2 was partially supported.

Table 5. Results of hypothesis 2

Independent Variable	Dependent Variable	F	p	R^2	β	p
Affective Commitment	Turnover Intentions	320.16***	<0.001	0.48	–0.70***	<0.001

Note. *** $p < 0.001$

The results show that affective commitment (β = –0.70, p <0.001) has a negative and significant effect on turnover intentions (Table 5). The model explains 48% of the variability of the dependent variable. The model is statistically significant (F (1, 343) = 320.16, p <0.001) (Table 5). Hypothesis 3 was supported.

Hypothesis 4 assumed a mediation effect, so the assumptions were followed according to [29]. The assumptions were confirmed in hypotheses 1, 2, and 3. For this reason, we will only test if affective commitment mediates the relationship between two dimensions of organizational culture (support and goals) and turnover intentions.

The results in Table 6 show that affective commitment has a total mediating effect on the relationship between organizational culture (support and goals) and turnover intentions since when the mediating variable was introduced into the regression equation, both the support culture and the goals culture no longer had a significant effect on turnover

Table 6. Results of the mediation effect

Independent Variable	Turnover Intentions	
	β Step1	β Step2
Support Culture	−0.37***	−0.05
Goals Culture	−0.15*	−0.05
Affective Commitment		−0.65***
F	47.51***	108.33***
$R^2{}_a$	0.21	0.48
$\Delta R^2{}_a$		0.27***

Note. * p <0.05; *** p <0.001

intentions. The model explains 48 per cent of the variability in turnover intentions. The increase in variability is statistically significant ($\Delta R2a = 0.27$, $p < 0.001$). Both models are statistically significant ($F1 = 47.51$, $p < 0.001$; $F2 = 108.33$, $p < 0.001$). Hypothesis 4 was partially supported.

5 Discussion

This study aimed to study the effect of organizational culture on intentions to leave and whether this relationship was measured by affective commitment.

Firstly, only hypothesis 1 was partially confirmed: only two organizational culture dimensions (support and goal cultures) negatively and significantly affect turnover intentions. These results are in line with those obtained by [30].

Secondly, hypothesis 2 was partially confirmed, as only supportive and goal-oriented cultures positively and significantly affect affective commitment. These results are partially in line with the literature since when employees identify with their organization's culture, they have higher levels of affective commitment [31].

Thirdly, hypothesis 3 confirmed that affective commitment significantly decreases turnover intentions. These results align with existing literature since employees' intentions to leave decrease when they feel emotionally committed to their work organization [32]. Some studies prove the great interest in studying affective commitment because it is the best reducer of turnover intentions [33, 34].

Finally, hypothesis 4 was partially proven. Affective commitment fully and significantly mediates the relationship between organizational culture (support and objectives) and turnover intentions. These results align with the literature because when employees identify with the organization where they work, their affective commitment to it strengthens, and their turnover intentions decrease [12].

It should be noted that the culture perceived with the highest average was the culture of rules, followed by the culture of support. In Portugal, there is still a high level of hierarchical distance in organizations, which may be why the culture of rules is perceived more highly. These results align with those obtained in the study carried out in Portugal [30].

5.1 Limitations

This exploratory study has some limitations. One of the main limitations is the data collection procedure.

The data was collected from various organizations and sectors, which may not fully represent the diversity of existing organizations. Therefore, the generalization of the results may be limited to other organizational realities. For future research, broader, more diverse samples, new sectors, and company sizes should be considered to broaden the applicability of the results.

5.2 Practical Implications

This study confirms that implementing strategies in organizations that promote a positive work environment is necessary to reduce turnover intentions. A positive work environment reinforces employees' emotional attachment to their work organization [17].

This study found that younger workers change workplaces more frequently. At the same time, those with a higher level of education tend to value professional development within the organization more [35]. In this way, retention strategies should be personalized according to the demographic characteristics of the job to be more effective.

The contributions of this study to human resources management are significant. They indicate that they should be concerned with maintaining a healthy and supportive work environment and promoting the appreciation of employees because only in this way will they be able to ensure that the organizations where they work have an organizational culture that promotes affective commitment and reduces turnover intentions. Among the various types of culture, according to the contrasting values model, the results of this study show us that the most important are the culture of support and the culture of goals.

6 Conclusions

The main objective of this study was to verify whether affective commitment mediated the relationship between organizational culture and turnover intentions.

Firstly, the study confirms that employees who feel supported, valued and respected by the organization have greater job satisfaction and a high level of commitment [8]. This positive perception of the work environment reduces turnover intentions, as employees develop a strong affective commitment to the organization [36].

Employee turnover results in significant costs for organizations, such as the costs associated with recruiting, selecting and training new employees, and indirect costs, such as those related to inexperience and unfamiliarity with operating procedures. This can lead to negative impacts, delays and errors, impacting the company's revenue and reputation [6].

References

1. Stone, P., Mooney-kane, C., Horan, E., Glance, T., Zwanziger, J., Dick, A.: Nurse working conditions and patient safety outcomes. Med. Care **45**, 571–578 (2007)

2. Rodrigues, D.L., Lopes, D., Pereira, M., De Visser, R., Cabaceira, I.: Sociosexual attitudes and quality of life in (non)monogamous relationships: the role of attraction and constraining forces among users of the second love web site. Arch. Sex. Behav. **48**, 1795–1809 (2019)
3. Quinn, R., Cameron, K.: Organizational life cycles and shifting criteria of effectiveness some preliminary evidence. Manage. Sci. **29**, 1–18 (1983)
4. Saripudin, Hubeis, M., Yusnita, N.: The influence of organizational culture on turnover through the quality of work life. J. Soc. Sci., **4**(1) (2023)
5. Said, H., Kamel, N., Abdel Aziz, S. Understanding the influence of organizational culture on turnover intention in Egyptian travel agencies. Int. Rev. Tourism, **20**(1) (2020)
6. Cotton, J.L., Tuttle, J.M.: Employee turnover: A meta-analysis and review with implications for research. Acad. Manag. Rev. **11**(1), 55–70 (1986)
7. Kumar, R., Ramendran, C., Yacob, P.: A study on turnover intention in fast food industry: employees' fit to the organizational culture and the important of their commitment. Int. J. Acad. Res. Bus. Soc. Sci. **2**(5), 2222–6990 (2012)
8. Mashile, D.A., Munyeka, W., Ndlovu, W.: Organizational culture and turnover intentions among academics: a case of a rural - based university. Stud. High. Educ. **46**(8), 1–9 (2019)
9. Meyer, J. P., Stanley, D. J., Herscovitch, L., Topolnytsky, L.: Affective, continuance and normative commitment to the organizational: A meta-analysis of antecedents, correlates and consequences. J. Vocational Behav., **61**(1) (2002)
10. Meyer, J.P., Smith, C.A.: HRM practices and organizational commitment: test of mediation model. Univ. Western Ontario **17**, 319–331 (2000)
11. Allen, N.J., Meyer, J.P.: Organizational commitment and career stage. J. Bus. Res. **26**(1), 49–61 (1993)
12. Nikolaou, I., Georgiou, K., Bauer, T. N., Truxillo, D. M.: Applicant reactions in employee recruitment and selection: the role of technology. In The Cambridge Handbook of Technology and Employee Behavior (pp. 100–130). Cambridge University Press (2019)
13. Haffar, M., Al-Karaghouli, W., Djebarni, R., Al-Hyari, K., Gbadamosi, G., Oster, F., Alaya, A., Ahmed, A.: Organizational culture and affective commitment to e-learning' changes during COVID-19 pandemic: The underlying affects of readiness for change. J. Bus. Res., **155** (2022)
14. Dunger, S.: Culture meets commitment: How organizational culture influences affective commitment. Int. J. Organ. Theory Behav., **26** (1 /2), 41–60 (2023)
15. Blau, P.M.: Exchange and Power in Social Life. Wiley, New York (1964)
16. Gouldner, A.W.: The norm of reciprocity: a preliminary statement. Am. Sociol. Rev. **25**, 161–178 (1960)
17. Zhu, K., Wang, X., Jiang, M.: The impact of organizational commitment on turnover intention of substitute teachers in public primary schools: Taking psychological capital as a mediator. Front Psychol., **13** (2022)
18. Dwivedi, S., Kaushik, S., & Luxmi. Impact of organizational culture on turnover intentions in BPO sector in India. Indian Journal of Industrial Relations, 48(4), 6 9–691 (2003)
19. Saeed, I., Waseem, M., Sikander, S., Rizwan, S.: The relationship of Turnover intention with job satisfaction, job performance, Leader member exchange, Emotional intelligence and organizational commitment. Int. J. Learn. Develop., **4** (2) (2014)
20. Trochim, W.: The Research Methods Knowledge Base. (2nd ed). Atomic Dog Publishing (2000)
21. Hill, M., Hill, A.: Investigação por Questionário. 2ª Ed. Edições Sílabo (2005)
22. Bryman, A., Cramer, D.: Análise de dados em ciências sociais. Introdução às técnicas utilizando o SPSS para windows (3ªed.) (2003)
23. Finney, S. J., DiStefano, C.: Non-normal and categorical data in structural equation modeling. In Hancock, G. R., R. Mueller, O. (eds.), Structural equation modeling: a second course (2nd ed., pp. 439–492). IAP Information Age Publishing (2013)

24. Neves, J. G.: Clima Organizacional, Cultura Organizacional e Gestão de Recursos Humanos. Lisboa: RH Editora, ISBN 972–6897–3–3 (2000)
25. Hair, J.F., Ringle, C.M., Sarstedt, M.: PLS-SEM: indeed, a silver bullet. J. Market. Theory Practice **19**(2), 139–152 (2011)
26. Bozeman, D., Perrewé, P.: The effect of item content overlap on organizational commitment questionnaire - turnover cognitions relationships. J. Appl. Psychol. **86**, 161–173 (2002)
27. Bártolo-Ribeiro, R.: Desenvolvimento e Validação de uma Escala de Intenções de Saída Organizacional. In: Pereira, M., et al. (eds.) Diagnóstico e Avaliação Psicológica: Atas do 10º Congresso da AIDAP/AIDEP, pp. 378–390. Coimbra; pp, AIDAP/AiDEP (2018)
28. Meyer, J.P., Allen, N.J.: Commitment in the workplace: theory, research and application, thousand Oaks. Sage Publication, CA (1997)
29. Baron, R.M., Kenny, D.A.: The moderator-mediator variable distinction in social psychological research: conceptual, strategic and statistical considerations. J. Pers. Soc. Psychol. **51**, 1173–1182 (1986)
30. Salvador, M., Moreira, A., Pistacho, L.: Perceived organizational culture and turnover intentions: the serial mediating effect of perceived organizational support and job insecurity. Soc. Sci., 11 (2022)
31. Niguse, G. T.: The effects of organizational culture on turnover intention: The mediating roleof job satisfaction, a case of Oromia Forest and Wild Life Enterprise. African J. Bus. Manage., **13** (2019)
32. Lee, J., Sohn, Y.W., Kim, M., Kwon, S., Park, I.-J.: Relative importance of human resource practices on affective commitment and turnover intention in South Korea and United States. Front. Psychol. **9**, 669 (2018)
33. Moreira, A., Sousa, M.J., Cesário, F.: Competencies development: the role of organizational commitment and the perception of employability. Social Sciences **11**, 125 (2022)
34. Palma-Moreira, A., Tomás, C., Antunes, A.: The mediating effect of affective commitment on the relationship between competence development and turnover intentions: does this relationship depend on the employee's generation? Admin. Sci. **14**, 97 (2024)
35. Hong, P. Y., Raymond, C. L., Gail, S. F.: Organizational Commitment and Turnover Intention in Union and Non-Union Firms. SAGE Open 1–11 (2014)
36. Hussain, T., Asif, S.: Is employees' turnover intention driven by organizational commitment and perceived organizational support. J. Qual. Technol. Manage. **8**, 01–10 (2012)

Psychological Capital as a Predictor of Perceived Performance: Study Applied to Portuguese Air Force Academy Students

Gonçalo Margarido[1](✉), Ana Gomes[1], and Sónia P. Gonçalves[2]

[1] Portuguese Air Force Academy, Centro de Investigação da Academia da Força Aérea, Instituto Universitário Militar, Sintra, Portugal
goncalomargarido@gmail.com, apgomes@academiafa.edu.pt

[2] Institute of Social and Political Sciences, Centre for Public Administration and Public Policies, Universidade de Lisboa, Lisbon, Portugal
spgoncalves@iscsp.ulisboa.pt

Abstract. The main objective of this study was to analyze the influence of Positive Psychological Capital on the performance of the Portuguese Air Force Academy students. Thus, the relation between these two variables was analyzed, adding work engagement as a mediator. In this way, Positive Psychological Capital promotes Work Engagement, which in consequence leads to better performance by students. The performance variable was analyzed through task performance, contextual performance, and counterproductive work behaviors. A cross-sectional correlational study was conducted with a quantitative methodology, through a questionnaire survey. The sample consisted of 242 students. Results confirmed the existence of a positive and significant relation between Positive Psychological Capital and Work Engagement, with Individual Work Performance. Regarding the three components of Individual Work Performance there was a positive and significant relationship between Positive Psychological Capital and Work Engagement with task and contextual performance, and negative and significant with counterproductive work behaviors. It was possible to confirm that Work Engagement plays an important role in the relationship between Positive Psychological Capital and Individual Work Performance. This research has expanded the knowledge of variables that lead to better performance levels in students of the Air Force Academy. Providing the necessary information so that, efforts can be invested to amplify these levels.

Keywords: Positive Psychological Capital · Individual Work Performance · Work Engagement · Air Force Academy · military students

1 Introduction

Today, more than ever, new ways of increasing competitiveness in the labor market are being sought. One of the ways found to increase this edge over other companies was to bet on the workers and their ability to improve. In this way, they started looking for new

M. J. Sousa (Ed.): UFHRD 2024, ISEM 52, pp. 156–171, 2025.
https://doi.org/10.1007/978-3-031-92548-1_10

fields to invest efforts in and thus increase organizations' productivity. With this, fields of study emerged that sought to increase the productivity of companies, paying special attention to individuals and their ability to improve (Lopes, Cunha, & Palma, 2005), as well as to their happiness and well-being, since they believe that this is an added value (Wright, 2003) As a result, new concepts emerged, such as Positive Psychological Capital and Work Engagement.

It is therefore relevant to understand the relationship between Positive Psychological Capital and Work Engagement because, as the literature states, Work Engagement is a predictor of performance and, as such, when analyzing this positive psychology component in employees, the aim will always be that work engagement levels are higher, which in turn will lead to better performance.

Due to its mission, performance is a central variable in any context, especially in a military organization. This was highlighted by Chambel et al. (2019), in a literature review on burnout and work engagement variables in a military context, who concluded that it is crucial to study the effects of these two variables on military behaviors, as well as on their performance.

In the organizational behavior literature, perceived performance has been conceptualized with the following dimensions, task performance, contextual performance and counterproductive work behaviors. These three types of perceived performance appear in 1990 with Campbell, who defines them as actions that contribute to the achievement of a company's objectives, the so-called Individual Work Performance.

This paper aims to study the relationship between Positive Psychological Capital and the perceived performance of students at the Portuguese Air Force Academy (PoAFA), with Work Engagement as a mediating variable. It is assumed that PPC is determinant for a psychological state of energy, enthusiasm and involvement with the study, which is embodied in better task and contextual performance and less counterproductive work behaviors in students of the Air Force Academy.

Based on these concepts, the following Research Question becomes pertinent:

What is the influence of Positive Psychological Capital on the perceived performance of Air Force Academy students?

Resulting of this Research Question, the need to deepen the theme arises, to understand the type of existing relationships between the variables. Thus, the following derived questions arise:

- *Students with high levels of Positive Psychological Capital show greater Work Engagement?*
- *Students with high levels of Positive Psychological Capital have higher perceived task and contextual performance and fewer counterproductive work behavior?*
- *Students with a high level of Work Engagement show higher perceived task and contextual performance and fewer counterproductive work behaviors?*
- *Does Work Engagement play a mediating role in the relation between Positive Psychological Capital and perceived performance, task performance, contextual performance and counterproductive work behaviors?*

A cross-sectional correlational study was conducted with a quantitative methodology, through a questionnaire survey. The sample consisted of 242 students of the PoAFA.

The paper is divided into five sections. After introduction, the second section provides an integrated overview of the main variables. Section 3 describes the method before setting out the results in Sect. 4. The paper then discusses the implications of the results and concludes with recommendations for further research.

2 Literature Review

2.1 Positive Psychological Capital

The term human capital refers to workers' skills, abilities and knowledge. Positive Psychological Capital, from now on referred to as psycap, reflects how these employee characteristics can be managed. Not only in what they represent in the present, but also in the idea that they can be optimized, and in the future, these same workers may achieve better results. Thus, reaffirming the advantages of investing in people (Lopes, Cunha & Palma, 2005).

To better understand these four fields of PPC, the following concepts emerge: self-efficacy, hope, optimism and resilience (Luthans, Youssef & Avolio, 2007). The term self-efficacy has been studied for several years, with Albert Bandura, with his sociocognitive theory, being the one who has focused on it the most (Stoykova, 2013). According to Bandura, self-efficacy can be defined as "the way people evaluate their abilities and how, through the perception of their own efficacy, this is reflected in their motivation and behavior" (Bandura, 1982, p. 122). The term hope, on the other hand, is seen by Snyder, Irving and Anderson (1991), as a positive motivational state, which leads the individual to achieve success. Something also important to mention, is that effectively a person who presents this characteristic often starts an action with a certain end, and is able to establish several paths in order to achieve that objective, even if some adversities arise (Snyder, 1994). According to Luthans (2002), optimism is an emotional and cognitive ability that is directly related to the expectations that an individual holds not only about the present, but also about the future, and that, in a way, will dictate the individual's determination to successfully achieve their goals (Nunes, 2015). According 2to Luthans and Youssef (2004) and Page and Donohue (2004), optimism associates internal and permanent causes with positive events, and on the other hand, any temporary and specific situation is derived from negative and undesirable events. Resilience is defined by Yunes (2003), and within the scope of psychology, as a term that is used to justify the ability of individuals, groups or organizations to overcome obstacles and crises. For Luthans et al.(2007), an individual with this characteristic is more likely to develop greater job satisfaction, an increase in organizational commitment and an improvement in social capital.

2.2 Work Engagement

Work Engagement is defined by Shaufeli et al. in 2002 as a positive, rewarding and work-related motivational state characterized by vigor, dedication and absorption. Vigor is when an individual displays "high levels of energy and mental resilience while working, the willingness to invest effort in one's work, and persistence even in the face of difficulties" (Schaufeli et al., 2002, p. 74). Dedication, "is characterized by a sense of significance, enthusiasm, inspiration, pride, and challenge" (Schaufeli et al., 2002, p. 74). And finally, absorption, as the name implies, is when an individual is so focused on work that time passes without the individual noticing, also leading to a difficulty in separating from it (Schaufeli et al., 2002).

Work Engagement has been highlighted in the literature because of the positive influence it has shown on outcome variables. For example, according to Kahn (1990), individuals with high levels of engagement show much greater commitment to their work because they effectively identify themselves with it. Those who feel more engaged in their work will have a much stronger bond and will therefore be much more motivated to give their best contribution to the success of the organization. It is also important to note that, because they have a greater commitment to the organization, it will be more difficult for them to break this bond with the organization (Bakker & Schaufeli, 2014). Schaufeli (2013) further argues that although engagement is directly related to the satisfaction and commitment of the individual in the work environment, it shows more significant evidence when associated with job performance.

2.3 Individual Work Performance

Perceived performance is one of the essential conditions to achieve success in different tasks/objectives. Therefore, organizations must take the necessary actions to motivate and satisfy individuals to enhance their performance (M. Cunha, Rego, R. Cunha, Cabral-Cardoso, & Neves, 2014). To study the concept of *perceived performance*, the variable *Individual Work Performance,* was used, which for this research proved to be the best fit. Thus, from this moment on, the perceived performance will be designated as IWP. In 1990, Campbell defines *IWP* as behaviors/actions relevant to an organization's goals. A paradigm shift is immediately noticeable. The importance is given not only to the results, but also to the workers' behaviors and actions.

For Campbell (1990), task performance becomes the most relevant component of IWP and he characterizes it as being "the proficiency with which workers perform the technical tasks essential to their work" (p. 708–709). Years later, Borman and Motowidlo (1993), 'divide' the performance sphere into *task performance* and *contextual* or *extra-role performance. Task performance* is defined as the effectiveness with which job holders carry out activities that contribute to the technical core, either directly by implementing a part of its technological process, or indirectly by providing it with the necessary materials and/or services (Borman & Motowidlo, 1997). Borman and Motowidlo (1993) and Motowidlo, Borman and Schmit (1997), also add, that in addition to all the so-called mandatory tasks of each individual, there is still an extra component, as the name indicates, which intends to expand even further the success of the organization, through actions at the psychological and social level. These behaviors go beyond duty, beyond task performance (Brandão, 2010).

Finally, Robinson and Bennett (1995), develop a third integral part of this model through their empirical analysis of deviant behaviors in the workplace. These deviant behaviors are all voluntary actions of individuals, which go against the rules of the organization, thus putting at risk other workers and even the organization itself. This new dimension is called *counterproductive work behaviors.*

2.4 Hypotheses and Research Model

According to Avey and colleagues (2011), individuals with high levels of *psycap*, are more likely to feel more energetic and to try harder, which is reflected in a superior performance over longer periods. This is because individuals with high levels of self-efficacy put effort into tasks they believe to be able to solve. It is also because they have the will to solve problems and can develop a variety of problem-solving solutions (hope), also because they accept that positive events are due to internal causes and actually expect positive outcomes (optimism), and finally because they respond positively and with great resilience when they encounter adversities and setbacks (resilience). This result is confirmed by a study conducted in Nigeria in 2011, applied to public sector workers, in which the effect of *psycap* on *work engagement* was assessed, by exploring the mediating role of emotional intelligence. With a sample of 557 workers, the results showed full mediation of the emotional intelligence variable, and that psycap has a positive and significant influence on work engagement (George & Okon, 2021). Thus, the research model is presented in Fig. 1 and the following hypotheses are formulated:

H1: *PsyCap* is positively associated with work engagement.

According to the study of Teixeira, Soares, and Lopes (2015), there is statistical evidence between *psycap* and *performance.* This study, applied to a sample of 100 salespeople from a Portuguese service provision organization, shows that there is a positive relationship between *psycap* and the two types of performance studied, task and contextual. According to the literature, several authors confirm the existence of a positive relationship between these two variables, namely Avey, Luthans and Youssef (2009), Luthans (2002), Luthans, Avolio, Walumbwa and Li (2005), and Luthans, Youssef and Avolio (2007). Thus, the following hypotheses are formulated:

H2: *PsyCap* is positively associated with *IWP.*

H2a: *PsyCap* is positively associated with *task performance.*

H2b: *PsyCap* is positively associated with *contextual performance.*

A study by Alqasmi, Manzoor and Hassan (2015), which aimed to explore the interference of *psycap* in controlling counterproductive work behaviors, showed a significant relationship, demonstrating that the spheres of *psycap* have the strength to limit deviant behaviors on workers in a particular organization. Studies such as Rodríguez-Cifuentes, Segura-Camacho, García-Ael and Topa (2020), Raaghul (2014), and T. Shah and S. Shah (2016) come to corroborate this assumption. The following hypothesis is thus formulated:

H2c: *PsyCap* is negatively associated with *counterproductive work behaviors.*

Christian, Garza and Slaughter (2011), developed a meta-analytic model that concluded that engagement is a predictor of *work performance*, either in the case of *task performance* or *contextual performance.* Schaufeli (2013) in his study also reinforces that the reason why engagement has such a strong relationship with performance, more than any other work attitude, is because it reflects an energetic force that guides the individual in the right direction, rather than a feeling of satisfaction, which is common for job satisfaction.

Based on the above literature, three further hypotheses are formulated:

H3: *Work Engagement* is positively associated with *IWP.*

H3a: *Work Engagement* is positively associated with *task performance.*

H3b: *Work Engagement* is positively associated with *contextual performance.*

According to Lee and Allen (2002), *counterproductive work behaviors* contain both affective and cognitive components. Therefore, negative emotions have direct and indirect effects on deviant behaviors, considering engagement as a positive emotional response. The authors conclude that the response to a positive affective emotion in an employee's tasks should have a negative relationship with *counterproductive work behaviors.* More specifically, there is a negative relationship between deviant behaviors and *work engagement.* Several studies confirm this theory (Rich, 2006; Colbert, Mount, Harter, Witt, & Barrick, 2004). Therefore, an additional hypothesis is formulated:

H3c: *Work Engagement* is negatively associated with *counterproductive work behaviors.*

Alessandri et al. (2018) conducted a study in a communications company, with a sample of 420 individuals, to test a dynamic model of the impact of *PsyCap* on *Work Engagement* and *Work Performance.* One of the hypotheses considered was precisely *work engagement* as a mediator of *psycap* levels and *work performance.* It should be noted that in this study the hypotheses concerning this mediation of work engagement were both confirmed. These results were also confirmed in other studies, not only for individual performance, but also for task and contextual performance (Chen, 2015; Zuberbühler, Calcagni, Martínez, & Salanova, 2021). Thus, the remaining hypotheses of the study are formulated:

H4: *Work Engagement* has a mediating effect on the relationship between *PsyCap* and *IWP.*

H4a: *Work Engagement* has a mediating effect on the relationship between *PsyCap* and *task performance.*

H4b: *Work Engagement* has a mediating effect on the relationship between *PsyCap* and *contextual performance.*

H4c: *Work Engagement* has a mediating effect on the relationship between *PsyCap* and *counterproductive work behaviors.*

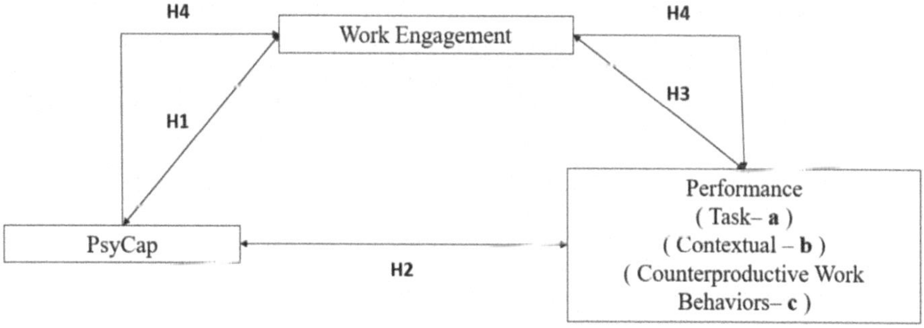

Fig. 1. Research Model

3 Method

This chapter will characterize the data collection instruments, the data collection and analysis procedures, and also the population and sample.

The method adopted was the quantitative method, used to test theories by analyzing the relationship between variables. These can be tested through instruments that later allow performing statistical analysis (Creswell & Creswell, 2018). Given the nature of this study, a questionnaire survey was used for data collection. It is also important to emphasize that this is a hypothetical-deductive study, in which, based on the previously analyzed literature, an analysis model was developed relating the three study variables, applied to the Air Force Academy.

3.1 Characterization of the Data Collection Instrument

To meet the defined objectives, a questionnaire was developed, which is divided into four distinct sections. The first section is composed of purely demographic questions (gender, age, specialty and rank), while the remaining sections focus on the three main variables of this study. These are the *PsyCap*, *Work Engagement* and *Individual Work Performance.*

To operationalize the *PsyCap* variable, the PCQ-24 questionnaire, developed by Luthans, Youssef, and Avolio (2007), was used. It consists of 24 questions assessing the four dimensions (self-efficacy, optimism, hope, resilience), each of them being assessed in six items on a *likert* scale from 1 to 6, where 1 corresponds to "I strongly disagree" and 6 to "I strongly agree". The respective translation into Portuguese was performed by Matias and D'Oliveira (2011), in an unpublished study taken from Moreira's (2012) Master's dissertation. The measurement instrument underwent a slight adjustment to adapt it to the military context, and was later revised and approved by the supervisors.

The Utrecht Work Engagement Scale adapted for students, developed by Schaufeli and Bakker (2004), was used to operationalize the variable Work Engagement. A *likert*-type scale was used, from 0 "never/not at all" to 6 "Always/every day". Finally, the Individual Work Performance Questionnaire, IWPQ by Koopmans, Bernaards, Hildebrandt, de Vet and van der Beek (2014). This questionnaire allowed measuring and rating an individual's perceived performance, in 18 items, through three dimensions, task performance, contextual performance and counterproductive work behaviors. This scale presents a closed response format, *likert* type, with 5 response options, from 0 "Rarely" to 4 "Always". The translation was taken from the master thesis of Neves (2015).

3.2 Data Collection and Data Analysis Procedure

To meet the defined objectives, a questionnaire was developed, which is divided into four distinct sections. The first section is composed of purely demographic questions (gender, age, specialty and rank), while the remaining sections focus on the three main variables of this study. These are the *PsyCap*, *Work Engagement* and *Individual Work Performance.*

To collect data, it was necessary to request the Commander of the Air Force Academy to accept the application of the questionnaire. After permission was granted for the application of the questionnaire, through the Google Forms platform, the questionnaire was sent by email to all students, with the respective introductory note and contextualization of the variables under study, and the answers were confidential and anonymous, being exclusively intended for scientific research purposes.

Subsequently, the Statistical Package for the Social Sciences (SPSS) version 28.0 platform was used for data processing. Initially, the psychometric qualities of the measurement instruments were analyzed through exploratory and confirmatory factor analysis, reliability through Cronbach's alphas, and sensitivity through the Kolmogorov-Smirnov tests, and then through asymmetry and kurtosis analysis. Then, descriptive analyses of the variables under study were performed, where it was possible to check and compare their means and standard deviations. Finally, the correlational analysis and the analysis of the mediation hypotheses were performed using the macro Process v3.1 tool.

3.3 Population and Sample

The target population of this study was all students of the Air Force Academy, in the current academic year 2021/2022.

As for the sampling, it is a non-probability or convenience sampling, applied to all students from 1st to 7th grade. A total of 242 questionnaires were collected in a population of 255 students, which is reflected in a response rate of 94.9%.

4 Results

This chapter will present the results obtained through the analysis of the data collected.

4.1 Descriptive Analysis

The target population of this study was all students of the Air Force Academy, in the current academic year 2021/2022.

After verifying the psychometric properties of the instrument under study, the mean and standard deviation of the different variables were analyzed. Table 1 shows that the variable with the highest mean value was contextual performance, with 2.93 on a scale from 0 to 4.

The variable with the lowest mean value was counterproductive work behaviors, with 1.69 on a scale from 0 to 4. It should be noted that a conversion to percentage was performed in order to confirm these results. To facilitate interpretation, when analyzing the IWP variable as a whole, the items in the negative, of the dimension counterproductive work behaviors were inverted, so that all dimensions of this variable are analyzed in the same 'direction'. On the other hand, when analyzing counterproductive work behaviors alone, the original items were used. Therefore, a lower mean, as is the case here, is positive, since the lower the value of this dimension, the lower the negative and harmful behaviors affecting the performance of an individual.

Table 1. Descriptive Statistics of the global variables

	Mean	Standard Deviation
PsyCap	4.56	0.48
Work Engagement	3.97	0.81
Individual Work Performance	2.73	0.40
Task Performance	2.83	0.53
Contextual Performance	2.93	0.52
Counterproductive Work Behaviors	1.69	0.80

4.2 Correlation Between Variables

A correlational analysis was performed between the different variables to measure the intensity and direction of the relationship between them. Pearson's correlation coefficient, which measures precisely this aspect, was used for this purpose.

Through the analysis of Table 2, it was possible to verify that most of the correlations are significant at the 0.01 level. We can also verify that the variables, PsyCap and *IWP* correlate with an intensity of 0.63 ($p < 0.01$), thus presenting the strongest correlation of the study. This means that these two variables correlate positively, that is, when one increases the other also increases. The relationship between *IWP* and contextual performance, despite presenting the strongest relationship, will not be considered, given that contextual performance is a dimension of *IWP*. The same applies in the reverse case, with the strongest and most negative relationship, between *IWP* and counterproductive work behaviors.

Thus, the two variables with the strongest and most negative correlation are *psycap* and counterproductive work behaviors, with a value of -0.35. This means that these two variables are negatively correlated, when one increases the other decreases.

Table 2. Correlational analysis between variables

	1	2	3	4	5	6
1 PsyCap	–					
2 Work Engagement	0.37*	–				
3 Individual Work Performance	0.63*	0.48*	–			
4 Task Performance	0.39*	0.17*	0.62*	–		
5 Contextual Performance	0.51*	0.42*	0.79*	0.49*	–	
6 Counterproductive Work Behaviors	−0.35*	−0.32*	−0.57*	0.05	−0.07	–

4.3 Testing the Mediation Hypotheses

To confirm the mediation hypothesis, the macro Process v3.1 tool for SPSS was used. This tool allows testing the mediation effects in an automated way, thus understanding whether there is a relationship between the independent variable, *PsyCap* and the dependent variable *IWP*, mediated by the mediating variable, *Work Engagement*.

Sobel's test was also applied, to provide further support to the mediation models of the study. However, due to the low reliability of this test, regarding the requirement that the ab relationship must have a normal distribution, the bootstrap method was also used to estimate the indirect effects of the mediation models (Hayes, 2009).

It was thereby possible to verify that *psycap* has a significant contribution in *work engagement*, explaining in 14% ($p < 0.01$, $R^2 = 0.14$), and also in IWP, explaining in 40% ($p < 0.01$, $R^2 = 0.40$). In the mediation hypothesis between *PsyCap* and *IWP*, a $p < 0.01$ was obtained in Sobel's test, and also a significant indirect effect through the bootstrap method (B = 0.877, 95% Boot IC [0.0469, 0.1411]). It is still possible to verify that the model has a significant contribution in *IWP*, explaining 47% of its variance ($p < 0.01$, $R^2 = 0.47$).

It was also observed that the model has a significant contribution in *task performance*, with a $p < 0.01$, explaining 15%. Also, in this model between *psycap* and *task performance*, mediated by *work engagement*, the mediation hypothesis was confirmed through Sobel's test, where a $p = 0.017$ was obtained. However, no significant relationship was found between *work engagement* and *task performance* ($p = 0.69$, $R^2 = 0.15$). Thus, there is no mediation, which is confirmed by the non-existence of a significant indirect effect (B = 0.0104, 95% Boot IC [−0.0477, 0.0775]).

In the model between *psycap* and *contextual performance*, mediated by *work engagement*, it was also found that *psycap* has a significant contribution in *contextual performance*, with a $p = 0.00$, explaining 26%. Also in this mediation model, the mediation hypothesis was confirmed through Sobel's test with a $p < 0.01$, and also through the bootstrap method (B = 0.1084, 95% Boot CI [0.0527, 0.1781]). It was still possible to verify that the model has a significant contribution in contextual performance, explaining 32% of its variance ($p < 0.01$, $R^2 = 0.32$).

Finally, in the model between *psycap* and *counterproductive work behaviors*, mediated by *work engagement*, it was found that *psycap* has a significant contribution in counterproductive work behaviors, with a $p < 0.01$, explaining 12%. Also in this model, the mediation hypothesis was confirmed through Sobel's test, where a $p < 0.01$ was obtained, and also through the bootstrap method (B = -0.1320, 95% Boot CI [−0.2274, −0.0510]). It is still possible to verify that work engagement has a significant contribution in counterproductive work behaviors, explaining 16% of its variance ($p = 0.00$, $R^2 = 0.16$).

5 Discussion

This chapter will analyze the results of the study. Firstly, in the descriptive analysis, it was possible to find that students generally perceive higher levels of contextual performance (M = 2.93) and lower levels of counterproductive work behaviors (M = 1.69). These results show that students perceive the extra tasks as being more important, thus highlighting the feeling of mission accomplishment, giving more focus on solving problems that affect the functioning of the Air Force Academy. The low levels perceived in the sphere of counterproductive work behavior show that students believe that they do not have very strong attitudes and behaviors that could compromise their objectives as students at the Academy. These results show a strong component of social desirability, as reflected in other studies of self-report scales (Crandall, Crandall, & Katrovsky, 1965; Crowne & Marlowe, 1960; Guedes, 2015).

As for the hypotheses of the study, it was possible to confirm all of them, except for hypothesis H4a, through the correlational analysis and the linear regressions performed with the support of the macro Process tool.

Testing of Hypothesis H1 - Hypothesis H1 aimed to test whether *psycap* is positively associated with *work engagement*. Through the correlational test, it was possible to verify a positive correlation ($r = 0.37$, $p < 0.01$), where the perceived *psycap* of the Air Force Academy students explains 14% of the variation in *work engagement* ($R^2 = 0.14$). Therefore, an increase in psycap levels is also associated with an increase in *work engagement* levels. These results are consistent with the literature previously analyzed (Avey, Reichard, Luthans, & Mhatre, 2011; George & Okon, 2021; Harris, 2012; McElravy, Matkin, & Hastings, 2017; Niswaty, Wirawan, Akib, Saggaf, & Daraba, 2021). Therefore, **H1 is confirmed**.

Test of Hypotheses H2 - H2, H2a, H2b and H2c aimed to test whether the *PsyCap* is positively associated with the *IWP* and respective dimensions. Through the correlational test, it was possible to verify that there is a positive correlation between *PsyCap* and *IWP* ($r = 0.63$), *task performance* ($r = 0.39$), *contextual performance* ($r = 0.51$) and negative with *counterproductive working behaviors* ($r = -0.35$), all of them significant ($p \leq 0.01$). The results show that 40% of the variation in students' *IWP* is explained by *psycap* levels ($R^2 = 0.40$). On the other hand, *psycap* levels explain 15% of the variations in task performance levels ($R^2 = 0.15$). It was also possible to verify that 26% of the variations in *contextual performance* are explained by *psycap* levels ($R^2 = 0.26$). The results also show that 12% of variations in *counterproductive work behaviors* are due to variations in *psycap* levels ($R^2 = 0.12$). These results are consistent with the literature previously analyzed (Avey, Reichard, Luthans, & Mhatre, 2011; Kappagoda, Othman, & Alwis, 2014; Luthans, 2002; Luthans, Avolio, Walumbwa, & Li, 2005; Luthans et al., 2007; Shah & Shah, 2016; Teixeira, Soares, & Lopes, 2015). Therefore, hypotheses H2, H2a, H2b and H2c were confirmed.

Hypotheses H3 and H4a- H3, H3a, H3b and H3c aimed to test whether *work engagement* is positively associated with *IWP* and its dimensions. Through the correlational test, it was possible to verify that there is a positive correlation between *work engagement* and *IWP* ($r = 0.48$), *task performance* ($r = 0.17$), *contextual performance* ($r = 0.42$) and negative correlation with *counterproductive work behaviors* ($r = -0.32$), all of them significant ($p = 0.01$). It was also found that 47% of the variations in *IWP* are due to variations in *psycap* and *work engagement* levels ($R^2 = 0.47$), that 32% of the variations in *contextual performance* are due to variations in the model levels ($R^2 = 0.32$), and that 16% of the variations in *counterproductive work behaviors* are due to variations in the model levels ($R^2 = 0.16$), however, the linear regression between *work engagement* and *task performance* could not be confirmed ($R^2 = 0.15$, $p = 0.69$). Therefore, hypotheses H3, H3a, H3b and H3c are all confirmed, except for H4a, as previously mentioned (Christian, Garza, & Slaughter, 2011; Colbert, Mount, Harter, Witt, & Barrick, 2004; Schaufeli, 2013). One justification for the results not being in line with what was expected may be the fact that the work method at the Air Force Academy is completely different from the work methods used by a worker or a student at a company/educational institution, respectively. A student at the Academy, for being at the same time a military, focuses his attention on other components, in this case on the concern with ensuring the proper functioning of his Academy.

Test of Hypotheses H4 - H4, H4a, H4b and H4c aimed to test whether psycap, mediated by work engagement, acted positively on the IWP and respective dimensions, and negatively on deviant behaviors. As seen above, H4a could not be confirmed, however, H4, H4b and H4c were all confirmed. As seen in Chapter 3.3, all regressions between the different variables were significant except for the linear regression between work engagement and task performance, and it was also possible to confirm the inexistence of a significant indirect effect in all models except for the model of hypothesis H4a. Thus, hypotheses H4, H4b and H4c were confirmed, which is in line with the literature previously analyzed (Alessandri, et al., 2018; Chen, 2015).

6 Conclusions

Initially, a set of objectives were proposed, as well as a starting question, "*What is the influence of Positive Psychological Capital, on the perceived performance of Air Force Academy students*?", which throughout this study were answered. Now, a balance of these answers will be elaborated.

Through the data analysis, it was possible to better understand the importance of *psycap* and *work engagement* variables in improving the performance of students of the Air Force Academy. It was found that these two variables, in addition to correlating with each other, also correlate positively with *IWP*, task and contextual performance, and negatively with counterproductive work behaviors. It should be noted that the strongest correlation was between the *Psycap* and *IWP* variables.

It was also possible to confirm the mediation hypotheses, thus ensuring the mediating role of work engagement in the relationship between *Psycap* and *IWP*, as well as with the respective dimensions of *IWP*, except for *task performance*. In fact, it wasn't possible to conclude that work engagement is a predictor of task performance, despite the existence of an association in this relationship. Through the literature, it was not possible to answer this conclusion given its specificity. However, a possible justification may be the fact that Air Force Academy students often give more emphasis to extra tasks, as observed in the correlational analysis, being more concerned with the resolution of internal problems, which lead to the organization's success.

This study allowed us to understand some of the most important variables that lead to higher levels of performance by the students. This was innovative research on the relationship between positive psychology and the perceived performance of the students of the Air Force Academy.

It was also possible to perceive that the main focus of the students is in fact to guarantee the correct functioning of the organization in which they work and which they value so much, evidence of the foundations of necessary basis as future managers and leaders of the Air Force.

As limitations of this study, we highlight the possible bias since it is a self-report measurement instrument, thus exposing the students' perceptions and not their actual performance. In addition, the fact that this was a cross-sectional study did not allow us to assess with confidence the evolution of the levels of the variables studied. Finally, the fact that the research was conducted in a period of adaptation due to the pandemic situation, the social relationships between students and, consequently, their motivation may have changed.

As future suggestions, it would be interesting to cross-reference the data on students' perceptions with their actual performance results (scores in the assessment strands). It would also be interesting to understand which dimensions of *psycap* and *work engagement* have more impact on performance, since this research did not focus on this aspect. As previously mentioned, this is a cross-sectional study, but in fact, a longitudinal study could contribute to the enrichment of this thematic, as well as an expansion of this study to the remaining Military Academies, so that with the final results experiences could be shared, always to promote better instruction. Finally, the application of open-ended questionnaires or interviews should be taken into account, to find new predictors of performance, such as *Learning Empowerment* or *Emotional Intelligence.*

Declaration of Interest Statement

The authors declare that they have no known competing financial interests or personal relationships that could have appeared to influence the work reported in this paper. It was an innovative research, which aimed to improve the student's performance at the Portuguese Air Force Academy.

References

Alessandri, G., Consiglio, C., Luthans, F., Borgogni, L.: Testing a dynamic model of the impact of psychological capital on work engagement and job performance. Career Dev. Int. **23**(1), 33–47 (2018). https://doi.org/10.1108/CDI-11-2016-0210

Alqasmi, T., Manzoor, S.R., Hassan, S.: Psychological capital intervention towards controlling counterproductive work behavior. Abasyn J. Soc. Sci. **8**(1), 62–73 (2015)

Avey, J.B., Reichard, R.J., Luthans, F., Mhatre, K.H.: Meta-analysis of the impact of positive psychological capital on employee attitudes, behaviors, and performance. Hum. Resour. Dev. Q. **22**(2), 127–152 (2011). https://doi.org/10.1002/hrdq.20070

Avey, J., Luthans, F., Youssef, C.: The additive value of positive psychological capital in predicting work attitudes and behaviors. J. Manag. **36**(2), 430–452 (2009). https://doi.org/10.1177/0149206308329961

Bakker, A.B., Schaufeli, W.B.: Work engagement. In: Flood, P.C., Freeney, Y. (eds.) Organizational behaviour. The Blackwell Encyclopedia of Management, vol. of, pp. 1–6. Wiley, West Sussex (2014)

Bandura, A.: Self-efficacy mechanism in human agency. Am. Psychol. **37**(2), 122–147 (1982). https://doi.org/10.1037/0003-066X.37.2.122

Borman, W.C., Motowidlo, S.J.: Expanding the criterion domain to include elements of contextual performance. In: Schmitt, N., Borman, W.C. (eds.) Personnel Selection in Organizations, pp. 71–98. Jossey-Bass, San Francisco (1993)

Borman, W., Motowidlo, S.J.: Task performance and contextual performance: the meaning for personnel selection research. Hum. Perform. **10**, 99–109 (1997)

Brandão, C. D.: O Desempenho Individual de dirigentes de topo da Administração Pública Portuguesa: um contributo para o desenvolvimento de uma taxionomia comportamental e motivacional. University of Psychology and Education Sciences of Porto's University (2010)

Campbell, J.P.: Modeling the performance prediction problem in industrial and organizational psychology. In: Dunnette, M.D., Hough, L.M. (eds.) Handbook of Industrial and Organizational Psychology, pp. 687–732. Consulting Psychologists Press (1990)

Chambel, M. J., Lopes, S., Castanheira, F., Silveira, C. R.: Militaries' Burnout and Work Engagement: A Qualitative Systematic Literature Review. In U. Kumar, *The Routledge International Handbook of Military Psychology and Mental Health* (pp. 283–316). Routledge (2019)

Chen, S.-L.: The relationship of leader psychological capital and follower psychological capital, job engagement and job performance: a multilevel mediating perspective. Int. J. Human Resour. Manage. **26**(18), 2349–2365 (2015). https://doi.org/10.1080/09585192.2015.1020443

Christian, M.S., Garza, A.S., Slaughter, J.E.: Work engagement: a quantitative review and test of its relations with task and contextual performance. Pers. Psychol. **64**(1), 89–136 (2011). https://doi.org/10.1111/j.1744-6570.2010.01203.x

Colbert, A.E., Mount, M.K., Harter, J.K., Witt, L.A., Barrick, M.R.: Interactive effects of personality and perceptions of the work situation on workplace deviance. J. Appl. Psychol. **89**(4), 599–609 (2004). https://doi.org/10.1037/0021-9010.89.4.599

Crandall, V.C., Crandall, V.J., Katrovsky, W.: A children's social desirability questionnaire. J. Consult. Psychol. **29**(1), 27–36 (1965). https://doi.org/10.1037/h0020966

Creswell, J. W., Creswell, J. D.: *Research Design: Qualitative, Quantitative, and Mixed Methods Approaches (5th ed.).* Los Angeles: SAGE Publications. ISBN 13: 9781506386706 (2018)

Crowne, D.P., Marlowe, D.: A new scale of social desirability independent of psychopathology. J. Consult. Psychol. **24**(4), 349–354 (1960). https://doi.org/10.1037/h0047358

Cunha, M., Rego, A., Cunha, R., Cardoso, C., Neves, P.: *Manual de Comportamento Organizacional e Gestão.* Editora Rh (2014)

George, O. J., Okon, S. E.: Psychological Capital and Work Engagement among Employees in the Nigerian Public Sector: The Mediating Role of Emotional Intelligence. *International Journal of Public Administration* (2021). https://doi.org/10.1080/01900692.2021.2001010.

Guedes, D.: Comparação da desejabilidade social entre uma amostra de pais em avaliação das competências parentais e uma amostra de controlo [Master Thesis, University of Psychology and Education Sciences of Coimbra's University] (2015). http://hdl.handle.net/10316/31336

Harris, C.: Relationships between psychological capital, work engagement and organizational citizenship behaviour in South African automative dealerships. *Doctoral dissertation, Nelson Mandela Metropolitan University, Port Elizabeth, South Africa* (2012)

Hayes, A.: Beyond Baron and Kenny: statistical mediation analysis in the new millennium. Commun. Monogr. **76**(4), 408–420 (2009). https://doi.org/10.1080/03637750903310360

Kahn, W.: Psychological conditions of personal engagement and disengagement at work. Acad. Manage. J., 692–772 (1990)

Kappagoda, S., Othman, H., Alwis, G.: Psychological capital and job performance: the mediating role of work attitudes. J. Human Resour. Sustain. Stud. **2**(2), 102–116 (2014). https://doi.org/10.4236/jhrss.2014.22009

Koopmans, L., Bernaards, C.M., Hildebrandt, V.H., de Vet, H.C., van der Beek, A.J.: Construct validity of the individual work performance questionnaire. J. Occup. Environ. Med. **56**(3), 331–337 (2014). https://doi.org/10.1097/JOM.0000000000000113

Lee, K., Allen, N.J.: Organizational citizenship behavior and workplace deviance: the role of affect and cognition. J. Appl. Psychol. **87**(1), 131–142 (2002). https://doi.org/10.1037/0021-9010.87.1.131

Lopes, M. P., Cunha, M. P., Palma, P. J.: Positive psychological capital: Distinguishing profiles and their impact on organizational climate. Manuscript Under Review (2005)

Luthans, F.: Positive organizational behaviour: developing and managing psychological strengths. Acad. Manag. Exec. **16**(1), 57–75 (2002). https://doi.org/10.5465/AME.2002.6640181

Luthans, F., Youssef, C.M.: Human, social and now positive psychological capital management: investing in people for competitive advantage. Organ. Dyn. **33**(2), 143–160 (2004). https://doi.org/10.1016/j.orgdyn.2004.01.003

Luthans, F., Avolio, B. J., Walumbwa, F. O., Li, W.: The psychological capital of Chinese workers: exploring the relationship with performance. Manage. Organ. Rev., 249-271 (2005). https://doi.org/10.1111/j.1740-8784.2005.00011.x

Luthans, F., Youssef, C.M., Avolio, B.J.: Psychological capital. Developing the human competitive edge. Oxford University Press, New York (2007)

Matias, M., D'Oliveira, T.C.: Adaptação para português da escala Psychological Capital Questionnaire (PCQ-24). Unpublished Work, Lisboa (2011)

McElravy, L., Matkin, G., Hastings, L.: How can service-learning prepare students for the workforce? exploring the potential of positive psychological capital. J. Leader. Educ., **16**(3), 97–117. https://doi.org/10.12806/V16/I3/T1

Moreira, C.: Quão perto é perto o suficiente: a influência do capital psicológico positivo, do envolvimento no trabalho e da segurança. *[Master Thesis, ISPA]. Repository of ISPA* (2012). http://hdl.handle.net/10400.12/2317

Motowidlo, S., Borman, W., Schmit, M.J.: A theory of individual differences in task and contextual performance. Hum. Perform. **10**(2), 71–83 (1997). https://doi.org/10.1207/s15327043hup1002_1

Neves, I.: A liderança e a sua importância na gestão de equipas comerciais [Master Thesis, University of Economy of Porto's University]. *Publications SASUP* (2015). https://sigarra.up.pt/sasup/pt/pub_geral.pub_view?pi_pub_base_id=37275

Niswaty, R., Wirawan, H., Akib, H., Saggaf, M. S., Daraba, D.: Investigating the effect of authentic leadership and employees' psychological capital on work engagement: evidence from Indonesia. CellPress, **7**(5) (2021). https://doi.org/10.1016/j.heliyon.2021.e06992

Page, L. F., Donohue, R.: Positive psychological capital: a preliminary exploration of the construct. *Monash University. Business and Economics*, Working Paper 51/04 (2004)

Raaghul, R.: An empirical study on psychological capital in relation to organizational citizenship behavior and counter-productive work behavior. *Department of Management Studies, School of Management* (2014)

Rich, B. L.: Job engagement: Construct validation and relationships with job satisfaction, job involvement, and intrinsic motivation. *University of Florida* (2006)

Robinson, S.L., Bennett, R.J.: A typology of deviant workplace behaviors: a multidimensional scaling study. Acad. Manag. J. **38**(2), 555–572 (1995). https://doi.org/10.2307/256693

Rodríguez-Cifuentes, F., Segura-Camacho, A., García-Ael, C., Topa, G.: The Mediating Role of Psychological Capital between Motivational Orientations and Their Organizational Consequences, 17(13). *Environmental Research and Public Health* (2020)

Schaufeli, W. B.: What is engagement?. In Truss, C., Alfes, K., Delbridge, R., Shantz, A., Soane, E. (eds.). *Employee Engagement in Theory and Practice*, 1–37 (2013)

Schaufeli, W.B., Salanova, M., González-Romá, V., Bakker, A.B.: The measurement of engagement and burnout: a two sample confirmatory factor analytic approach. J. Happiness Stud.: Interdisc. Forum Subject. Well-Being **3**(1), 71–92 (2002)

Schaufeli, W., Bakker, A.: Utrecht work engagement scale. Preliminary Manual, Version 1.1. *Utrecht University* (2004)

Shah, T.A., Shah, S.Z.: Combined effects of psychological capital and psychological contract on employees' job level outcomes. Euro. J. Bus. Manage. **8**(26), 5–19 (2016)

Snyder, C.R.: Hope and optimism. Encyclopedia Hum. Behav. **2**, 535–542 (1994)
Snyder, C. R., Irving, L. M., Anderson, J. R.: Hope and Health. In: Snyder, C. R., Forsyth, D. R. (eds.), Handbook of social and clinical psychology (1991)
Stoykova, P. S.: *Developing Psychological Capital: Test of a Training Intervention with Bulgarian Students and Professionals.* ISCTE Business School (2013)
Teixeira, A. P., Soares, A. E., Lopes, M. P.: Impacto do Capital Psicológico Positivo no Desempenho de uma Equipa Comercial. *Psique Magazine*, 9–33 (2015)
Wright, T.A.: Positive organizational behavior: an idea whose time has truly come. J. Organ. Behav. **24**(4), 437–442 (2003). https://doi.org/10.1002/job.197
Yunes, M.A.: Psicologia positiva e resiliência: O foco no indivíduo e na família. Psicologia em Estudo **8**, 75–84 (2003)

Competencies Development and Turnover Intentions: The Mediating Effect of Affective Commitment

Beatriz Pereira and Ana Palma-Moreira(✉)

Universidade Europeia, Quinta do Bom Nome, Estr. da Correia 53, 1500-210 Lisbon, Portugal
ana.moreira@universidadeeuropeia.pt

Abstract. The main objective of this study was to investigate the relationship between organizational practices of competencies development and turnover intentions and whether this relationship is mediated by affective organizational commitment. To this end, we hypothesized: (1) organizational practices of competencies development have a negative and significant relationship with turnover intentions; (2) organizational practices of competencies development have a positive and significant relationship with affective organizational commitment; (3) affective organizational commitment has a negative and significant relationship with turnover intentions; (4) affective organizational commitment mediates the relationship between organizational practices of competencies development and turnover intentions. The sample of this study consisted of 345 participants working in organizations based in Portugal. The results indicate that training, individualized support, and affective commitment negatively and significantly affect turnover intentions. Training and individualized support have a positive and significant effect on affective commitment. Affective commitment directly mediates the relationship between organizational practices of competencies development (training and individualized support) and turnover intentions.

This study points out to Human Resources Management professionals the importance of training and individualized support in boosting affective commitment, reducing turnover intentions, and keeping the human capital that is so important for the organization's competitiveness.

Keywords: Organizational practices of competence development · turnover intentions · affective commitment · quantitative study

1 Introduction

In recent years, the global economic situation, the arrival of new technologies in the labor market and consequently, the various changes that have taken place have meant that the labor market today is increasingly competitive (De Vos et al., 2011). What is more, organizations need to be more and more able to guarantee the security of their employees, which often leads to them leaving voluntarily in search of stability and security (Benson, 2006).

M. J. Sousa (Ed.): UFHRD 2024, ISEM 52, pp. 172–185, 2025.
https://doi.org/10.1007/978-3-031-92548-1_11

Organizations must thus modify their approaches to keep their top talent and preserve the competencies necessary for their strategic plans (Afiouni, 2007). When a company invests in its workers, they will return the favor by acting and thinking positively. This relationship can be interpreted using the social exchange premise, a theory developed by Blau (1964). It states that we can understand the patterns formed to start, maintain, or end a relationship by observing how employees and organizations interact through mutual and contingent exchanges. When positive experiences are positive, employees develop a strong affective commitment to the organization (Meyer & Allen, 1991; Meyer et al., 2002). In the view of Meyer and Allen (1991), there is great interest in studying affective organizational commitment mainly because it is the best reducer of intentions to leave the organization.

The main objective of this research is to study the relationship between organizational skills development practices and intentions to leave the organization and whether affective commitment is the mechanism explaining this relationship.

2 Literature Review

2.1 Organizational Practices of Competencies Development

The primary goal of competency-based management is to plan, develop, and evaluate the competencies required to accomplish the suggested goals, whether organizational, group, or individual (Brandão & Guimarães, 2001).

Skills management techniques have gained popularity in recent years. Ceitil (2016) asserts that this is mostly because of the acknowledgement of the significance of this field, which supports the successful operation of organizations.

According to De Vos et al. (2011), competence development-focused HRM systems help us determine the competencies required to complete a task and, as a result, assign the work to people who have those competencies. In other words, these systems assist businesses in assigning work based on demonstrated competencies rather than positions held to improve performance (De Vos et al., 2011).

Additionally, Knapik et al. (2020) contend that competence emphasizes the results of human work in these systems and institutes changes in organizations' structures, systems, policies, and practices.

According to De Vos et al. (2011), training, functional rotation, and individualized support are examples of competencies-development techniques. One of the most popular methods for developing abilities in business is training. Organizations It is 'an organizational practice that aims to develop employees' positive feelings towards the company through a favorable perception of organizational support' (Wayne et al., 1997, p.85).

Ravikumar et al. (2020) claim that functional rotation enables staff members to experience and hold various roles and responsibilities inside the company. It is fundamental for developing competencies, increasing motivation, performance, and employee knowledge, and reducing employee fatigue. Arasi and Aghdam (2016) also argue that functional rotation helps increase employee commitment.

Lastly, coaching, mentoring and career development are all included in individualized support.

2.2 Turnover Intentions

We can define turnover intentions as employees' willingness to leave the organization where they work (Saeed et al., 2014). Turnover intentions cause serious problems for organizations, such as the costs of recruiting and selecting new employees and the loss of highly qualified employees who are important for the organization's strategic development. Thus, turnover intentions present a significant challenge for organizations (Lai & Kapstad, 2009).

Kumar et al. (2012) identified five factors that influence turnover intentions in an organization: satisfaction with pay, satisfaction with working conditions, satisfaction with supervision, affective commitment, and the job itself. It is also important to note that trust in those who manage the organization also reduces intentions to leave. Lai and Kapstad (2009) also consider motivation and skills mobilization as variables influencing turnover intentions.

2.2.1 Organizational Practices of Competencies Development and Turnover Intentions

The importance attached to the study of skills development is a fundamental tool for the strategic management of organizations because it is related to the high turnover of highly competent employees (Mulder et al., 2007). Organizations must concern themselves with developing the skills of their employees, as this is the only way they will be able to stand out in a highly competitive job market. When an organization invests in developing the skills of its employees, its turnover intentions decrease (Benson, 2006; Chambel, 2012).

In addition, individuals should consider the pros and cons of an organization. For example, if an organization invests more in its employees, knowledge, competencies, and growth, it will be more attractive to the employees, causing them to change organizations (Benson, 2006).

Hypothesis 1: Organizational practices of competencies development have a negative and significant relationship with turnover intentions.

2.3 Affective Commitment

Organizational commitment can be understood as the bond between employees and the organization where they work (Meyer & Herscovitch, 2001). According to Meyer et al. (2004), commitment is a difficult concept to define, which may be because it can take various forms, including affective commitment, normative commitment, and continuity commitment.

In this sense, according to Meyer and Allen's (1991) Three Component Model of Organizational Commitment, affective commitment relates to the emotional part, in other words, the emotional connection and identification that the individual feels with the organization where they work. Continuance commitment is related to the perception of the consequences of leaving the organization, namely the loss of some benefits associated with seniority, among others. Lastly, normative commitment is correlated with the moral obligation that the individual feels to remain in the organization (Meyer et al., 2004).

Affective commitment is a fundamental social tool that allows us to understand how the relationship between an individual and their work organization influences their performance, job satisfaction and turnover intentions (Mercurio, 2015).

2.3.1 Organizational Practices of Competencies Development and Affective Commitment

According to Shore et al. (2006), when an organization invests in developing the skills of its employees, it feels motivated and tries to repay this investment by developing a more outstanding affective commitment to the organization. We can interpret this relationship considering the norm of reciprocity (Gouldner, 1960) and the social exchange theory (Blau, 1964). According to the norm of reciprocity, employees' affective commitment to the organization where they work depends on their perception of what it offers them (Benson, 2006). According to Moreira et al. (2024), employees feel emotionally attached when they perceive that the organization cares about developing their skills. For Hermansyah et al. (2024), developing competencies positively and significantly affects affective commitment.

Hypothesis 2: Organizational practices of competencies development have a positive and significant relationship with affective organizational commitment

2.3.2 Affective Commitment and Turnover Intentions

In the study by Meyer and Allen (1991), the authors state great interest in studying affective commitment because it is considered the best reducer of turnover intentions. Among the three components of organizational commitment for these authors, affective commitment is the greatest reducer of turnover intentions. More recent studies have confirmed that affective commitment is the best reducer of turnover intentions (Moreira & Cesário, 2021; Nguyen et al., 2020).

Hypothesis 3: Affective organizational commitment has a negative and significant relationship with turnover intentions.

2.3.3 Mediating Effect

According to Moreira et al. (2024), affective commitment is fundamental to understanding the relationship between organizational skills development practices and turnover intentions.

According to Benson (2006), organizational competences development practices, especially training, are perceived by employees as helping them to develop and improve the skills necessary for their professional activity, which increases their emotional attachment to the organization and reduces turnover intentions. In this way, the more the organization cares about employee development, the more employees reciprocate by increasing their emotional commitment and reducing their intentions to leave the organization (Moreira et al., 2024).

Martini et al. (2022) also confirmed this relationship. Several theories, including the social exchange theory, can justify this relationship. This theory states that the more an organization invests in developing its employees and thus satisfying their needs, the greater the affective commitment is and the lower the intention to leave for another organization (Blau, 1964).

Hypothesis 4: Affective organizational commitment mediates the relationship between organizational practices of development competencies and turnover intentions.

The literature review justifies formulating the hypotheses summarized in the following research model (Fig. 1).

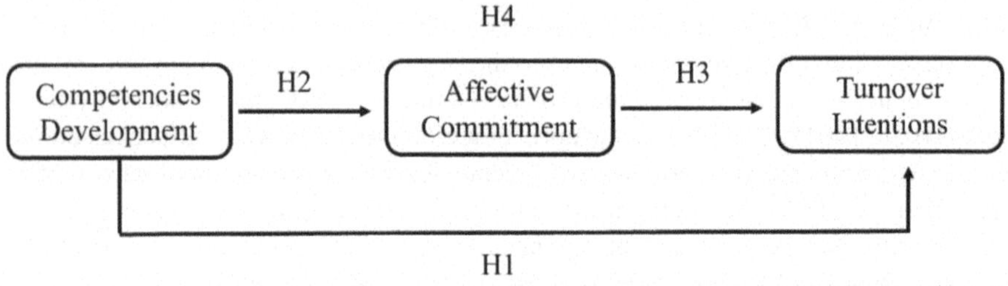

Fig. 1. Research Model

3 Method

3.1 Data Collection Procedure

Three hundred forty-five participants participated in this study voluntarily and needed to work in organizations based in Portugal. Snowball sampling was a convenient, purposeful, non-probabilistic sampling technique (Trochim, 2000). Since the data was collected simultaneously, it was also an exploratory and cross-sectional study.

The questionnaire was created on the Google Forms platform, and its link was shared via email and LinkedIn. Before answering the questionnaire, the participants read the informed consent form. Then they were asked if they would consent to complete the questionnaire. Participants were directed to the next portion of the questionnaire if they selected "yes" and to the end if they selected "no." 345 out of the 349 volunteers agreed to take part in the study.

To describe the sample, the questionnaire asked sociodemographic questions on age in years, gender, seniority within the company, marital status, educational background, industry, and contract type. The participants also responded to three scales: turnover intentions, affective commitment, and organizational practices of competency development.

3.2 Participants

Three hundred forty-five individuals working in organizations based in Portugal participated in the questionnaire. Of the 345 participants, 55.4% were female and 44.6% male. The average age of the participants was 37.95, ranging from 19 to 70 years old.

Concerning educational qualifications, most of the sample had a degree (47.5%), followed by a master's degree or higher (30.1%), and 22.3% had studied up to the 12th grade. As for seniority in the organization, 18% had been in the organization for less than one year, 16.8% for 1 to 2 years, 24.3% for 2 to 5 years, 15.1% for 5 to 10 years, 8.1% for 10 to 15 years, 4.6% for 15 to 20 years and 13% of the sample for more than 20 years. Regarding employment, 64.1% were in the private sector, 29.3% in the public sector and 6.7% in the public/private sector. Regarding contracts, 13.9% of participants had an open-ended contract, 12.5% had a fixed-term contract, 65.8% had an open-ended contract, and 7.8% had another contract. Finally, 43.2% of participants are single, 48.4% are married, 7.8% are divorced, and 0.6% are widowed.

3.3 Data Analysis Procedure

Once the data had been collected, it was imported into the SPSS Statistics 29 software to be processed. The metric qualities of the instruments used in this study were then tested. Confirmatory factor analyses were carried out using the AMOS Graphics 29 software to test the validity of the instruments. Reliability was tested by calculating Cronbach's alpha, which should be higher than .70, the minimum acceptable in organizational studies (Bryman & Cramer, 2003).

As for the sensitivity study, the different measures of central tendency, dispersion and shape were calculated for the different items of the scales used, thus carrying out the normality study for all the items and the different scales. According to Finney and DiStefano (2013), the absolute values of asymmetry and kurtosis should be less than two and seven, respectively, and they should have answers at all points and not skew towards any of the extremes.

Next, descriptive statistics were carried out on the variables under study by performing Student's t-tests for one sample. The link between the variables under study was examined using Pearson's correlations. Multiple linear regressions were used to test the hypotheses, with a significance level of.05.

3.4 Instruments

We used the instrument created by De Vos et al. (2011) and adapted for the Portuguese population by Moreira and Cesário (2022) to measure organizational practices of competency development. It has 12 items distributed across three dimensions: training, individualized support, and functional rotation. These items are categorized on a 5-point Likert rating scale from 1 ('Never') to 5 ('Always'). Upon doing the CFA, it was discovered that every adjustment index was sufficient. (χ^2/df = 2.17; GFI = .96; CFI = .94; TLI = .91; RMSEA = .058; SRMR = .093). As for internal consistency, Cronbach's alpha values ranged from .66 (individualized support) to .73 (training).

The instrument Bozeman & Perrewé (2001) created, translated and modified for the Portuguese population by Bártolo-Ribeiro (2018), was used to measure turnover intentions. It had six items categorized on a 5-point rating scale, ranging from 1 ('Strongly Disagree') to 5 ('Strongly Agree'). (χ^2/df = 2.65; GFI = .99; CFI = .99; TLI = .98; RMSEA = .069; SRMR = .027). Cronbach's alpha for internal consistency was .93.

The tool created by Meyer and Allen (1997), which consists of six items categorized on a 7-point Likert rating scale (from 1 'Strongly Disagree' to 5 'Strongly Agree'), was used to measure affective commitment. Upon doing the CFA, it was discovered that every adjustment index was sufficient. (χ^2/df = 3.03; GFI = .99; CFI = .99; TLI = .98; RMSEA = .077; SRMR = .037). The Cronbach's alpha value for internal consistency is .91.

Only items 11 and 12 of the organizational practices of competencies development scale have a median close to the lower extremes in terms of sensitivity. Every item has answers at every stage, and its kurtosis and asymmetry absolute values are less than 7 and 2, respectively.

4 Results

4.1 Descriptive Statistics of the Variables Under Study

Several one-sample Student t-tests were carried out to understand the position of the answers given by the participants in this study to the various constructs.

The participants' answers in this study were significantly below the central point (3) regarding training, individualized support, functional rotation, and turnover intentions. However, regarding affective commitment, the answers were significantly above the scale's center point (4) (Table 1).

Table 1. Descriptive statistics of the variables under study

Variable	t	gl	p	d	Mean	SD
Training	–4.08***	344	< .001	.22	2.81	.85
Individualized Support	–8.69***	344	< .001	.47	2.51	1.05
Functional Rotation	–6.34***	344	< .001	.34	2.57	1.26
Turnover Intentions	–9.19***	344	< .001	.50	2.47	1.06
Affective Commitment	7.11***	344	< .001	.38	4.58	1.52

Note. *** p < .001

4.2 Association Between Variables

The participants' age correlates significantly with all the variables under study (Table 2). Older participants perceive training, job rotation, and greater affective commitment more highly, but they perceive individualized support less positively and have fewer turnover intentions.

Organizational practices of competencies development (training, individualized support and functional rotation) are negatively and significantly associated with turnover intentions and positively and significantly associated with affective commitment (Table 2). In turn, affective commitment is negatively and significantly associated with intentions to leave (Table 2).

Table 2. Association between variables

Variable	1	2.1	2.2	2.3	3	4
1. Age	--					
2.1. Training	.15**	--				
2.2. Individualized Support	-.15**	.32***	--			
2.3. Functional Rotation	.12*	.42***	.33***	--		
3. Affective Commitment	.20***	.31***	.33***	.26***	--	
4. Turnover Intentions	-.20***	-.26***	-.24***	-.21***	-.70***	--

Note. * $p < .05$; ** $p < .01$; *** $p < .001$

4.3 Hypotheses

The hypotheses were tested using simple and multiple linear regressions. Age was used as a control variable.

Age ($\beta = -.20$, $p < .001$), training ($\beta = -.14$, $p = .014$), and individualized support ($\beta = -.20$, $p < .001$) negative and significantly affect turnover intentions (Table 3). These results indicate that older participants with a higher perception of training and individualized support have fewer intentions to leave the organization.

The model explains 13% of the variability on dependent variable (Table 3). The model is statistically significant ($F(4, 340) = 13.31$, $p < .001$) (Table 3).

Hypothesis 1 was partially supported.

Table 3. Effect of organizational practices of competencies development on turnover intentions

Independent Variable	Dependent Variable	F	p	R^2_a	β	p
Age	Turnover Intentions	13.31***	< .001	.13	-.20***	< .001
Training					-.14*	.014
Individualized Support					-.20***	< .001
Functional Rotation					-.06	.316

Note. * $p < .05$; ***$p < .001$.

Age ($\beta = .21$, $p < .001$), training ($\beta = .16$, $p = .005$), and individualized support ($\beta = .29$, $p < .001$) are positive and significantly associated with affective commitment (Table 4). These results indicate that older participants with a higher perception of training and individualized support have higher levels of affective commitment.

The model explains 19% of the variability on the dependent variable (Table 4). The model is statistically significant ($F(4, 340) = 21.75$, $p < .001$) (Table 4).

Hypothesis 2 was partially supported.

Table 4. Effect of organizational practices of competencies development on affective commitment

Independent Variable	Dependent Variable	F	p	R^2_a	β	p
Age	Affective Commitment	21.75***	< .001	.19	.21***	< .001
Training					.16**	.005
Individualized Support					.29***	<. 001
Functional Rotation					.07	.207

Note. ** p < .01; ***p < .001.

Affective commitment (β = -.68, p = < .001)) is negative and significantly associated with turnover intentions (Table 5). These findings suggest that participants' turnover intentions are lower when their affective commitment levels are higher.

The model explains 48 per cent of the variability in turnover intentions (Table 5). The model is statistically significant (F (2, 342) = 161.81, p < .001) (Table 5).

Hypothesis 3 was supported.

Table 5. Effect of affective commitment on turnover intentions

Independent Variable	Dependent Variable	F	p	R^2_a	β	p
Age	Turnover Intentions	161.81***	< .001	.48	-.06	.133
Affective Commitment					-.68***	< .001

Note. *** p < .001

Hypothesis 4 presupposes a mediating effect, so it was only tested after checking the three conditions proposed by Baron and Kenny (1986), which were confirmed in hypotheses 1, 2, and 3.

A two-step multiple linear analysis was carried out to test this hypothesis. The first step introduced age, training, and individualized support as independent variables. The second step introduced the mediating variable (affective commitment).

Affective commitment has a total mediating effect on the relationship between organizational practices of competencies development (training and individualized support) and turnover intentions (Table 6). When the mediating variable (affective commitment) was introduced into the regression equation, training and individualized support no longer significantly affected turnover intentions (Table 6).

The model explains 48 per cent of the variability in turnover intentions (Table 6). The increase in variability is statistically significant (ΔR2 = .35, p < .001) (Table 6). Both models are statistically significant (Table 6).

This hypothesis was supported.

Table 6. Mediating effect results

Independent Variables	Turnover Intentions	
	β Step 1	β step 2
Age	-.20***	-.06
Training	-.16**	-.04
Individualized Support	-.22***	-.01
Affective Commitment		-.67***
F	17.41***	81.05***
R^2_a	.13	.48
ΔR^2		.35***

Note. * p < .05; *** p < .001

5 Discussion and Conclusions

This study aimed to investigate the effect of organizational practices of competencies development (training, individualized support, and job rotation) on turnover intentions and whether this relationship is mediated by affective commitment.

Firstly, hypothesis 1 was partially confirmed since only training and individualized support negatively and significantly affect turnover intentions. These results align with literature since, in the view of Syed et al. (2023), training negatively and significantly affects turnover intentions, i.e. the higher the perception of training, the lower the turnover intentions. Indra et al. (2022) conducted a study in which they found that the support provided by the organization and the manager reduces turnover intentions. As for functional rotation not significantly affecting turnover intentions, the results go against what most of the literature says. In three studies by Moreira et al. (2020, 2022, 2024) in which this same instrument was used, functional rotation has a negative and significant effect on turnover intentions despite being, among the three dimensions, the dimension with the weakest effect. However, in the Hendry et al. (2022) study, these authors did not find a significant effect of functional rotation on turnover intentions.

Secondly, Hypothesis 2 was partially supported since only training and individualized support positively and significantly affected affective commitment. Concerning training and individualized support, the results align with literature. Chambel (2012) argues that when the organization invests in skills development practices, they increase affective commitment to the organization. As for functional rotation, the results go against what the literature says. The study by Moreira et al. (2024) tells us that training, individualized support and functional rotation positively and significantly impact affective commitment. Arasi and Aghdam (2016) also found that functional rotation positively affects affective commitment, although they consider this weak.

Thirdly, Hypothesis 3 was confirmed. Affective commitment is negatively and significantly associated with turnover intentions. This result is in line with the literature. According to Indra et al. (2023), when employees are affectively committed to the organization where they work, their turnover intentions decrease. To Meyer and Allen (1991), affective commitment is the greatest reducer of turnover intentions since the more an employee identifies with the organization's values, the less likely they are to want to leave (Benson, 2006).

Finally, the mediating effect of affective commitment on the relationship between organizational practices of competencies development (training and individualized support) and turnover intentions was confirmed. When an organization invests in developing the skills of its employees, their emotional commitment to the organization increases, and their turnover intentions decrease. These results align with the literature since, according to Moreira et al. (2024), the more the organization is concerned with employee development, the greater their affective commitment and the lower their turnover intentions. These relationships can be interpreted in the light of the social exchange theory developed by Blau (1964) or the norm of reciprocity (Gouldner, 1960).

Regarding descriptive statistics, the organizational practices of competencies development (training, individualized support and functional rotation) are significantly below the central point, although training is the closest to it. The fact that, in this study, training is the competencies development practice closest to the central axis can be explained by the fact that training is one of the competencies development practices that organizations use most frequently (Ludwikowsha, 2018). On the other hand, coaching and mentoring are the practices with the lowest level of perception among the participants in this study. Coaching is indeed a practice that few human resources managers use, but employees may perceive mentoring differently than the organization would like it to be. According to Whitener (2001), there are instances where procedures are in place but are not viewed by staff members as the company would prefer. The low perception of competencies development among the participants in this study aligns with other studies by Moreira et al. (2022, 2024), which also show employees' low perception of skills development practices.

5.1 Limitations

The main limitation of this study is the non-probabilistic, intentional, snowball data collection procedure. another limitation is that this is a cross-sectional study.

Finally, the fact that the questionnaire was closed-ended could lead to biased results due to social desirability.

5.2 Practical Implications

According to Afiouni (2007), organizations must invest in their employees, adjust their strategies to retain their talents, and keep critical competencies to fulfil their strategic plans. Meyer and Allen (1991) argue that the main interest in studying affective commitment is that it is the most significant reducer of turnover.

This study reinforces the importance of organizations investing in their employees. The greater the employees' perception of competencies development practices, the greater the affective commitment and the lower the intentions to leave the organization.

Organizations should invest more in individualized support, particularly in practices such as mentoring and coaching. Individualized support is the competence development practice that most boosts affective commitment and reduces turnover intentions.

References

1. Afiouni, F.: Human resource management and knowledge management: a road map toward improving organizational performance. J. Am. Acad. Bus. Camb. **11**, 124–130 (2007)
2. Arasi, D.A., Aghdam, M.B.A.: The relationship between job rotation and duty commitment of employees: a case study among employees of Islamic Azad University, District 13. Int. J. Organ. Leadersh. **5**(2), 87–95 (2016)
3. Baron, R.M., Kenny, D.A.: The moderator-mediator variable distinction in social psychological research: conceptual, strategic, and statistical considerations. J. Pers. Soc. Psychol. **51**, 1173–1182 (1986)
4. Bártolo-Ribeiro, R.: Desenvolvimento e Validação de uma Escala de Intenções de Saída Organizacional. In Diagnóstico e Avaliação Psicológica: *Atas do 10º Congresso da AIDAP/AIDEP.* Edited by Pereira, M., Alberto, I. M., Costa, J. J., Silva, J. T., Albuquerque, C. P. A., Santos, M. J. S., Vilar, M. P., & Rebelo, T. M. D. AIDAP/AiDEP, Coimbra; pp. 378–90, ISBN:978–989–20–9329–1/978–989–20–9341–3 (2018)
5. Benson, G.S.: Employee development, commitment and intention to turnover: a test of "employability" policies in action. Hum. Resour. Manag. J. **16**, 173–192 (2006)
6. Blau, P.: Exchange and Power in Social Life. Wiley, New York (1964)
7. Bozeman, D., Perrewé, P.: The effect of item content overlap on organizational commitment questionnaire–turnover cognitions relationships. J. Appl. Psychol. **86**, 161–173 (2002)
8. Bryman, A., Cramer, D. *Análise de dados em ciências sociais. Introdução às técnicas utilizando o SPSS para windows* (3ª ed.). Oeiras: Celta (2003)
9. Ceitil, M.: *Gestão e Desenvolvimento de Competências* (2ª ed.). Lisboa: Edições Sílabo, Lda (2016)
10. Chambel, M.J.: Práticas de recursos humanos e duplo comprometimento afetivo por parte dos trabalhadores terceirizados. Revista Psicologia: Organizações e Trabalho **12**, 267–282 (2012)
11. Brandão, H.P., Guimarães, T.A.: Gestão de competências e gestão de desempenho: tecnologias distintas ou instrumentos de um mesmo construto? Revista de Administração de Empresas **41**(1), 8–15 (2001)
12. De Vos, A., De Hauw, S., Van der Heijden, B.I.J.M.. Competency development and career success: the mediating role of employability. J. Vocat. Behav. **79**, 438–447 (2011)
13. Finney, S. J., Cristine DiStefano, C. *Nonnormal and categorical data in structural equation modeling. In Structural Equation Modeling: A Second Course.* Gregory, R. (ed.) Hancock and Ralph O. Mueller. Charlotte: IAP Information Age Publishing, pp. 439–92 (2013)
14. Gouldner, A.W.: The norm of reciprocity: a preliminary statement. Am. Sociol. Rev. **25**, 161–178 (1960)
15. Hendry, A., Jones, E.J.H., Charman, T.: Executive function in the first three years of life: precursors, predictors and patterns. Dev. Rev. **42**, 1–33 (2016)
16. Hill, M., Hill, A.: *Investigação por Questionário.* Lisboa: Edições Sílabo (2002)
17. Indra, I.P., Sudiro, A., Rofiq, A.: Organizational and supervisor support on turnover intention mediated by affective commitment. Interdisc. Soc. Stud. **2**(4), 1816–1828 (2023)
18. Knapik, J., Fernandes, B.H.R., Sales, S.S.: Modelos de Gestão por Competências: um estudo longitudinal em uma empresa automobilística. Revista Psicologia: Organizações e Trabalho **20**(3), 1122–1131 (2020)
19. Kumar, R., Ramendran, C., Yacob, P.: A study on turnover intention in Fast food industry: - employees' fit to the organizational culture and the important of their commitment. Int. J.-Acad. Res. Bus. Soc. Sci. **2**(5), 9–42 (2012)

20. Lai, L., Kapstad, J. C.: Perceived competence mobilization: an explorative study of predictors and impact on turnover intentions. Int. J. Hum. Resour. Manage., **20**(9), 1985–1998 (2009)
21. Ludwikowski, W.M.A., Armstrong, P.I., Lannin, D.G.: Explaining gender differences in interests: the roles of instrumentality and expressiveness. J. Career Assess. **26**(2), 240–257 (2018)
22. Martini, M., Gerosa, T., Cavenago, D.: How does employee development affect turnover intention? ex-ploring alternative relationships. Int. J. Train. Dev. **27**, 39–56 (2023)
23. Mercurio, Z.A.: Affective commitment as a core essence of organizational commitment: an integrative literature review. Hum. Resour. Dev. Rev. **14**(4), 389–414 (2015)
24. Meyer, J.P., Allen, NJ.: A three-component conceptualization of organizational commitment. Hum. Resour. Manage. Rev., **1**, 61–89 (1991)
25. Meyer, J.P., Allen, N.J.: Commitment in the workplace: Theory, research, and application. Sage Publications, Thousand Oaks, CA (1997)
26. Meyer, J.P., Becker, T.E., Van den Berghe, C.: Employee commitment and motivation: a conceptual analysis and integrative model. J. Appl. Psychol. **89**, 991–1007 (2004)
27. Meyer, J.P., Herscovitch, L.: Commitment in the workplace: toward a general model. Hum. Resour. Manag. Rev. **11**, 299–326 (2001)
28. Meyer, J.P., Stanley, D.J., Herscovitch, L., Topolnytsky, L.: Affective, continuance, and normative commitment to the organization: a meta-analysis of antecedents, correlates, and consequences. J. Vocat. Behav. **61**, 20–52 (2002)
29. Moreira, A., Cesário, F.: Organizational Commitment as a reducer of Turnover Intentions:which component (affective, calculative and normative) is the best predictor? *Academia Letters*, Article 3636 (2021)
30. Moreira, A., Cesário, F.: Organizational practices of competencies development: adaptation and validation of an instrument. Acad. Lett., Article 5032 (2022)
31. Moreira, A., Cesário, F., Chambel, M.J., Castanheira, F.: Competences development and turnover intentions: the serial mediation effect of perceived internal employability and affective commitment. Euro. J. Manage. Stud. **25**(1), 5–22 (2020)
32. Moreira, A., Sousa, M.J., Cesário, F.: Competencies development: the role of organizational commitment and the perception of employability. Soc. Sci. **11**, 125 (2022)
33. Moreira, A., Tomás, C., Antunes, A.: The mediating effect of affective commitment on the relationship between competence development and turnover intentions: does this relationship depend on the employee's generation? Admin. Sci. **14**, 97 (2024)
34. Mulder, M., Weigel, T., Collins, K.: The concept of competence in the development of vocational education and training in selected EU member states: a critical analysis. J. Vocational Educ. Train., **59**(1), 67–88 (2007)
35. Nguyen, H., My Thi Tran, D., Ba Vu, T., Thuy Thi Le, P.: An empirical study of affective commitment: the case of machinery enterprises in Hochiminh City. Organ. Markets Emerg. Econ., **11**(2), 429–455 (2020)
36. Ravikumar, B., Mohan, K., Ram, V. S.: The influence of job rotation practices on employee performance, mediation effect of human factor and motivation. Int. J. Manage. Human., 2394–0913 (2020)
37. Saeed, I., Waseem, M., Sikander, S., Rizwan, M.: The relationship of turnover intention with job satisfaction, job performance, leader member exchange, emotional intelligence and organizational commitment. Int. J. Learn. Dev. **4**, 242–256 (2014)
38. Shore, L. M., Tetrick, L. E., Lynch, P., Barksdale, K.: Social and economic exchange: construct development and validation. J. Appl. Soc. Psychol., **36**(4), 837–867 (2006)
39. Syed, M., Khan, U. R., Faisal, M.: Impact of training and development on turnover intention. Pak. J. Int'L Aff., **6** (1) (2023)
40. Trochim, W.: *The Research Methods Knowledge Base*. 2nd Edition. Atomic Dog Publishing, Cincinnati, OH (2000)

41. Wayne, S.J., Shore, L.M., Liden, R.C.: Perceived organizational support and leader-member exchange: a social exchange perspective. Acad. Manag. J. **40**, 82–111 (1997)
42. Whitener, E.M.: Do, "high commitment" human resource practices affect employee commitment? A cross-level analysis using hierarchical linear modeling. J. Manag. **27**, 515–535 (2001)

A Well-Being Culture Influencing the Retention of Talent Generation Y Case Study of Top50 Companies in Thailand 2023

Nishima Rensep and Jutamas Kaewpijit(✉)

National Institute of Development Administration, 148 Serithai Road, Khlong Chan Subdistrict, Bangkapi District 10240, Bangkok, Thailand
juthamas@nida.ac.th

Abstract. The Study of Well-being Culture Influencing the Retention of Talent Generation Y Case Study of Top50 Companies in Thailand 2023 has two objectives: (1) to study wellbeing factors to retain talent Generation Y in organization; and (2) to study retention factors to retain talent Generation Y in organization. This research was conducted using a qualitative approach by utilizing semi- structured interviews and analyzing data through the content analysis method with 31 participants of Generation Y from five different business sectors including (1) Food and Beverages; (2) Food Production; (3) Construction Materials; (4) Commerce and Real Estate; and (5) Energy and Utilities. The results indicated that 13 factors influence Well-being contributing to the retention of Generation Y talent in the organization, respectively as follows: (1) Healthy Relationships at Work; (2) Health and Flexible Benefits; (3) Pay; (4) Work-Life Harmony; (5) Strategic Leadership; (6) Autonomy; (7) Work Environment; (8) Facilities; (9) Challenging Work; (10) Opportunity for Growth; (11) Work Process; (12) Mattering at Work; and (13) Work Location. Findings reveal that the top three factors are: Healthy Relationships at Work (17.32%), Health and Flexible Benefits (16.54%), and Pay (10.24%) of the total elements that affect Well-being, comprising 44.10% of all factors. In addition, there are 11 factors that influence the retention of talented Generation Y in Organization, respectively as follows: (1) Healthy Relationships at Work; (2) Opportunity for Growth; (3) Work-Life Harmony; (4) Benefits; (5) Mattering at Work; (6) Learning Opportunity; (7) Pay; (8) Strategic Leadership; (9) Corporate Image; (10) Work Location; and (11) Company Culture. Findings reveal that the top three factors are: Healthy Relationships at Work (18.99%); Opportunity for Growth (12.66%), and Work-Life Harmony (12.66%) of the total elements that affect Retention, comprising 44.31% of all factors. It can be concluded from the study results of the two variables that Talent Generation Y is most dependent on Healthy Relationships at Work. The Study will help organizations understand the factors that contribute to employee retention and well-being in the Thai business environment, where Generation Y is playing a significant role in replacing Baby Boomer retirement and Generation X in the workforce. These factors will help the organization develop a competitive edge and set goals for the future. In order to enable talent generation Y to remain with the organization, it is necessary to build a solution for managing a well-being culture that influences talent retention. Due to the time periods or variations across the years,

M. J. Sousa (Ed.): UFHRD 2024, ISEM 52, pp. 186–200, 2025.
https://doi.org/10.1007/978-3-031-92548-1_12

the study's conclusions might not fully capture the context or circumstances surrounding labor and organizations that existed during that time. Nevertheless, when using research findings, one should take the importance of the business context into account. Furthermore, research on the connection between Generation Y and other Generations in organization is essential for a comprehensive understanding. Also, design appropriate guidelines for fostering harmonious coexistence among individuals in the organization from a holistic perspective.

Keywords: Well-being · Retention · Talent Generation Y · Millennials · 3C

1 Background and Rationale

Following the spread of Covid-19, the trends in economy, health and geopolitics continued to fluctuate the labor markets around the world in 2023. Although the labor markets in high income economies have returned to prosperity, low to medium income economies have continued to encounter unemployment rates higher than those before the Covid-19 pandemic. Together with the changing labor expectations and concerns pertaining to labor quality arising from the pandemic, these have led to an important issue that the world currently pays attention to (World Economic Forum, 2023). The Great Resignation has resulted from employees who were affected during the Covid-19 pandemic. During such time, those employees reviewed their work life or preferred lifestyle. As a result, it gave rise to much of talent uprising. When cultures and organizational goals do not match, talented people willingly resign in pursuit of a new role or workplace that provides flexibility, decent pay or a culture that meets their expectations. Organizations should be aware of the importance of organizational wellbeing as it is essential for their cultures and systems. If employees' well-being is positive, it gives rise to effectiveness and competitive businesses (Adecco Group Thailand, 2021; Adair et al., 2021; Tanachaporn Putiyanun, 2022). Trend 2024 (2022) by Creative Thailand stated that Millennials are those born in 1980-1995. They give importance to mental health and good quality of life. Among them, 41% always suffer from cumulative stress. Thus, they seek activities that help to give them work-life balance. This is in accordance with Deloitte Global (2023) who surveyed data from 40% of Generation Z and 24% of Millennials or 22,856 people in 44 countries around the world. It was found that they resigned from work within two years. One third decided to resign despite not having a new job secured. This was because the factors around salaries, burnout, negative working environments affecting mental health and assigned tasks did not match their norms or values. Further, the findings support what PwC (2023) reported in Global Workforce Hopes and Fears Survey 2023. The survey was based on opinions of 52,000 people from 44 countries around the world including Thailand. It was found that employees were concerned they would not receive any up-skilling or adequate physical and metal care. All of which directly impacted their effectiveness. When employees feel they are not adequately taken care of, they start looking for a new job. Consequently, organizations lose capable employees in the end.

In the survey of PwC (2020), it was revealed that Millennials accounted for 50% of the world population and a large number were entering into the workforce. While this is happening, the Baby Boomers are gradually entering into the retirement age. Younas and Bari (2020) reported that Generation Y demonstrated a need for a different format of management and organizational cultures from the past. Following global situations and rapid economic growth, they have led to an increasing need for capable individuals. Hence, it has become a challenge for organizations to create strategies to attract and retain talented employees in order to create, add values and increase effectiveness of services through the use of skills and existing knowledge to achieve organizational goals. The Future of Job Report 2023 revealed the survey results from perspectives of more than 803 companies, more than 11.3 billion employees covering 27 industries from 45 countries around the world. This was to prepare for an economic shift in the timeframe of 2023–2027. It was found that 48% of these companies stated that career progression and promotion processes for highly talented employees were practices resulted in preparation and competition among organizations (World Economic Forum, 2023)

Based on the background and the importance of the issues above, along with research findings relating to retention of Generation Y talents or organizational well-being , they are still limited in terms of not focusing on the culture of well-being in combination with talent Generation Y retention, which both play essential roles in moving forward an organization and being a successor in an important role of the organization in the future. As a result, the researcher conducted research organizational cultures focusing on well-being that influenced retention of talents. It was a case study of five private organizations in Thailand, which sit in the Top 50 Companies in Thailand 2023. This research would allow organizations to understand factors relating to well-being and retention of Generation Y talents. This would lead to designing management of wellbeing cultures influencing retention of Generation Y talents, promoting emotions and mental health, and allowing talents to become relaxed and energized to work and stay with their organization.

2 Literature Review

2.1 Definitions and Elements of Well-Being Organizations

Well-being means a way that a person decides on the quality of life, happy emotions, satisfaction with the body and the mind, working conditions, relationships with the environment and satisfaction with work, all of which impact upon the effectiveness and efficiency of organizations (Garg and Rastogi, 2009; Rasool et al., 2021; Guest and Conway (2004 as cited in Jain et al., 2008). Based on the literature review of researchers and scholars, it was found that there were 14 factors influencing well-being of employees in similar or different ways as follows:

Positive relationships with others, in Guest and Conway (2004, as cited in Jain et al., 2008); Gelencsér et al. (2023); Garg and Rastogi (2009); Seligman (2012, as cited in Madeson 2017); Office of the U.S. Surgeon General (2022); GALLUP PRESS et al. (2010), it means support, readiness to be involved together and being valued by other people, all of which are important for employees to work better, contributing positively to employees in the long run and helping to create strong social networks. **Meaning,** in Simone (2014); Garg and Rastogi (2009); Seligman (2012, as cited in Madeson 2017); Office of the U.S. Surgeon General (2022), it means being respected and seeing value in people, having a clear role, goals and life directions from the past, and the present in order to reach future goals that correspond to organizational goals. **Opportunity for Growth,** in Gelencsér et al. (2023); Garg and Rastogi (2009); Office of the U.S. Surgeon General (2022); GALLUP PRESS et al. (2010), it means receiving continuous performance development, opportunities to be exposed to new experiences in order to increase knowledge, having a career path plan that corresponds to the needs of employees and communicating growth in order to achieve people and organizational goals.

Nature of Work, in Guest and Conway (2004, as cited in Jain et al. 2008); Gelencsér et al. (2023); Garg and Rastogi (2009), it means work that is meaningful, challenging, receiving collaboration from colleagues, autonomous, being able to think and self-decide without being influenced by organizations or external causes that are beyond control.

Positive Emotion, in Simone (2014) and Seligman (2012, as cited in Madeson 2017), it means levels of energy, enthusiasm, attention and appreciation of people that lead to positive interactions with the society, being essential to being an armor protecting any potential incoming negativity.

Support from Colleagues and Supervisors, in Guest and Conway (2004, as cited in Jain et al. 2008) and Gelencsér et al. (2023), it means relationships between employees and bosses in a sense that the role of bosses can support their subordinates to be able to better work and utilize their skills.

Contingent Rewards and Pay, in Gelencsér et al. (2023) and GALLUP PRESS et al. (2010), it means the way that organizations value employees, pay wages, have a system to increase wages as per performance, are fair and nondiscriminatory when it comes to roles and responsibilities for performance, and provide welfare benefits or flexible gains to employees.

Feeling Involved in Making Organizational Changes, in Guest and Conway (2004, as cited Jain et al. 2008) and Seligman (2012, as cited in Madeson 2017), it means the use of skills, abilities and strengths of people to initiate involvement in life activities.

Environment, in Garg and Rastogi (2009) and GALLUP PRESS et al. (2010), it means an ability to manage an environment and choose or create the context that is suitable for oneself as well as an opportunity to help the community and interact with the society.

Safety, in Office of the U.S. Surgeon General (2022) and GALLUP PRESS et al. (2010), it means a workplace that is safe and hygienic including an organization that focuses on investing in the health of its employees so they are healthy and energized to work toward achieving goals.

Operating Procedures, in Gelencsér et al. (2023), it means standard procedures for work with clear criteria in order to promote effectiveness and provide opportunities of creativity and innovation for employees.

Communication, in Gelencsér et al. (2023), it means opening up and trusting each other, communicating goals and organizational cultures from leaders to employees. **Self-acceptance,** in Garg and Rastogi (2009), it means having positive attitudes toward oneself, leading to positive work.

Work-life Balance, in Office of the U.S. Surgeon General (2022), it means working as professionals in a role of responsibility and being able to integrate work and lifestyle.

2.2 Talent Generation Y Retention Talent Generation Y

In this research, the researcher aimed to study Talent Generation Y that matched with talented employees as valued by organizations. This was in line with the definition of an objective of the research. Talent as Mastery, it means talented people who train themselves to become competent. They become knowledgeable, capable and highly adept at operational skills. They are creative in their work and able to solve issues and demonstrate high performance and high potential which is different from other people (Collings et al., 2019; Juthamas Kaewpijit, 2020; Wasita Ritbumroong, 2005). They display the 3C qualities as follows: (1) Competency - knowledgeable, capable and skilled at working. They are sought by organizations, which leads to determination and decision-making in order to create effectiveness toward outcomes; (2) Commitment - being involved or bonded with an organization, being responsible, willing, dedicated and determined to deliver values to the organization; and (3) Contribution - being ready to dedicate oneself to the organization and work as if being the owner of the organization, leading to positive productivity and success to the organization (Ulrich and Smallwood, 2009)

Talent Generation Y are those born in 1980–2000 or aged 23–43 years old. They are the core workforce of the present and the future. They first joined the workforce when the Baby Boomers were beginning to retire. Generation Y are offsprings of the Baby Boomers or Generation X. They are enthusiastic, skilled at technologies, grow up in diverse cultures, are able to adapt to circumstances, and value learning to succeed professionally. Generation Y or Millennials value mental health and well-being. They look for activities that allow work-life balance and are prepared to risk uncertainty in their professional life (Cennamo & Gardner, 2008 and Weyland, (2011 as cited in Younas & Bari, 2020); Bussin (2014, as cited in Ting 2017); Rasdi et al., 2018; Creative Thailand, 2022).

Talent Retention

Juthamas Kaewpijit (2020) stated that organizational talents could be seen as twofold - cost or investment. Even when these talents leave an organization, they continue to be talents in Thailand. Talent retention refers to retaining talents in the organization for as long as the organization wants them. This is because a growing organization needs to prioritize talent management against other resources so as to sustainably increase the potential of competitiveness advantages (Wasita Ritbumroom, 2005; Rasdi et al., 2018). According to literature review through the perspectives of researchers and scholars, 16 factors were found to contribute to the retention of Generation Y in similar or different ways as follows:

Creative Communication, in Glanz, B. (2002); Petcharat et al. (2013); Ting (2017); Weerarathne et al. (2022), it means creating a positive communication setting, giving importance to team building or arranging activities that are conducive to relationships and allowing opportunities for direct involvement in order to build motivation to work. **Strategic Leadership,** in Younas and Bari (2020); Petcharat et al. (2013); Weerarathne et al. (2022), it means an ability of leaders to demonstrate strategic visions and persuade others to make possible a future together in order to achieve organizational goals.

Mentoring, in Younas and Bari (2020) and Weerarathne et al. (2022), it means connecting people who have more skills and experience than those who have less in order to advice about growth and learning support, and give encouragement about life and work.

Respect and Reason for Being, in Glanz, B. (2002); Ting (2017); Rasdi et al. (2018); Weerarathne et al. (2022), it means respecting rights of others, honoring everyone equally, allowing opportunities, freedom and roles to employees so they are responsible for decision-making and discovering options in order to create achievements for the organization.

Achievement, in Ting (2017); Rasdi et al. (2018); Weerarathne et al. (2022); Gelencsér et al. (2023), it means employees feeling they are valuable, making achievements for themselves and the organization. This influences their motivation and involvement.

Work-Life Balance, in Ting (2017); Rasdi et al. (2018); Weerarathne et al. (2022), it means an assigned role that balances life and work, be it work flexibility and convenient commuting.

Atmosphere and Appreciation for All, in Glanz, B. (2002) and Ting (2017), it means creating a fun atmosphere, promoting positive relationships among colleagues, having mentors to give advice, feeling being part of the organization and having a boss who gives creative appreciations.

Empathy and Enthusiasm, in Glanz, B. (2002) and Gelencsér et al. (2023), it means receiving assistance and empathy among members in the organization, having activities that stimulate employees to be enthusiastic and allowing opportunities for employees to get involved.

Opportunities for Advancement and Use of Potential, in Petcharat et al. (2013) and Ting (2017), it means allowing opportunities for employees to develop skills, knowledge and capabilities, and to perform important work that is meaningful for themselves and the organization in order to stimulate employees to become determined to grow in the organization.

Learning and Development, in Petcharat et al. (2013) and Ting (2017), it means receiving opportunities to learn and train, as well as having clear human resources plans to develop and increase capabilities to work with the organization.

Rewards and Pay, in Petcharat et al. (2013) and Ting (2017), it means several rewards that an organization gives to employees either in cash or in kind to reward them for their roles as well as to stimulate employees to work effectively.

Challenging and Meaningful Assignments, in Ting (2017) and Rasdi et al. (2018), it means being trusted to perform an important and meaningful assignment in order to increase capabilities from the value gained from working.

Knowledge Sharing, in Younas and Bari (2020) and Weerarathne et al. (2022), it means a process or a tool to exchange knowledge among individuals or organizations by knowledgeable people in order to pass on such knowledge to colleagues in organizations.

Social Media, in Younas and Bari (2020), it means the use of social media to be a technology for online communication among people and organizations.

Responsibility, in Ting (2017), it means clear objectives and strategies as set by an organization for employees to know their roles and responsibility.

Turnover Intention, in Gelencsér et al. (2023), it means an intention to resign willingly on the part of employees due to external and internal factors including those from workplace settings, which set satisfaction of the employees.

2.3 General Data of Top50 Companies in Thailand 2023

WorkVenture is a consultancy business that recruits personnel in the Asian region. It is a leader in employer brand building for leading countries in Thailand. It was found in 2015. One of the projects that WorkVenture managed was the Top50 Companies in Thailand campaign. Its objectives were (1) to build employer brands for top companies in Thailand, (2) to survey dream companies that received much interest and were accepted by employees and mass media across the country. The project surveyed employees aged 21-35 years, new graduates, first-time employees and those who had been employed for a while. The survey used open-ended questions including "What is the company you want to work for most? and Why do you choose this company?" The project allowed freedom to respond to the questions while the company names remained anonymous to avoid being guided. The responses received were used to assign rankings for Top50 Companies in Thailand in a given year. Toward the end of 2022, WorkVenture conducted an online survey on the target group of more than 12,372 subjects as a result of Covid-19 that came to play a role in the changing work lifestyles of employees. This led them to seek something that facilitated living, stability and valuing. It was found that the respondents deemed important "the image of employer brands" whether they were stable or not particularly the hybrid work arrangements, which were the approach that they felt worked well professionally (Workventure, 2023).

3 Research Methodology

The population group for this research was Talent Generation Y who worked in Top50 Companies in Thailand 2023 from five businesses including (1) food and beverages, (2) food production and sales, (3) real estate and construction, (4) retail and immovable property development, and (5) energy and public utility. These were organizations with new generation management structures or talent management in order to develop and prepare them to become successors. This was qualitative research using in-depth interviews to obtain in-depth data. The researcher applied the criterion sampling technique by selecting the samples based on the criteria according to the literature review. They included the group of high performance and high potential employees (Juthamas Kaewpijit, 2020; Wasita Ritbumroon, 2005). The snowball sampling technique was also used. This was done by having human resources personnel and Talent Generation Y who were interviewed recommend or tell other Talent Generation Y in a special program of an organization. The researcher relied on the number of sample size in the study by Hagaman and Wutich (2017) who stated that if the samples were diverse to achieve a wider scope, the subject size of 20-40 people should reach the saturation level. Since this research selected Talent Generation Y from various private organizations, the samples included 31 people.

The researcher developed the questions based on the review of the articles Well-being in the Workplace through Interaction between Individuals Characteristics and Organizational Context by Biggio and Cortese (2013) and The Relationship between Talent Management Practices and Retention of Generation 'Y' Employees: Mediating Role of Competency Development by Younas and Bari (2020). The scope of questions was set around the keywords and adjusted to suit the Thai context. Prior to using the questions in the interviews with the Talent Generation Y samples, the construct validity, content validity and appropriate wording were assessed by the researcher's advisor who was an expert in research studies. This was to ensure that the objectives of the research, both in the factors around well-being and retention of Talent Generation Y were met.

4 Findings

4.1 Demographic Findings of the Samples

Following data collection using the in-depth interview technique (semi-structure interview) from 31 Talent Generation Y who worked in five of the Top50 Companies in Thailand 2023, the demographic findings of the samples were as follows: (1) gender: 18 females or 58.06%; 10 males or 32.25%; and three undisclosed genders or 9.67% (2) aged 23-43 years old, divided into the 23-29 year-old group of 6 persons or 19.35%; the 30-36 year-old group of 15 persons or 48.38%; and the 37-43 year-old group of 10 persons or 32.25% (3) year of birth 1980-2000: 10 persons or 32.25% born in 1980-1986; 15 persons or 48.38% born in 1987-1993; six persons or 19.35% born in 1994-2000 (4) average length of employment in the current organization was six years, the shortest service was one year and the longest was 17 years: 17 persons or 54.83% had served 1-6 years; 13 persons or 41.93% had served 7-12 years; one person or 3.22% had served 7–12 years (5) positions in the organization ranged from: one person or 3.22% was a supply chain officer; one person or 3.22% was a management associate; two persons or 6.45% were engineers; two persons or 6.45% architects; three persons or 9.67% were marketing officers; three persons or 9.67% were information technology officers; and 15 persons or 48.38% were human resources officers.

4.2 Findings of Well-Being Factors Influencing Retention of Talent Generation Y

The researcher used the content from Part 2 of the interviews about factors influencing organizational well-being in Talent Generation Y. There were two questions including 2.1) What is organizational well-being for you? and 2.2) Which factors influenced your organizational well-being? Because the interviewees responded to the questions in the same direction, the researcher used the content analysis technique to search the keywords that the interviewees stated most frequently. It was found that there were 13 wellbeing factors influencing retention of Talent Generation Y. They were (1) Healthy Relationships at Work at 17.32%; (2) Health and Flexible Benefits at 16.54%; (3) Pay at 10.24%; (4) Work-Life Harmony at 8.66%; (5) Strategic Leadership at 7.87 (6) Autonomy at 6.30%; (7) Work Environment at 5.51%; (8) Facilities at 5.51% (9) Challenging Work at 5.51%; (10) Opportunity for Growth at 5.51%; (11) Work Process at 4.72%; (12) Mattering at Work at 3.94% and (13) Work Location at 2.36% respectively.

4.3 Findings of Factors Influencing Retention of Talent Generation Y

The researcher used the content from Part 3.4 of the interviews: Today what are the factors influencing your desire to work for the organization everyday without resigning?

The researcher used the content analysis technique to search the keywords. It was found that there were 11 factors influencing retention of Talent Generation Y. There were (1) Healthy Relationships at Work at 18.99%; (2) Opportunity for Growth at 12.66%; (3) Work-Life Harmony at 12.66%; (4) Benefits at 11.39%; (5) Mattering at Work at 10.13%; (6) Learning Opportunity at 8.86%; (7) Pay at 7.59 %; (8) Strategic Leadership at 6.33%; (9) Corporate Image at 5.06%; (10) Work Location at 3.80%; and (11) Company Culture at 2.53% respectively (Fig. 1).

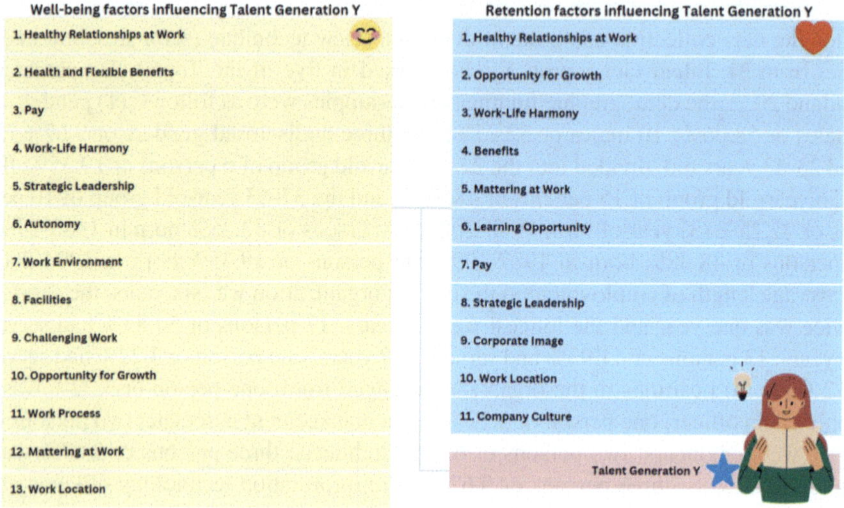

Fig. 1. Findings of 13 well-being factors influencing Talent Generation Y and 11 retention factors influencing Talent Generation Y

4.4 Findings of Consistency Between Well-Being Factors and Retention Factors Influencing Talent Generation Y

The researcher conducted an analysis in order to consider the consistency of the factors Talent Generation Y prioritized between the well-being factors and the factors influencing retention of Talent Generation Y. It was found that there were eight consistent factors including (1) Healthy Relationships at Work; (2) Benefits; (3) Pay; (4) WorkLife Harmony; (5) Strategic Leadership; (6) Opportunity for Growth; (7) Mattering at Work; and (8) Work Location (Fig. 2).

According to the study on well-being factors and retention factors influencing Talent Generation Y, it was found that the Healthy Relationships at Work factor was the one that Talent Generation Y prioritized.

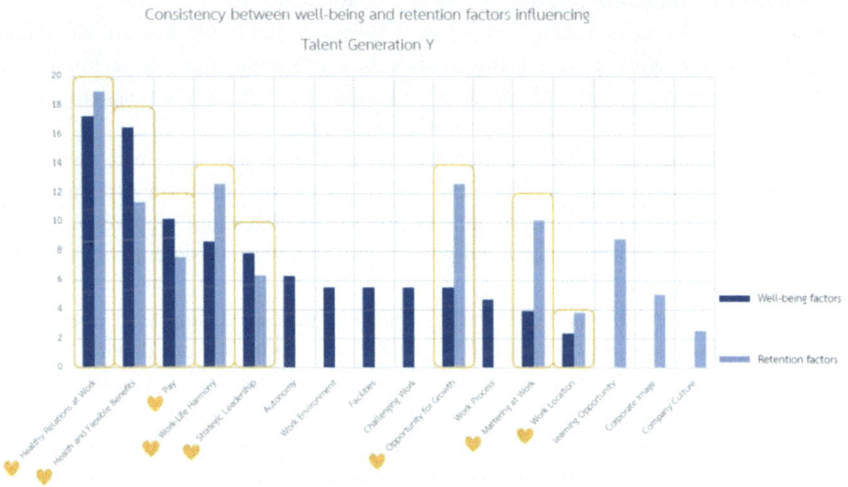

Fig. 2. Findings 8 factors of consistency between well-being and retention factors influencing Talent Generation Y

5 Discussion

5.1 Well-Being Factors that Retained Talent Generation Y with the Organization

Based on the findings of well-being factors influencing retention of Talent Generation Y, the discussion around the 13 factors that were consistent, similar or different from the concepts, theories and related studies was as follows:

(1) **Healthy Relations at Work** referred to having colleagues who were ready to provide assistance, mentoring, encouragement and positive attitudes; open up to others and exchange opinions; were prepared to overcome obstacles by allowing each individual's ability to create teamwork in order to find a solution and drive for outcomes. This was consistent with the studies of Gelencsér et al. (2023); Garg and Rastogi (2009); Seligman (2012, as cited in Madeson, 2017); GALLUP PRESS et al. (2010).

(2) **Health and Flexible Benefits** referred to having standardized health benefits for employees including coverage over family members or being able to buy such befits for family members with employee prices. This factor extended to flexible benefits whereby employees were able to choose them according to their individual preferences allowing equal rights to benefits. This was in line with the studies of Gelencsér et al. (2023); GALLUP PRESS et al. (2010).

(3) **Pay** referred to pay commensurate with assigned work and competitive with external markets. Additionally, it had to reflect the effort of talents who dedicated themselves to the organization. In other words, if work was pressing or difficult, it would not be an obstacle provided that the pay was decent and such pay could be used to acquire happiness through other dimensions. This was consistent with the studies of Gelencsér et al. (2023); GALLUP PRESS et al. (2010)

(4) **Work-Life Harmony** referred to being able to create balance between work and personal life because Talent Generation Y believed that work time should be dedicated to work while personal time should be spent with family or living in order to create positive energy or happiness for themselves. This was consistent with the study of Of- fice of the U.S. Surgeon General (2022)

(5) **Strategic Leadership** referred to executives or supervisors who were able to give clear directions in order to reach the outcomes the organization expected. They were able to give advice, recommend or provide feedback and prepared to support employees both professionally and personally. This was consistent with the study of Gelencsér et al. (2023).

(6) **Autonomy** referred to being trusted to important tasks; learning by trial and error, learning and deciding by oneself to prove one's abilities; and driving assigned work to its goals. This was in line with the study of Garg and Rastogi (2009), but different from Seligman (2012, as cited in Madeson, 2017) in terms of the definition, which meant using skills, abilities and strengths of a person to become involved in life activities. (7) **Work Environment** referred to having a work environment with standardized wellbeing including cleanliness, noise, smell and safety. Additionally, there was a corner for employees to relieve themselves from stress. This was in line with the study of Office of the U.S. Surgeon General (2022), but different from the study of Garg and Rastogi (2009), which stated that environment mastery referred to an ability to manage complex environments and choose or create appropriate context.

(8) **Facilities** referred to having equipment and facilities in support of work, allowing effective work. This was consistent with the study of GALLUP PRESS et al. (2010). (9) **Challenging Work** referred to valuing challenging work other than routines as it was deemed as a way for self skill development and growth. This was consistent with the studies of Gelencsér et al. (2023); Garg and Rastogi (2009), but different in terms of the definition provided in Seligman (2012, as cited in Madeson, 2017); Simone (2014), which meant positive emotional state as a result of work assessment or work experience for achievement.

(10) **Opportunity for Growth** referred to having an opportunity to grow in a career whether vertically or horizontally. This was in line with the studies of Gelencsér et al. (2023); Garg and Rastogi (2009); Office of the U.S. Surgeon General (2022); GALLUP PRESS et al. (2010)

(11) **Work Process** referred to having data and work processes that clearly supported usage and reduced unnecessary processes in order to reduce work process delays while allowing business competitiveness. This was consistent with the study of Gelencsér et al. (2023).

(12) **Mattering at Work** referred to knowing that work mattered and created value for the organization and individuals. This was in line with study of Seligman (2012, as cited in Madeson, 2017); Office of the U.S. Surgeon General (2022).

(13) **Work Location** referred to having a workplace close to home, allowing convenient commuting, avoiding fatigue and allowing time with family. This was consistent with the study of Garg and Rastogi (2009), but different in terms of the definition provided in GALLUP PRESS et al. (2010), which meant an organization with a

culture and workplace conducive to well-being and stimulating healthy behaviors for the benefits of both the organization and individuals.

5.2 Retention Factors that Retained Talent Generation Y with the Organization

Based on the findings of factors influencing retention of Talent Generation Y, the discussion around the 11 factors that were consistent, similar or different from the concepts, theories and related studies was as follows:

(1) **Healthy Relationships at Work** referred to team members sharing knowledge and experience with one another, having positive attitudes, opening up to other people while being reasonable, and ready to support any implementation to reach goals. This was consistent with the studies of Wasita Ritbumroon (2005); Ting (2017); Rasdi et al. (2018); Weerarathne et al. (2022)

(2) **Opportunity for Growth** referred to a role of responsibility that was important and challenging, the organization seeing the potential and providing an opportunity to learn and develop skills as well as being prepared to support growth. This was consistent with the studies of Petcharat et al. (2013); Ting (2017); Rasdi et al. (2018), but different from the study of Gelencsér et al. (2023) in terms of the definition, which meant employees feeling free from obligations because of each individual experience and cultural factors setting levels of determination.

(3) **Work-Life Harmony** referred to understanding and allowing autonomy for management both professionally and personally. The focus was for outcomes as key. This was consistent with the studies of Ting (2017); Rasdi et al. (2018); and Weerarathne et al. (2022).

(4) **Benefits** referred to having benefits that were useful for making one's way including giving health benefits for employees and family members. This was in line with the study of Ting (2017).

(5) **Mattering at Work** referred to the value from assigned tasks, being a part of creating the value for oneself and the organization, and being able to make other people happy from using products or services from the organization. This was consistent with the studies of Wasita Ritbumroong (2005) and Ting (2017).

(6) **Learning Opportunity** referred to an organization providing opportunities to learn and develop skills from various work tasks including creating an environmental setting to always allow self-development. This was consistent with the studies of Younas and Bari (2020); Ting (2017); and Petcharat et al. (2013).

(7) **Pay** referred to pay commensurate with assigned tasks and adequate for the needs to cover expenses and take care of the family. This was in line with the study of Petcharat et al. (2013); and Ting (2017).

(8) **Strategic Leadership** referred to leaders or executives who could give directions appropriate for the organizational goals; understand and manage the teams under their care appropriately both for professional and personal matters; open up to new ideas; see the potential and be willing to support growth. This was consistent with the studies of Younas and Bari (2020), Petcharat et al. (2013); and Weerarathne et al. (2022). **9) Corporate Image** referred to the reputation of being a good organization with stability, an ability to take care of employees despite critical times, leading to

pride and willingness to communicate the corporate image to others. This was in line with the study of Gelencsér et al. (2023).

(9) **Work Location** referred to a workplace close to home or convenient for commuting, allowing time to spend with family. This was consistent with the studies of Wasita Ritbumroong (2005) and Ting (2017).

(10) **Company Culture** referred to cultivating employees to hold the same value, be candid and see goals in the same direction. This was consistent with the studies of Petcharat et al. (2013); and Ting (2017), but different from the study of Wasita Ritbumroong (2005), which meant receiving assistance and that the organization creating an environment or activities stimulating employees to be enthusiastic and providing opportunities to be involved in such activities.

6 Limitation and Future Research

6.1 Research Limitations

(1) This was qualitative research on Talent Generation Y. One of its limitations was the factors relating to the organizational well-being of Generation X and Generation Z, which affect the environment conducive to the overall organizational well-being and retention.

(2) Due to the case study and the samples of the current research being the group that came from a variety of businesses and careers, the limitation on the cluster of businesses and careers of similar samples applied. This would be particularly true for a case study that aimed to study factors influencing Talent Generation Y of each business group.

6.2 Research Recommendations and Applications

(1) A study on cultures focusing on well-being of retention of Talent Generation Y in five business groups including (1) food and beverages; (2) food production and sales; (3) immovable properties and construction; (4) retail and immovable property development; and (5) energy and public utility. The findings ought to be considered in the business context that is similar in nature.

(2) A case study: Top50 Companies in Thailand 2023. The findings ought to be considered in relation to the periods to repeat the research. This is because different years of study may render results that do not cover the labor context or circumstances that occur in an organization in those times.

6.3 Recommendations for Future Research

(1) Future research should study further the relationships between Generation Y and Generation Z so that Talent Generation Y may study, understand and develop Generation Z to play a role in the organization. This is in preparation for Generation X for their retirement.

(2) Future research should conduct a study on each generation by integrating organizations as a whole. This can be a comparison between needs or relationships among three generations including Generation X, Generation Y and Generation Z in order to design appropriate guidelines for co-existence, effectiveness and efficiency of the organization.
(3) Future research should study in order to find clusters of specific business groups and sample groups. This is to design a culture that focuses on well-being for retention of Talent Generation Y that is appropriate for each context and business group.

References

Juthamas Kaewpijit (2020). Effective Talent Management with "Work-Come-GetStay" Concept. https://www.scb.co.th/th/personal-banking/stories/businessmaker/ibe-talent-managment.html

Tanachaporn Putiyanun (2022). Why Well-Being of Employees "Should" be Prioritized?. Brightside People

Wasita Ritbumroong: Talent Management for Developing Organizational Competitiveness Advantage (2005)

Adair, C., Donohue, J., Friedrich, T., Hebbar, A., Kissack, J., Wich, J. G.: The Talent Uprising - The Revolution Driving Evolution (2021)

Adecco Group Thailand.: Knowing The Great Resignation (2021). https://adecco.co.th/en/knowledge-center/detail/hr-insight/the-great-resignation

Biggio, G., Cortese, C.: Well-being in the workplace through interaction between individual characteristics and organizational context [Health and Well-being]. Taylor and Francis Group. (2013, 18) https://doi.org/10.3402/qhw.v8i0.19823

Coe, E., Enomoto, K., Gupta, A., Lewis, R.: National employer survey reveals behavioral health in a COVID-19 era as a major concern. McKinsey and Company (2020). https://www.mckinsey.com/industries/healthcare/our-insights/nationalemployer-survey-reveals-behavioral-health-in-a covid-19-era-as-a-major-concern/

Collings, D. G., Scullion, H., Caligiuri, P. M.: Global Talent Management. Routledge Global Human Resource Management (2019)

Creative Thailand. (2022). Trend 2022 : READY SET GO. Creative Economy Agency (Public Organization)

Deloitte Global.: 2023 Gen Z and Millennial Survey Waves of change: acknowledging progress, confronting setbacks (2023)

Gallup Press, Rath, T., Harter, J.: Wellbeing: the five essential elements (2010)

Garg, P., Rastogi, R.: Effect of Psychological Wellbeing on Organizational Commitment of Employees [The Icfai University Journal of Organizational Behavior] (2009)

Gelencsér, M., Szabó-Szentgróti, G., Komüves, Z. S., Hollósy-Vadász, G.: The Holistic Model of Labour Retention: The Impact of Workplace Wellbeing Factors on Employee Retention [administrative sciences]. MDPI (2023). https://doi.org/10.3390/admsci13050121

Hagaman, A. K., Wutich, A.: How Many Interviews Are Enough to Identify Metathemes in Multisited and Cross-cultural Research? Another Perspective on Guest, Bunce, and Johnson's (2006) Landmark Study. Sage Journals (2017). https://doi.org/10.1177/1525822X16640447

Jain, A. K., Giga, S. I., Cooper, C. L.: Employee wellbeing, control and organizational commitment [Leadership and Organization Development Journal] Emerald Group Publishing Limited (2008). https://doi.org/10.1108/01437730910949535

Madeson, M.: Seligman's PERMA+ Model Explained: A Theory of Wellbeing (2017). https://positivepsychology.com/perma-model/#google_vignette

Office of the U.S. Surgeon General. (2022). Workplace Mental Health and WellBeing
Petcharat, W., Sadangharn, P., Kruapong, A.: Talent retention and intention to stay of talented people in an industrial estate, Chonburi Province. WMS J. Manage. (2013)
PwC. (2020). Millennials at work Reshaping the workplace
PwC. (2023). Is your workforce reinvention ready? PwC. https://www.pwc.com/gx/en/issues/workforce/hopes-and-fears.html
Rasdi, R. M., Kusnin, N., Chen, Y. S.: Predictors and intervening variables of talent retention. Int. J. Acad. Res. Bus. Soc. Sci. (2018). https://doi.org/10.6007/IJARBSS/v8-i12/5007
Rasool, S. F., Wang, M., Tang, M., Saeed, A., Iqbal, J.: How Toxic Workplace Environment Effects the Employee Engagement: The Mediating Role of Organizational Support and Employee Wellbeing [International Journal of Environmental Research and Public Health]. MDPI (2021). https://doi.org/10.3390/ijerph18052294
Said, D. S., Agarwal, S., Alzoubi, Y. I., Momin, M. M., Naqvi, A.: Occupational stress among Generation-Y employees in the era of COVID-19: cases from universities in India. Emerald Publishing Limited (2022). https://doi.org/10.1108/ICT-03-2022-0017
Simone, S. D.: Conceptualizing Wellbeing in the Workplace [International Journal of Business and Social Science]. IJBSS, 5 (2014)
Ting, X.: Identifying Motivation Preferences of Generation Y Talent in China [Business Economics]. VAASAN AMMATTIKORKEAKOULU International Business (2017)
Ulrich, D., Smallwood, N.: What is talent? [Executive Education]. T. R. Group. (2009)
Weerarathne, R. S., Walpola, M. D. C. P., A. D. W. D. Piyasiri, Jayamal, I. A. U. M., Wijenayaka, T. H. P. C., Pathirana, G. Y.: 'Leave or remain': intentions of Gen X and Y employees. Springer (2022). https://doi.org/10.1007/s11135-022-01456-z
Workventure. (2023). Poll Realeased: Top 50 Companies New Generations Want to Work For Most 2023. https://www.workventure.com/top50-companies-2023
World Economic Forum. (2023). Future of Jobs Report 2023. W. E. Forum. https://www.weforum.org/reports/the-future-ofjobs-report-2023/
Younas, M., Bari, M. W.: The relationship between talent management practices and retention of generation 'Y' employees: mediating role of competency development [Economic Research-Ekonomska Istraživanja]. Routledge Taylor and Francis Group (2020). https://doi.org/10.1080/1331677X.2020.1748510

Employee Churn Prediction in Commercial Airlines in Portugal

António Pimenta de Brito[1](✉), Maria José Sousa[2], and Ana Moreira[3]

[1] BRU-ISCTE / ESCAD-IP LUSO, Lisbon, Portugal
acpbo@iscte-iul.pt
[2] ISCTE-IUL, Lisbon, Portugal
maria.jose.sousa@iscte-iul.pt
[3] Universidade Europeia, Lisbon, Portugal
ana.moreira@universidadeeuropeia.pt

Abstract. Employee churn occurs when an employee leaves voluntarily from a company. It's essential to measure this indicator due to the financial costs a company deals with, especially in the commercial aviation sector. "Turnover intention" is the best antecedent to actual turnover. Job satisfaction encompasses many dimensions and is the best antecedent to actual turnover. There are few studies about employee churn in the commercial aviation sector, and no study has yet been conducted to correlate employee satisfaction with turnover intentions in the commercial aviation secttheor in Portugal. The methodology of this study is performing a literature review on the factors commonly associated with employee churn and a review of the principal concepts involved. The evaluation result retrieved most common factors associated with employee churn. "Pay satisfaction" (PS), "safety satisfaction" (SS), "work/ family satisfaction" (WFS), "career satisfaction" (CS), "leadership satisfaction" (LS) and, finally, Job Performance (JB), were validated to test the hypotheses about which factors are more associated with employee churn and the relative contribution of the variables for the dependent variable "turnover intentions". Multiple Linear Regression was used to study a sample of 370 valid responses from people working in the commercial aviation sector in Portugal. As for the effect of satisfaction on turnover intentions, the dimensions of CS SS, LS, and PS, have a negative and significant effect on turnover intentions. Other conclusions were drawn. Namely, relationships between demographic variables and the main variables were presented.

Keywords: airlines · employeechurn · hranalytics

1 Introduction

Employee churn is an HR Analytics indicator that aims to measure the voluntary abandonment of employees. Companies study and try to understand this phenomenon because of the high costs that the loss of a valuable professional means. The second reason is related to the willingness to act in the face of evidence of professional abandonment to avoid this cost for the company and ensure the best possible experience for the employee.

M. J. Sousa (Ed.): UFHRD 2024, ISEM 52, pp. 201–229, 2025.
https://doi.org/10.1007/978-3-031-92548-1_13

In the commercial aviation sector, the demands on the technical and non-technical qualifications of professionals are very high and expensive; therefore, the concern about voluntary turnover is all too justified. There is evidence that in the commercial aviation sector, this problem is faced by HR departments (Efthymiou et al. 2021; Kiernan, 2018), but it is made up of several realities. There are several types of professionals and airlines.

This research intends to understand what factors influence the turnover intentions of commercial aviation professionals in Portugal. It also intends to understand the relationships between the variables that impact turnover intentions and their relative contribution to the phenomenon. Finally, it intends to issue management recommendations for HR decision-making in this sector.

The exploratory study will begin with a literature review of the concepts involved in this research, namely the factors that appear in the literature as important to influence turnover intentions of professionals in general, in aviation, in particular. This paper aims to study the factors that influence employee churn, as well as the moderating effect of turnover intentions in the relationship between satisfaction and performance.

2 Theoretical Framework

2.1 Employee Churn

"Churn", differently from other concepts as "turnover", refers to the voluntary rather than involuntary abandonment of the worker and deals more with the financial consequences of an employee loss (Pirrolas, Correia & Nascimento, 2022). The concept of churn is commonly used in the realm of marketing and customer relationship management (Hassouna et al., 2015), "Customer churn". It is not so used in the HRM realm, but recently, the amount of scientific and industry applications increased as HR Analytics (HRA) became an important HR trend. HRA is the area of HRM that uses employee data to generate insights for decision-making (Margherita, 2022; Marler & Boudreau, 2016; Tursunbayeva et al., 2018).

2.2 Turnover Intentions

The costs of an employee's departure are various and are not only the training costs but the costs of finding another employee of the same value, the knowledge lost to a competitor, and the loss of value in general. Most research on predictive models of employee churn and turnover intention is based on surveys or real company data (Ekawati, 2019). Alternatively, the construct "turnover intention" is used, i.e., instead of the employee's departure when it happens, the variable "turnover intentions" represents the employee's propensity to leave his organization (Ali et al, 2022; Rombaut & Guerry, 2018; Hussain et al. 2020). Turnover intention is found to be the best antecedent to actual turnover (Griffeth, Hom & Gaertner, 2000; Joseph et al. 2007).

2.3 Turnover in Airlines

The topic of turnover is complex and nuanced in the aviation sector. In some cases, turnover is high (Chen, 2006; Efthymiou et al. 2021; Kiernan, 2018), while in others, it is not so high. Before and after the COVID-19 pandemic, the shortage of pilots didn't stop (Brito, 2020). CAE (2023) updated the estimate of the shortage of pilots for the next 10 years and doubled it by 252,000 new commercial pilots needed.

The cost of turnover and absenteeism of flight crew in the operation is high and a critical problem in aviation (Chen, 2006). In the case of pilots and aeronautical maintenance technicians, the impact is also high not only in the operation and in the company's revenue but also due to the cost of hiring, training, and having to comply with strict aviation regulations on training and safety of these professionals (for example the aircraft type certifications).

Efthymiou et al. (2021) studied the factors of retention in pilots of the low-cost company Ryanair. Ryanair has high turnover levels, probably due to its business model based on low wages, cost containment and contentious management of pilot claims. It concluded that among the factors studied, lifestyle, economics, and attractive rosters led to the choice of respondents, Ryanair pilots, as the main factors for retention.

The class of pilots has always been seen as privileged in terms of bargaining power for negotiating job conditions as salaries, and therefore, one can eventually talk about generalised dissatisfaction and turnover intentions in some context, but not high rates of voluntary turnover as a consummate act of leaving. There is pressure and stress derived from operational and safety complexity, as well as the challenges and possibility of conflicts resulting from proximity to the customer, such as cabin crew (Cheng et al., 2018; Karatepe & Eslamlou, 2017; Karatepe & Kim, 2020; Vatankhah & Darvishi, 2018). As with any service job, dissatisfaction may exist, but since aviation is a heavily unionized sector, the voice of dissatisfaction can be channelled and heard, and this negativity is not embodied in an action to leave the company (Freeman & Medoff, 1984). However, turnover intentions can exist perfectly and with all the consequences that this can have, because of dissatisfaction and give rise for example to the phenomenon of "boreout" (Karatepe & Kim, 2020) and other consequences analyzed in this study, which have been little addressed by science. "Boreout" is defined as "is a negative psychological state of low arousal (…)" (Stock, 2015, p. 574, cited by Karatepe & Kim, 2020). In short, there aren't many studies on the retention and turnover intentions of aviation professionals.

2.4 Job Satisfaction

The literature on the topic of job satisfaction is extensive; however, despite its development, there is no unanimous definition. Nevertheless, there are elements consistently present in the most widely disseminated conceptions in the literature (Wyrwa & Kaźmierczyk, 2020). The dimensions of job satisfaction most addressed in the literature are emotional and cognitive, and the degree of subjectivity that is originated by individual perceptions and feelings regarding job satisfaction. Besides these differences in the evaluation of job satisfaction, there are situational factors. The most popular definition of job satisfaction is "that of an attitude that determines the degree to which employees find their work favorable or unfavorable" (Wyrwa & Kaźmierczyk, 2020, p. 152). However, this definition poses problems regarding the subjectivity of individual cognitions and emotions and the interaction between both.

According to the conclusions of a recent systematic review of the literature on job satisfaction by Wyrwa & Kaźmierczyk (2020), there is not a single type of job satisfaction, but several dimensions of job satisfaction; likewise, there is a degree of job satisfaction rather than a single perspective on the various aspects of work. Regarding the factors contributing to job satisfaction, there is consensus in the literature that they are divided into two main groups: individual factors and group factors, that is, intrinsic and extrinsic (Pujol-Cols & Dabos, 2018; Wyrwa & Kaźmierczyk, 2020), respectively. There are reasons for satisfaction/dissatisfaction that depend on the perceptions, attitudes, emotions, and cognitions of the individual, and factors generated by the external environment, namely the organization and the broader business and social environment in which the worker operates. The interaction between these factors is also complex and can generate mixed influences on job satisfaction that are difficult to disentangle (Wyrwa & Kaźmierczyk, 2020).

Positive correlations were found in literature between job satisfaction and performance, organizational citizenship behavior and customer satisfaction, and negative correlations between job satisfaction and absenteeism, turnover and workplace deviance (Robbins & Judge, 2009).

Pay satisfaction significantly impacts turnover intentions (Danish et al. 2017; Motshegwa, 2010; Singh & Loncar, 2010; Vandenberghe & Tremblay, 2008). Effective leadership styles, particularly transformational, entrepreneurial, and coaching leadership, are associated with higher satisfaction and lower turnover intentions (Gan & Voon, 2021; Wee, Bang & Park, 2020; Wells & Peachey, 2011; Yang, Pu, & Guan, 2019). Career-related factors significantly impact job satisfaction (Gu et al. 2010; Huang et al., 2017; Mosadeghrad, Ferlie & Rosenberg, 2008). Work/family satisfaction significantly affects turnover intentions, mediated through various factors such as job satisfaction, work-to-family conflict, and the specific context of the employment (e.g., family businesses, healthcare) (Chen et al., 2015; Lim et al. 2021; Mihelic & Tekavcic, 2014; Shu et al., 2018). The literature illustrates a clear link between safety satisfaction and turnover intentions across various industries, with a particular focus on the transport sector in general (Huang et al. 2016; Smith, 2018; Siu, Cheung & Lui, 2015; Zhang et al., 2023). Literature shows a consistent link between job satisfaction and turnover intentions in the commercial aviation sector (Chen, 2006; Jou, Kuo, & Tang, 2013; Suifan, Diab & Abdallah, 2017). Finally, job dissatisfaction is one of the strongest predictors turnover intentions in an organization (Mowday, Porter & Steers, 1982; Mulki, Jaramilo & Lokander, 2006).

Literature demonstrates a nuanced relationship between turnover intentions and job performance. Overall, there is a negative relationship between turnover intentions and job performance (Bishop, 1990; Hancock et al., 2013; Trevor et al., 1997; Wynen & Kleizen, 2019).

Job satisfaction is a critical determinant of employee performance in the aviation industry. Improving job satisfaction through positive work environments, recognition, and support can lead to enhanced employee performance and organizational success (Isnanto, 2021; Petty et al., 1984 Mansour et al., 2021; Hoole & Vermeulen, 2003; Firdaus et al., 2022; Lin-jing, 2010; Ng et al., 2009; Bakhsh, 2020). However, it was also found that job satisfaction can have a negative impact on performance in aviation

(Supriyanto, 2018). Historically, the academic literature on the relationship between job satisfaction and performance is not unanimous and clear (Iaffaldano and Muchinsky, 1985; Judge et. al., 2001; Petty, McGee, and Cavender, 1984; Whitman, Van Rooy, and Viswesvaran, 2010) and is very dependent in the various dimensions of job satisfaction and different types of job satisfaction and job performance relationships (Edwards et. al 2008).

2.5 Factors Influencing Employee Churn

There are many factors that contribute to job satisfaction and to employee churn/ turnover intentions. The academic literature has studied a few. Below is a rundown of the factors that most influence churn/attrition. From the literature review, bellow (Table 1) are presented the most used approaches. In this study, a literature review was conducted with the following criteria. It was searched in the SCOPUS and WEB of SCIENCE academic repositories for the most recent scientific publications in Business/ Management on employee churn predictive models, written in English, with no time frame of publication, with the search term "EMPLOYEE CHURN" in the title and the results were 13 papers, 8 were validated and presented in the table were returned after debugging the search (Table 1).

Table 1. A literature review on employee churn predictive models (2011–2022)

	Authors	Factors influencing Employee churn
1	Saradhi & Palshikar (2011)	Age, designation, gender, department, qualification, past experience in years, employee location, experience in parent organization, experience in client organization, billed or not billed, on-site/off-site, designation in client organization
2	Yiğit & Shourabizadeh (2017)	Education, Education Field, Environment Satisfaction, Gender, Job Involvement, Job Level, Job Role, Job Satisfaction, Marital Status, Over-Time, Performance Rating, Relationship Satisfaction, Stock Option Level, Years Since Last Promotion, Years With Current Manager
3	Dolatabadi & Keynia (2017)	Age, Marital status, Level of Education, Work experience (total), Area of expertise, Average number of services given per year, Average time assigned to each service, Sex, Relating certificate to employee service, Duplicate service or not, Miss Call service or not, Number of duplicate service, Resolved service on the first day or not, Service Class, Request Channel, Scope of service, Service time, Service Type, Referring to the software development team or not, Year, Resolved service on the first calling or not, several features of customers and employees should be examined

(*continued*)

Table 1. (*continued*)

	Authors	Factors influencing Employee churn
4	Sisodia Vishwakarma & Pujahari (2017)	employee's salary, current position, promotion, (? ineligible)
5	Yahia, Hlel & Colomo-Palacios (2021)	Age, Marital status, Tenure, Grade, Rewards, Job involvement, Training, Business Travel, Job satisfaction, Job performance, Environment satisfaction
6	Srivastava & Eachempati (2021)	employee_satisfaction, appraisal_rating, employee_CTC_level, number of projects/tasks assigned per quarter, time spent per project per quarter, safety measure, promotion
7	Jain, Tomar, & Jana (2021)	satisfaction level, last evaluation, number project average monthly hour time spent, work accident, left, promotion last 5year, department, salary
8	Pirrolas, Correia & Nascimento (2022)	work environment, salary, recognition, leadership, work schedule

The table summarizes the most recent studies on predictive models of employee churn between 2011 and 2022. Few studies in the literature consider employee churn prediction and analysis (Yiğit & Shourabizadeh, 2017) and few with the dependent variable "turnover intentions".

The most recurrent factors associated with turnover intentions across these studies seem to be salary (Sisodia, Vishwakarma & Pujahari, 2017; Yahia, Hlel & Colomo-Palacios, 2021; Jain, Tomar, & Jana, 2021; Pirrolas, Correia & Nascimento, 2022), job satisfaction (Yiğit & Shourabizadeh, 2017; Yahia, Hlel & Colomo-Palacios, 2021; Jain, Tomar, & Jana, 2021; Srivastava & Eachempati, 2021), and promotion opportunities (Sisodia, Vishwakarma & Pujahari, 2017; Srivastava & Eachempati, 2021; and Jain, Tomar, & Jana, 2021). Other factors like age, marital status, job performance, leadership quality, and work environment also appear significant, but they are mentioned less frequently or are more study specific. These factors might be interconnected, influencing churn in combination rather than isolation.

It is concluded that the factors that influence turnover intentions most used in the models are essentially extrinsic job satisfaction variables, and the following dimensions were retrieved and created for the model to be tested: "Pay satisfaction" (PS), "safety satisfaction" (SS), "work/ family satisfaction" (WFS), "career satisfaction" (CS), "leadership satisfaction" (LS) and, finally, Job Performance (JB). The goal is to test how each one contributes separately to the churn phenomenon. According to the research carried out in the literature, this is not a usual approach; it is normal to address the topic of factors that influence job satisfaction. Satisfaction has many dimensions and is the most determining factor to influence turnover intentions (Mowday, Porter & Steers, 1982; Mulki, Jaramilo & Lokander, 2006).

Srivastava & Eachempati (2021) pointed out two limitations to most churn predictive models in the literature. Most of them claim to know which factors motivate and predict churn, but do not calculate the relative contribution of each of these factors to the model.

Multiple linear regression models are proposed to make them more robust and beyond the simple linear prediction of churn.

3 Method

3.1 Research Model and Hypotheses

The literature review carried out leads to formulate the following hypotheses, summarised in the research model (Fig. 1):

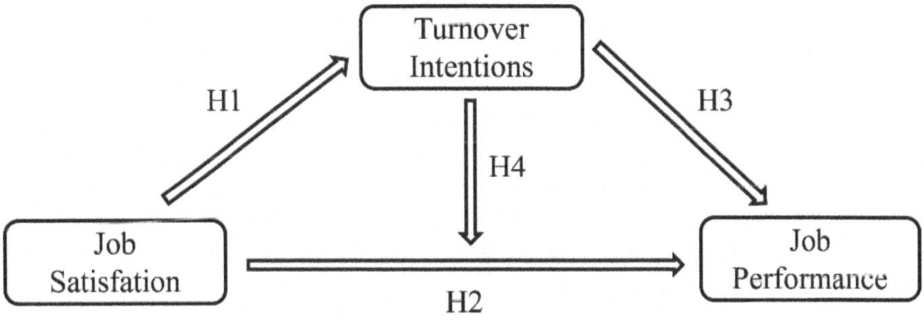

Fig. 1. Research conceptual model.

Hypothesis 1: Job satisfaction has a significant negative effect on turnover intentions.

Hypothesis 2: Job satisfaction has a positive and significant effect on performance.

Hypothesis 3: Turnover intentions have a significant negative effect on performance.

Hypothesis 4: Turnover intentions have a moderating effect on the relationship between job satisfaction and performance.

3.2 Data Collection Procedure

A total of 369 subjects voluntarily took part in this study, all of whom worked for commercial airlines. Although 418 responses to the questionnaire were posted online on the Google Forms platform, 49 responses were not considered valid because the subjects needed to meet the conditions for working in the commercial air transport sector. The questionnaire link was sent to members of the Portuguese Airline Pilots' Association and via LinkedIn to all types of airline professionals. Data collection took place between September 2023 and January 2024. The sample was non-probabilistic, intentional and snowball (Trochim, 2000). The first part of the questionnaire informed participants of the purpose of the study and guaranteed the confidentiality of their answers. As for informed consent, after the initial instructions, the participants answered a question about whether they wanted to take part in the study voluntarily. If they said no, they were sent to the final page (thank you). The questionnaire consisted of 7 sociodemographic questions and 15 items that make up an instrument designed to measure employee churn prediction.

Participants

This sample consisted of 369 participants, 139 (37.7%) female and 230 (62.3%) male. In terms of age, 59 (16%) were aged between 20 and 29, 99 (26.8%) were aged between 30 and 39, 143 (38.8%) were aged between 40 and 49, 59 (16%) were aged between 50 and 59 and 9 (2.4) were aged between 60 and 69. As for their profession, 78 (21.1%) belong to the Technical Crew (TC), 66 (17.9%) belong to the Cabin Personnel (CP), 67 (18.2%) belong to the Maintenance, Repair and Overhaul (MRO), 112 (30.4%) to the Ground Personnel (GP) and 46 (12.5%) perform other functions. In terms of academic qualifications, 88 (23.8%) had up to the 12th grade, 46 (12.5%) had a bachelor's degree, and 235 (63.7%) had a degree or higher. Among the participants, 95 (25.7%) had been with the organization for three years or less, 100 (27.1%) between 4 and 9 years, 95 (25.7%) between 10 and 19 years and 79 (21.4%) 20 years or more. About working hours, 93 (25.2%) work between 40 and 80 h a month, 160 (43.4%) between 81 and 160 h, 106 (28.7%) between 161 and 240 h and 10 (2.7%) more than 240 h. When asked if they had been promoted in the last five years, 202 (54.7%) said no, 157 (42.5%) said yes, and 10 (2.7%) said not applicable.

When seniority is cross-referenced with profession, the highest percentage for TC is between 5 and 9 years, for CP between 20 and 29 years, for MRO, GP, and others between 10 and 15 years (Table 2).

Table 2. Seniority of the participants by profession

		Profession					Total
		TC	CP	MRO	GP	Others	
Seniority	Less than a year	3	10	5	1	4	23
		3.8%	15.2%	7.5%	0.9%	8.7%	6.2%
	1 year	4	6	6	11	7	34
		5.1%	9.1%	9.0%	9.8%	15.2%	9.2%
	2 years	4	7	5	7	4	27
		5.1%	10.6%	7.5%	6.3%	8.7%	7.3%
	3 years	4	4	2	0	1	11
		5.1%	6.1%	3.0%	0.0%	2.2%	3.0%
	4 years	6	2	1	5	1	15
		7.7%	3.0%	1.5%	4.5%	2.2%	4.1%
	5 to 9 years	28	12	11	27	7	85
		35.9%	18.2%	16.4%	24.1%	15.2%	23.0%
	10 to 19 years	15	10	21	41	8	95
		19.2%	15.2%	31.3%	36.6%	17.4%	25.7%

(continued)

Table 2. (*continued*)

		Profession					Total
		TC	CP	MRO	GP	Others	
	20 to 29 years	9	13	11	16	8	57
		11.5%	19.7%	16.4%	14.3%	17.4%	15.4%
	30 to 39 years	5	2	5	4	6	22
		6.4%	3.0%	7.5%	3.6%	13.0%	6.0%
Total		78	66	67	112	46	369
		100.0%	100.0%	100.0%	100.0%	100.0%	100.0%

When cross-referencing the profession with the number of hours worked per month is performed, the study concludes that the highest percentage of TC and CP work between 81 and 160 h per month. In contrast, the highest percentage of MROs, GPs and others work between 161 and 240 h a month (Table 3).

Table 3. Profession of the participants crossed with hours worked per month

		Profession					Total
		TC	CP	MRO	GP	Others	
Hours per month	40 to 80 h	31	17	8	24	13	93
		39.7%	25.8%	11.9%	21.4%	28.3%	25.2%
	81 to 160 h	45	41	21	39	14	160
		57.7%	62.1%	31.3%	34.8%	30.4%	43.4%
	161 to 240 h	1	8	35	46	16	106
		1.3%	12.1%	52.2%	41.1%	34.8%	28.7%
	Over 240 h	1	0	3	3	3	10
		1.3%	0.0%	4.5%	2.7%	6.5%	2.7%
Total		78	66	67	112	46	369
		100.0%	100.0%	100.0%	100.0%	100.0%	100.0%

Data Analysis Procedure

The first step was to import the data into SPSS Statistics 29 software (IBM Corp., Armonk, NY., USA). Factor analyses were carried out to test the validity of the instruments. Confirmatory factor analyses were carried out for the employee churn and performance scales, and an exploratory factor analysis was carried out for the turnover intentions scale since this instrument consists of only three items. The KMO value was calculated, which should be greater than 0.70 (Sharma, 1996). Bartlett's sphericity test must be significant to indicate that the data comes from a multivariate normal population (Pestana & Gageiro, 2003). The total variance explained was also calculated, which should be greater than 50%. As for the factor weights of each item, all items with factor weights greater than 0.50 were considered.

The confirmatory factor analyses were carried out using AMOS Graphics software for Windows (IBM Corp., Armonk, NY, USA). The procedure followed a "model generation" logic (Jöreskog & Sörbom, 1993). Following the established recommendations (Hu & Bentler, 1999), six fit indices were combined: Chi-square ratio/degrees of freedom (χ^2/gl); Tucker-Lewis Index (TLI); Goodness-of-fit Index (GFI); Comparative et al. (CFI); Root Mean Square Error of Approximation (RMSEA); Root Mean Square Residual (RMSR). The chi-square/degrees of freedom ratio (χ^2/gl) is considered an acceptable value if it is below 5. For the CFI, GFI, and TLI, values above 0.90 indicate a good fit and above 0.80 indicate an acceptable fit. As for RMSEA, values below 0.08 indicate a good fit (McCallum et al., 1996). The lower the RMSR, the better the fit (Hu & Bentler, 1999). We then tested the construct reliability for each scale's dimensions, whose value should be higher than 0.70. Convergent validity was tested by calculating the average variance extracted (AVE), which should be greater than 0.50 (Fornell & Larcker, 1981). As for discriminant validity, it was tested by comparing the root value of the AVE value with the correlation values between the factors. The square root of the AVE value must be greater than the correlation between the factors whose discriminant validity is being analyzed.

Internal consistency was tested for each of the dimensions that make up the instrument by calculating Cronbach's alpha, which assesses the ratio between the variance of each item and the entire dimension. Its values vary between 0 and 1, with no negative values (Hill & Hill, 2002) and, in organizational studies, it should be higher than 0.70 (Bryman & Cramer, 2003).

Finally, with all the participants, the sensitivity of the items and their respective dimensions was tested. The items must have responses at all the response points, the median must not be close to one of the extremes, and the absolute values of asymmetry and kurtosis must be below 2 and 7, respectively (Finney & DiStefano, 2013).

To study the descriptive statistics of the variables under study, one-sample Student's t-tests were carried out. The effect of sociodemographic variables on the variables under study was tested using parametric t-student tests and One-Way ANOVA. The association between the variables under study was analysed using Pearson's correlations. Hypotheses 1, 2, and 3 were tested by multiple linear regressions using the stepwise method. Hypothesis 4, which presupposes a moderating effect, was tested using the Macro Process developed by Hayes (2013).

Instruments

To measure job satisfaction, we used the instrument developed by the researchers in this study, which is currently being validated. This instrument is made up of 15 items anchored on a five-point Likert scale (from 1 "strongly disagree" to 5 "strongly agree"). The 15 items are divided into five dimensions: career satisfaction (items 1, 2 and 3);

safety satisfaction (items 4, 5 and 6); leadership satisfaction (items 7, 8 and 9); work/life balance satisfaction (items 10, 11 and 12); and pay satisfaction (items 13, 14 and 15). A five-factor CFA was carried out, and the fit indices obtained were adequate ($\chi 2/gl$ = 1.27; CFI = 0.99; GFI = 0.95; TLI = 0.99; RMSR = 0.05; RMSEA = 0.03). These results indicate that this instrument is made up of five factors. All the items had factor weights greater than 0.50. The construct reliability ranged from 0.77 (career satisfaction) to 0.91 (leadership satisfaction). About convergent validity, the AVE values vary between 0.53 (career satisfaction) and 0.77 (leadership satisfaction). These results indicate that all the dimensions have good convergent validity (Fornell & Larcker, 1981). As for discriminant validity, all the AVE square root values are higher than the correlation between the respective dimensions, indicating discriminant validity.

To measure performance, we used a reduced version of the instrument developed by (Williams, 1991, consisting of five items anchored on a five-point Likert scale (from 1 "strongly disagree" to 5 "strongly agree"). A one-factor confirmatory factor analysis was carried out to test the validity of this instrument. The adjustment indices obtained are adequate ($\chi 2/gl$ = 2.83; CFI = 0.99; GFI = 0.99; TLI = 0.98; RMSR = 0.01; RMSEA = 0.07). This instrument has a construct reliability of 0.88, which can be considered good. As for convergent validity, it has an AVE value of 0.64, indicating that it has good convergent validity (Fornell & Larcker, 1981).

Turnover intentions were measured using the instrument developed by Bozeman and Perrewé (2001). This instrument consists of 3 items anchored on a five-point Likert scale (from 1 "strongly disagree" to 5 "strongly agree"). As this instrument consists of only three items, its validity was tested using an exploratory factor analysis. A KMO of 0.74 and an explained variance of 86.12 per cent were obtained. The results of Bartlett's test of sphericity were significant at $p < 0.001$, indicating that the data comes from a normal, multivariate population. All the items had factor weights greater than 0.90. The internal consistency showed a Cronbach's alpha of 0.92.

It was also found that none of the items that make up the instruments grossly violate normality since their absolute values of asymmetry and kurtosis are below 2 and 7, respectively (Finney & DiStefano, 2013).

4 Results

4.1 Descriptive Statistics of the Variables Under Study

The first step was to carry out descriptive statistics on the variables under study to understand the position of the answers given by the participants in this study (Table 4).

The participants in this study revealed that they were very satisfied with safety and leadership, as well as a high perception of performance, as their answers were significantly above the center point (3) (Table 4). On the other hand, they were not very satisfied with their career, work-life balance and pay, as their answers were significantly below the center point (3) (Table 4). Concerning turnover intentions, these do not differ significantly from the center point, which indicates that the participants' intentions to leave their organisation in this study are at the neutral point (3).

Table 4. Resume of the scores of the answers on the variables

Dimension	t	df	p	d	Mean	SD
Career Satisfaction	–3.22	368	<0.001	0.17	2.82	1.05
Safety Satisfaction	18.57	368	<0.001	0.97	3.90	0.93
Leadership Satisfaction	6.55	368	<0.001	0.34	3.39	1.15
WLB Satisfaction	–6.75	368	<0.001	0.35	2.64	1.03
Pay satisfaction	–3.96	368	<0.001	0.21	2.75	1.21
Performance	59.33	368	<0.001	3.08	4.59	0.52
Turnover intentions	0.15	368	0.439	0.01	3.01	1.36

Note. *** p <0.001

4.2 Effect of Sociodemographic Variables on the Variables Under Study

To test the effect of sociodemographic variables on the variables under study, several Student's t-tests for independent samples and One Way ANOVA were carried out.

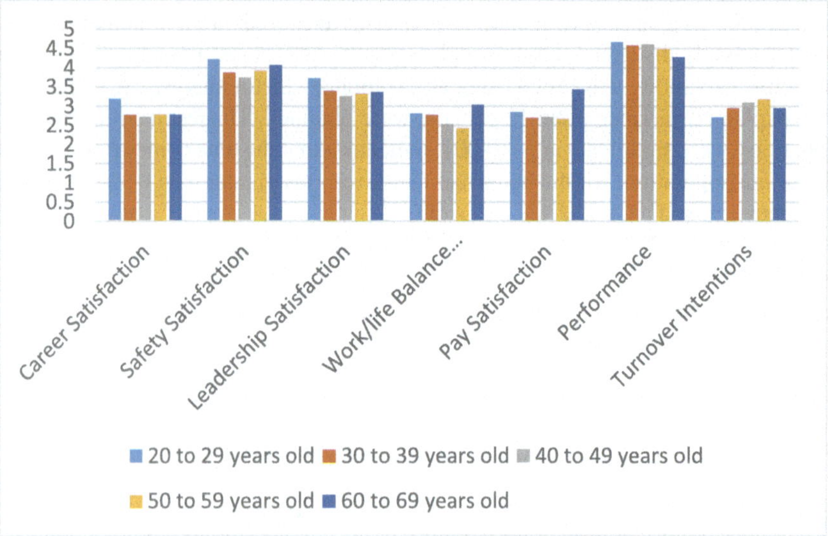

Fig. 2. Answer scores on the variables per age

Age only has a statistically significant effect on safety satisfaction (F (4, 364) = 2.97, p = 0.019) (Fig. 2). Participants between 20 and 29 were the most satisfied with safety, followed by those between 60 and 69. Participants aged between 40 and 49 were the least satisfied with safety.

Table 5. Effect of Gender in Satisfaction variables

Variable	t	df	p	Female		Male	
				Mean	SD	Mean	SD
Career Satisfaction	–4.86***	367	<.001	2.49	1.02	3.02	1.02
Safety Satisfaction	–5.91***	367	<.001	3.54	1.08	4.11	.75
Leadership Satisfaction	–3.98***	367	<.001	3.09	1.25	3.57	1.04
WLB Satisfaction	–4.16***	367	<.001	2.36	1.05	2.81	.97
Pay satisfaction	–4.15***	367	<.001	2.42	1.22	2.95	1.17
Performance	–1.60	367	.111	4.54	.58	4.63	.47
Turnover intentions	3.64***	367	<.001	3.34	1.39	2.81	1.30

Gender has a significant effect on all the variables under study, except for performance (Table 5). Male participants have a significantly higher perception of satisfaction with career, safety, leadership, work/life balance, and pay than female participants. However, female participants have the greatest turnover intentions across the aviation organizations.

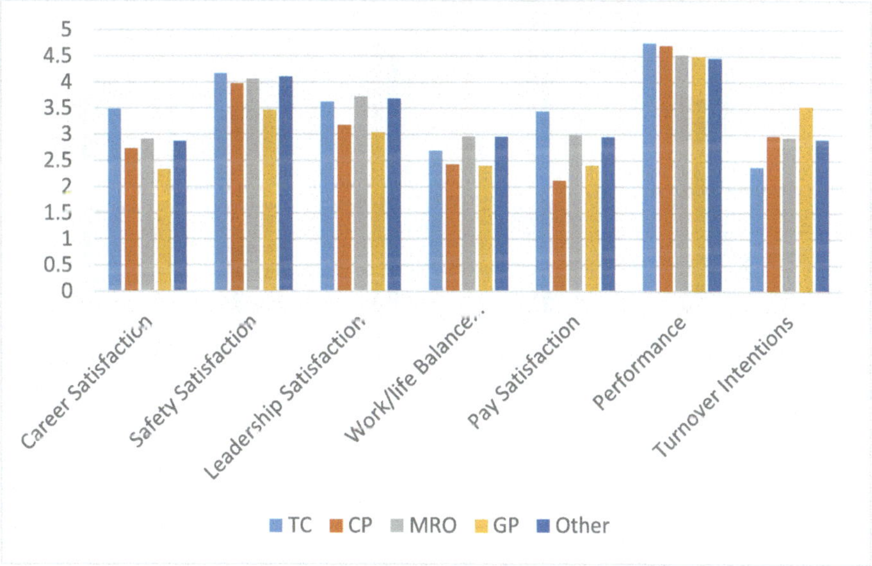

Fig. 3. Answer scores on the variables per profession

The results show that profession has a significant effect on career satisfaction (F (4, 364) = 16.56, p <0.001), safety satisfaction (F (4, 364) = 9.80, p <0.001), leadership satisfaction (F (4, 364) = 6.34, p <0. 001), work/life balance satisfaction (F (4, 364) = 5.03, p <0.001), pay satisfaction (F (4, 364) = 16.08, p <0.001), performance (F (4, 364) = 4.56, p = 0.001) and turnover intentions (F (4, 364) = 9.67, p <0.001) (Fig. 3). Regarding career satisfaction, the participants in TC roles were the most satisfied with their careers, differing significantly from all the others (Fig. 4). Regarding safety satisfaction, participants in GP were the least satisfied with safety, differing significantly from all the others (Figure). Participants working in CP and on land are the least satisfied with leadership, differing significantly from those working in TC, MRO, and others (Fig. 4). Those working in CP and onshore also feel least satisfied balancing work and personal life (Fig. 4). Finally, about pay satisfaction, participants working in TC and MRO are the most satisfied with pay, differing significantly from all the others (Fig. 4).

Participants working in CP have the highest perception of performance, differing significantly from ground staff and others. GP staff have the highest turnover intentions, differing significantly from all the others. Those with the lowest turnover intentions are those working in TC.

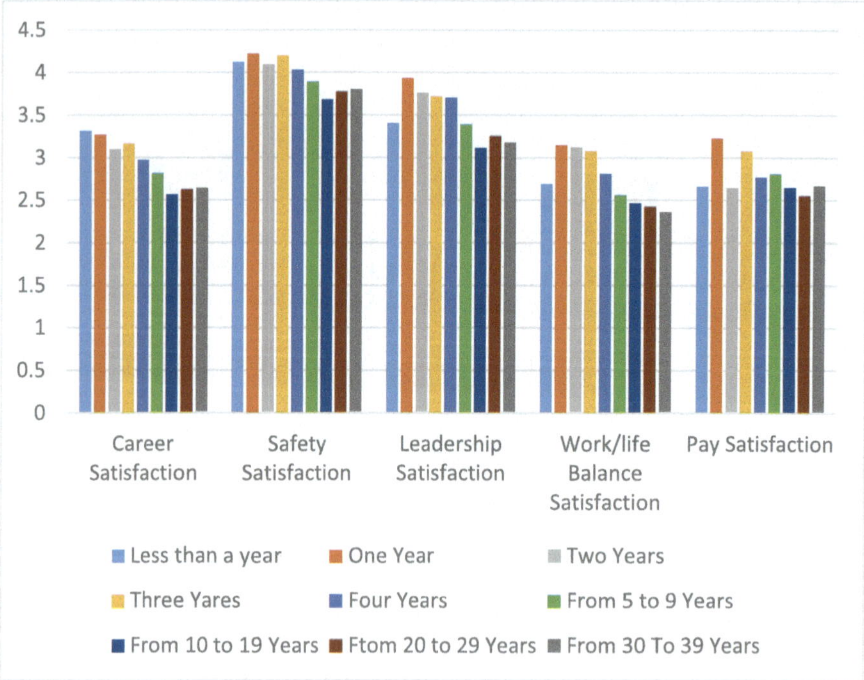

Fig. 4. Answer scores on the variables per seniority

The results show (Fig. 4) that seniority in the organization has a significant effect on career satisfaction (F (4, 364) = 3.08, p = 0.002), leadership satisfaction (F (4, 364) = 2.52, p = 0.011) and work/life balance satisfaction (F (4, 364) = 3.15, p = 0.002).

In safety satisfaction (F (4, 364) = 1.81, p = 0.074) and pay satisfaction (F (4, 364) = 1.12, p = 0.347), the effect was not statistically significant. As can be seen in the figure, all types of satisfaction decrease as seniority in the organization increases.

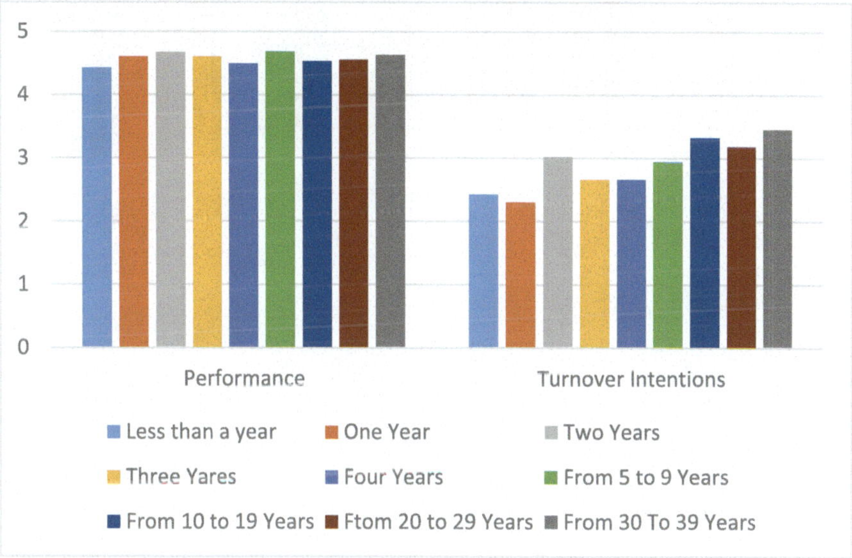

Fig. 5. Answer scores on performance and turnover intentions per seniority

Regarding the effect of seniority in the organization on performance and turnover intentions (Fig. 5), there were only statistically significant differences in turnover intentions (F (4, 364) = 3.11, p = 0.002). As seniority in the organization increases, so do turnover intentions.

Regarding the effect of the number of working hours on the variables under study (Fig. 6), there are only statistically significant differences in safety satisfaction (F (3, 365) = 5.12, p = 0,002). Participants who work more than 240 h a month feel significantly less satisfied with safety than all the other participants.

Regarding the effect of agreement with the last performance appraisal on the variables under study (Fig. 7), there are statistically significant differences in satisfaction with career satisfaction (F (4, 364) = 14.44, p <0.001), safety satisfaction (F (4, 364) = 18.62, p <0. 001), leadership satisfaction (F (4, 364) = 16.68, p <0.001), work/life balance satisfaction (F (4, 364) = 7.81, p <0.001), pay satisfaction (F (4, 364) = 10.77, p <0.001), perception of performance and turnover intentions (F (4, 364) = 13.25, p <0.001). Participants who strongly agree with their performance appraisal feel significantly more strongly about their career, safety, leadership, work/life balance, and remuneration. They have a higher perception than all other participants but have lower turnover intentions.

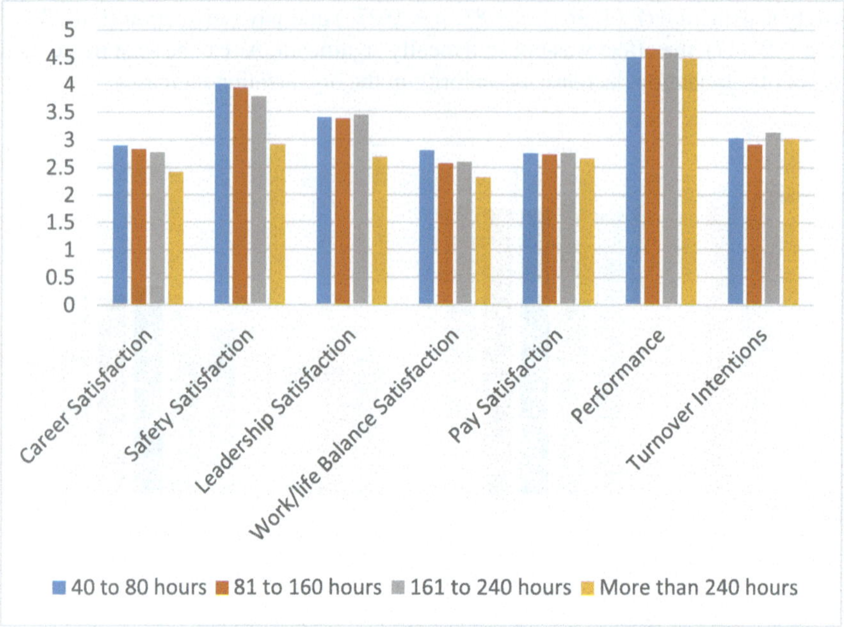

Fig. 6. Answer scores on the variables per number of working hours

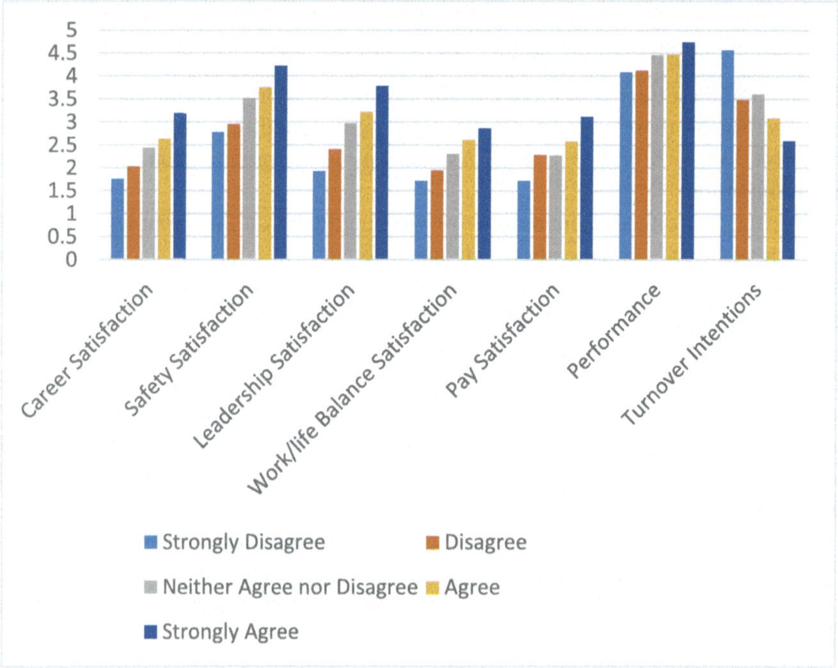

Fig. 7. Answer scores on the variables per agreement on the last performance appraisal

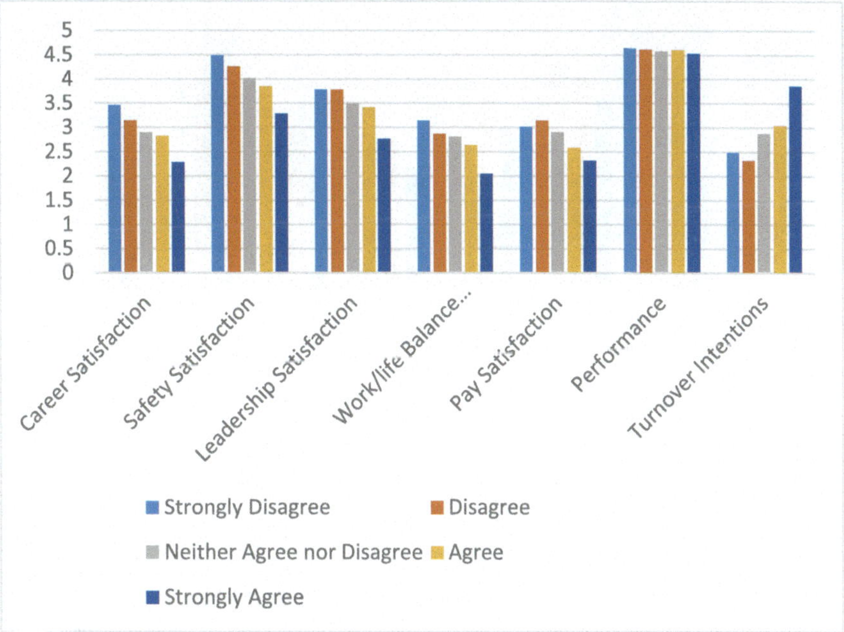

Fig. 8. Answer scores on the variables per number of projects/tasks assigned per quarter

The number of projects/tasks assigned per quarter has a significant effect (Fig. 8) on career satisfaction (F (2, 366) = 35.86, p <0.001), safety satisfaction (F (2, 366) = 9.88, p <0.001), leadership satisfaction (F (2, 366) = 14. 46, p <0.001), with work-life balance satisfaction (F (2, 366) = 13.92, p <0.001), pay satisfaction (F (2, 366) = 12.89, p <0.001) and turnover intentions (F (2, 366) = 17.65, p <0.001) (Fig. 8). Participants who strongly agree that the number of projects/tasks assigned per quarter is too high are the least satisfied and have more turnover intentions than all the other participants (Fig. 8).

The results (Fig. 9) show that career progression has a significant effect on career satisfaction (F (2, 366) = 35.86, p <0.001), safety satisfaction (F (2, 366) = 9.88, p <0.001), leadership satisfaction (F (2, 366) = 14.46, p <0.001), work/life balance satisfaction (F (2, 366) = 13.92, p <0. 001), pay satisfaction (F (2, 366) = 12.89, p <0.001) and turnover intentions (F (2, 366) = 17.65, p <0.001). Participants who have not progressed are less satisfied with their careers, safety, leadership, work/life balance, and pay and have more turnover intentions than all the other participants (Fig. 9).

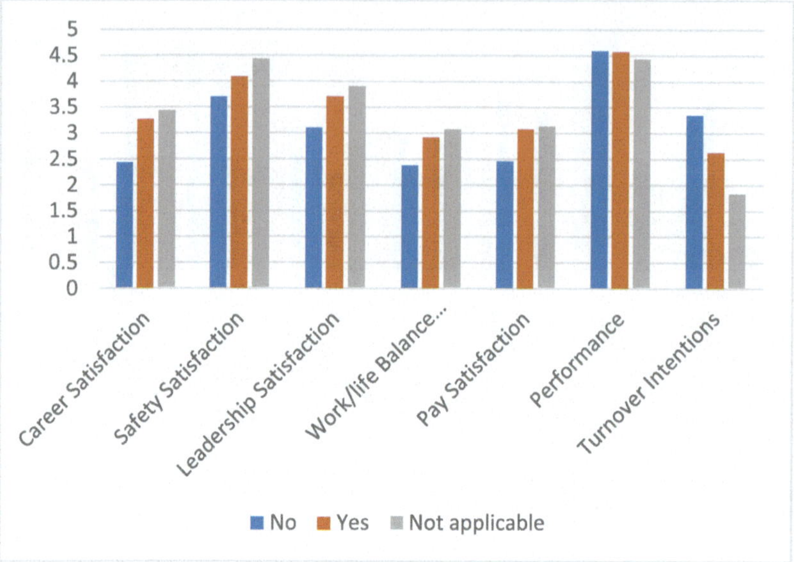

Fig. 9. Effect of career progression on the variables

4.3 Association Between the Variables Under Study

To study the association between the variables under study, Pearson's correlations were used.

Table 6. Correlation analysis between variables

Career Satisfaction	1.1	1.2	1.3	1.4	1.5	2	3
	--						
Safety Satisfaction	.54***	--					
Leadership Satisfaction	.53***	.63***	--				
WLB Satisfaction	.50***	.52***	.55***	--			
Pay Satisfaction	.59***	.41***	.51***	.46***	--		
Performance	.07	.18***	.07	.01	.02	--	
3. Turnover intentions	.63***	.49***	.58***	.50***	.55***	-.09	--

The results show that performance is only significantly associated with satisfaction with safety (Table 6). Turnover intentions are negatively and significantly associated with career satisfaction, safety satisfaction, leadership satisfaction, work/life balance satisfaction and pay satisfaction (Table 6). The most significant association is with career satisfaction (Table 6).

4.4 Hypotheses

The hypotheses formulated in this study were tested with simple and multiple linear regressions using the stepwise method.

Hypothesis 1: Job satisfaction negatively and significantly affects turnover intentions.

Table 7. Results of the fourth step of multiple linear regression using the stepwise method (H1)

Independent Variable	Dependent Variable	F	p	R2a	β	p
Career Satisfaction	Turnover Intentions	93.68***	<.001	.50	.34***	<.001
Leadership Satisfaction					.25***	<.001
WLB. Satisfaction					-.12*	.012
Pay satisfaction					.17***	<.001

Career satisfaction (β = -.34; p <.001), leadership satisfaction (β = -.25; p <.001), work/life balance satisfaction (β = -.12; p = -.012) and pay satisfaction (β = -.17; p <.001) have a significant and negative effect on turnover intentions. The model explains 50 per cent of the variability in turnover intentions. The model is statistically significant (F (4, 364) = 93.68, p <.001). This hypothesis was partially supported (Table 7).

Hypothesis 2: Job satisfaction has a positive and significant effect on performance.

Table 8. Results of the fourth step of multiple linear regression using the stepwise method (H2)

Independent Variable	Dependent Variable	F	p	R2a	β	p
Safety Satisfaction	Performance	8.59***	<.001	.04	.25***	<.001
Work/life B. Satisfaction					−.13*	.035

The results show that safety satisfaction positively and significantly affects performance (β = .25; p <.001). On the other hand, work/life balance satisfaction has a negative and significant effect on performance (β = −.13; p = .035). The model explains 4 per cent of the variability in performance. The model is statistically significant (F (2, 366) = 8.59, p <.001). This hypothesis was partially supported (Table 8).

Hypothesis 3: Turnover intentions have a negative and significant effect on performance.

Turnover intentions negatively and marginally significantly affect performance (β = -.09; p = .090). The model explains 1% of the variability in performance. The model is marginally significant (F (1, 367) = 2.88, p = .090). This hypothesis was partially supported (Tables 9 and 10).

Hypothesis 4: Turnover intentions have a moderating effect on the relationship between job satisfaction and performance.

Table 9. Results of the fourth step of multiple linear regression using the stepwise method (H3)

Independent Variable	Dependent Variable	F	p	R2a	β	p
Turnover Intentions	Performance	2.88*	.090	.01	.09*	.090

Table 10. Results of the moderating effect of turnover intentions (H4).

Variables	B	SE	t	*p*	95% CI
	Career Satisfaction → Performance (R^2 = .10; *p* = .282)				
Constant	4.58***	.03	142.41***	<.001	[4.51, 4.64]
Career Satisfaction	.01	.03	.43	.669	[–.05, .08]
Turnover intentions	–.03	.03	–1.00	.316	[–.08, .02]
CS*TI	–.02	.02	–.90	.371	[–.06, .02]
	Safety Satisfaction → Performance (R^2 = .04; *p* = .005)				
Constant	4.58***	.03	152.04***	<.001	[4.52, 4.64]
Safety Satisfaction	.12	.04	3.13**	.002	[.04, .19]
Turnover Intentions	.01	.02	.19	.849	[–.04, .05]
SS*TI	–.02	.02	–.83	.407	[–.07, .03]
	Leadership Satisfaction → Performance (R^2 = .01; *p* = .189)				
Constant	4.57***	.03	144.95***	<.001	[4.51, 4.63]
Leadership Satisfaction	.02	.03	.71	.478	[–.04, .08]
Turnover Intentions	–.02	.02	–1.01	.314	[–.07, .02]
LS*TI	–.02	.02	–1.34	.182	[–.06, .01]
	Work/life balance Satisfaction → Performance (R^2 = .01; *p* = .233)				
Constant	4.60***	.03	153.37***	<.001	[4.54, 4.66]
Work/life B. Satisfaction	–.03	.03	–.94	.346	[–.04, .08]
Turnover Intentions	–.05	.02	–1.97*	.050	[–.09, .00]
WLBS*TI	.01	.02	.73	.469	[–.02, .05]
	Pay Satisfaction → Performance (R^2 = .02; *p* = .092)				
Constant	4.57***	.03	146.69***	<.001	[4.50, 4.63]
Pay Satisfaction	–.02	.03	–.90	.364	[–.07, .03]
Turnover Intentions	–.04	.02	–1.75*	.081	[–.08, .00]
PS*TI	–.03	.02	–1.73*	.083	[–.07, .00]

The results showed that turnover intentions have a marginally significant moderating effect on the relationship between pay satisfaction and performance (ŋ 10).

For participants with low turnover intentions, pay satisfaction becomes relevant to boosting their performance than for participants with high turnover intentions (Figure 10). When turnover intentions are high, the positive effect of pay satisfaction on performance might be weaker, because these employees are less motivated to perform well despite being satisfied with their pay. The interaction effect is noticeable but not strong enough to be definitively confirmed by the data. There is some evidņence that turnover intentions moderate the relationships between pay satisfaction and performance, but this evidence is not conclusive. This hypothesis was partially supported.

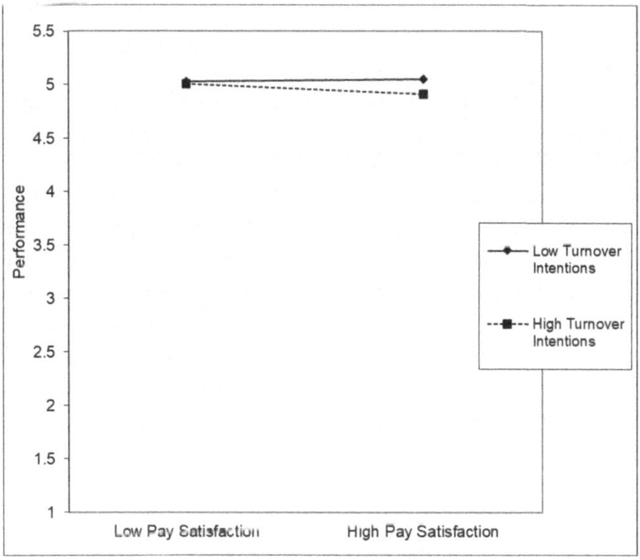

Fig. 10. interaction plot with illustration of the moderating effect of turnover intentions on the relationship between pay satisfaction and performance

5 Discussion and Conclusions

This study aimed to investigate the factors influencing turnover intentions and their impact on job performance among commercial aviation professionals in Portugal. This research contributes to the limited literature on employee churn in the commercial aviation sector by specifically examining job satisfaction dimensions and their moderating effects on turnover intentions and performance. Also, this study is innovative as it depicts an in-depth picture of the commercial airline's HR reality in Portugal.

As expected, hypothesis 1 has been partially confirmed because only career satisfaction, leadership satisfaction, work/life balance satisfaction and pay satisfaction have a negative and significant effect on turnover intentions, indicating that the higher the satisfaction levels, the lower the turnover intentions. These results are in accordance with the literature (Chen, 2006; Mowday, Porter & Steers, 1982; Jou, Kuo, & Tang, 2013; Mulki, Jaramilo & Lokander, 2006; Suifan, Diab & Abdallah, 2017). The reason why the perception of safety satisfaction has no effect on turnover intentions is uncertain. However, there are two possibilities. 1) This lack of a significant relationship between the variables confirms the hypotheses of the literature (Huang et al. 2016; Smith, 2018; Siu, Cheung & Lui, 2015; Zhang et al., 2023), in the sense that there is evidence of the relationship between variables in the transport sector, but not in the aviation sector. (2) On the other hand, the issue of aviation safety is something that is taken for granted as it is the means of transport with the most advanced technology, with more specific regulation and more subject to scrutiny, at least in most airlines, so its effect on turnover intentions may be marginal.

As expected, hypothesis 2 was partially confirmed, as only satisfaction with safety has a positive and significant effect on performance and satisfaction by work/life balance has a negative and significant impact on performance. These results indicate that the higher levels of satisfaction with safety and work/life balance, the higher the performance and the lower the performance, respectively. These results are consistent with the literature, since the positive relationship between job satisfaction and performance has never been fully proven in academic literature (Iaffaldano and Muchinsky, 1985; Judge et. al., 2001; Petty, McGee, and Cavender, 1984; Whitman, Van Rooy, and Viswesvaran, 2010). On the other hand, there have even been results in which job satisfaction has a negative relationship with performance, for example, in the case of air traffic controllers (Supriyanto, 2018). In particular, the negative relationship between work/life balance satisfaction and performance found in our study is counterintuitive and a contribution to the science on employment satisfaction in the aviation sector. The relationship between these variables is not much studied in this sector, so this study occupies an important research gap.

As expected, hypothesis 3 was partially confirmed, as turnover intentions have a negative and marginally significant effect on performance, indicating that the higher the turnover intentions, the lower the performance. These results are in line with the literature (Bishop, 1990; Trevor et al., 1997; Wynen & Kleizen, 2019) and are a relationship found in sectors such as transport (Hancock et al., 2013).

Finally, hypothesis 4 was partially confirmed. The results regarding the moderating effect of turnover intentions on the relationship between job satisfaction and performance were mixed. While turnover intentions marginally moderated the relationship between pay satisfaction and performance, it did not significantly moderate the relationships for other satisfaction dimensions. This suggests that turnover intentions might not be a strong moderator in this context, although it does play a role in how pay satisfaction influences performance.

With this study was filled the limitation pointed out by Srivastava & Eachempati (2021), which noted that most studies point to the factors that influence employee churn, but not the relative contribution between the variables.

In addition to these indicators, the questionnaire contains a rich set of descriptive and socio-demographic data that characterize the diverse workforce of the commercial aviation sector in Portugal. Participants are highly satisfied with safety and leadership and have a high perception of performance, with averages significantly above the midpoint

(3). There is dissatisfaction with work-life balance, career, and pay, with averages below the midpoint (3). Participants' turnover intentions are neutral, with averages around the midpoint (3).

Regarding the effect of sociodemographic variables. On the effect of the variable age, satisfaction with safety is significantly higher among participants aged 20–29 and 60–69, and lower among those aged 40–49. On the effect of gender in job satisfaction, male participants have higher satisfaction in all aspects (career, safety, leadership, work-life balance, pay) compared to female participants. Female participants have higher turnover intentions. On the effect of type of profession in the commercial aviation sector, profession significantly affects all aspects of satisfaction. Participants in technical crew (TC) roles are the most satisfied with their careers and pay. Ground personnel (GP) participants have the highest turnover intentions. It was concluded that the flight technical staff, namely pilots, are the most satisfied in the various dimensions and with less turnover intentions and with a high perception of performance. The most dissatisfied, exhausted and with significantly higher turnover intentions are ground personnel.

On the effect of seniority on the satisfaction variables, satisfaction decreases with increasing seniority, except for safety and pay. Turnover intentions increase with seniority.

Regarding the effect of working hours on job satisfaction, participants working more than 240 h/month are less satisfied with safety.

On the effect of performance appraisal perception on job satisfaction dimensions, participants who strongly agree with their last performance appraisal are more satisfied in all aspects and have lower turnover intentions.

When analyzed the effect of the variable "number of projects/ tasks" (too high? y/n) on satisfaction dimensions, the perception on the number of projects/tasks negatively impacts satisfaction and increases turnover intentions.

Finally, the effect of career progression on satisfaction: the lack of career progression reduces satisfaction in all aspects and increases turnover intentions.

The results indicate that participants' satisfaction is influenced by factors such as age, gender, profession, seniority, working hours, performance appraisals, number of tasks, and career progression. This highlights the complexity and interdependence of these factors in job satisfaction and turnover intentions.

5.1 Practical Implications

For HR managers and policymakers in the commercial aviation sector, these findings underscore the importance of enhancing job satisfaction to reduce turnover intentions and improve performance. Strategies should focus on:

- To focus the attention of human resources policies on the role of ground personnel of commercial airlines, namely to review the pay and benefits packages, to reduce/ restructure the demanding hours, to seek to mitigate the recurring dangers of the profession of direct contact with the customer and with ground tasks, to improve career plans, to ensure empathic, effective and motivating leadership, to increase the flexibility of schedules, to implement plans for reconciling family with work and offer mental health support;

In general, for all occupations in the commercial air transport sector:

- Improving career development opportunities: providing clear career paths and advancement opportunities can enhance career satisfaction and reduce turnover intentions.
- Ensuring workplace health/ safety: maintaining high safety standards not only ensures compliance with regulations but also boosts employee performance and satisfaction.
- Effective leadership: leadership training and development programs can improve leadership satisfaction and reduce turnover intentions.
- Work-life balance initiatives: implementing flexible work arrangements and support programs can address the negative impact of work-life balance satisfaction on performance;
- Competitive compensation packages: regularly reviewing and adjusting pay structures to meet market standards can increase pay satisfaction and mitigate turnover intentions.

5.2 Limitations

The study has several limitations that should be considered. The sample was restricted to employees in the commercial aviation sector in Portugal, which may limit the generalizability of the findings to other sectors or regions. The study's cross-sectional nature also means that causal relationships cannot be definitively established. Future research could employ longitudinal designs to better capture the dynamics of job satisfaction, turnover intentions, and performance over time, or large sample sizes.

In addition, the questionnaire can be improved, namely, the division of commercial airline professions and the specification of better professions like "ground personnel" and "others" as they encompass a plethora of activities. Also, it's questionable to include "Maintenance and engineering" in the commercial aviation activity, even if some airlines have that vertical integration. Still, it enriches the study.

More factors can be found that influence turnover intentions, and there is little academic literature in the field of aviation.

5.3 Future Research

Further research is needed to explore other potential moderating variables that might influence the relationship between job satisfaction and performance. The factors "marital status" and "work environment" that appear in the literature as associated with turnover intentions were left out of the churn predictive model for a reason of simplification. In future studies, they can be included in a new model.

Additionally, examining the impact of specific interventions to improve job satisfaction on turnover intentions and performance could provide valuable insights for HR practices. Studies could also expand to different geographic regions and sectors to enhance the generalizability of the findings.

As aviation is constituted by various activities, commercial aviation is only one of them. There is still a lack of studies on the relationship between job satisfaction, turnover intentions and performance of other aviation professionals like ground handling, maintenance & engineering, aircraft manufacturing, defence, ground handling, airport professionals, ATC (air traffic controllers), pilot training activity, aviation professionals in firefighting, rescue and medical aircraft operations, flight catering, safety and security professionals, regulators, cleaning professionals, etc.

The relationships between several variables and with the variable job satisfaction are partitioned by dimensions, like the "relationship between pay satisfaction and performance". However, there are several relationships between job satisfaction dimensions and other variables. It could be a good avenue of research to narrow the study of relationships between job satisfaction and other variables to more specific dimensions of job satisfaction.

5.4 Conclusions

In conclusion, this study highlights the critical role of job satisfaction in influencing turnover intentions and job performance among commercial aviation professionals in Portugal. By focusing on enhancing various dimensions of job satisfaction, organizations can reduce turnover intentions and improve overall performance. The findings provide valuable insights for HR managers seeking to develop effective retention strategies and create a more satisfied and productive workforce.

References

Ali, S., et al.: Abusive supervision and turnover intentions: a mediation-moderation perspective. Sustainability **14**(17), 10626 (2022)

Armstrong, M.: A Handbook of Human resource Management Practice, 10th edn., p. 264. Kogan Page Publishing, London (2006)

Aziri, B.: Job satisfaction: A literature review. Manage. Res. Practice, **3**(4) (2011)

Bakhsh, A.: Factors affecting employee job satisfaction of a major Middle Eastern airline. Transp. Modes eJournal (2020)

Bishop, J.H.: Job performance, turnover, and wage growth. J. Law Econ. **8**(3), 363–386 (1990)

Brito, A. P.: 'O Multi-Crew Pilot License (MPL), no contexto de aumento da procura do transporte aéreo, das necessidades de safety e os novos desafios da formação de pilotos', Provas públicas para a obtenção de título de Especialista em Gestão e Administração, Instituto Politécnico de Beja/ ISCTE-IUL/ Universidade do Algarve, Portugal (2020)

Bryman, A., Cramer, D.: Análise de dados em ciências sociais. Introdução as técnicas utilizando o SPSS para Windows. Celta (2003)

CAE.: *The global pilot shortage is a challenge to the world's airlines.* CAPA. Retrieved from (2024, Jan. 11) (2023, Aug 21): https://centreforaviation.com/analysis/reports/the-global-pilot-shortage-is-a-challenge-to-the-worlds-airlines-658033

Chen, C.F.: Job satisfaction, organizational commitment, and flight attendants' turnover intentions: a note. J. Air Transp. Manag. **12**(5), 274–276 (2006)

Chen, I. H., Brown, R., Bowers, B. J., Chang, W. Y.: Work-to-family conflict as a mediator of the relationship between job satisfaction and turnover intention. *Journal of advanced nursing*, *71*(10), 2350–2363.

Cheng, T.M., Chang, S. Y., Chan, Y.-Y.: I know you are suffering from burnout: the moderated mediation effects of "leisure benefits" and leisure coping" on the burnout model of flight attendants. J. Air Transp. Manag., **71** (August), 119–129 (2018)

Danish, R. Q., Gohar, A., Ahmad, A., Shaukat, S.: Mediating Role of work engagement in the relationship between job satisfaction and turnover intention: an empirical evidence from hospitals in Lahore, Pakistan. Int. J. Manage. Res. Emerg. Sci., **7**(1) (2017)

Dolatabadi, S. H., Keynia, F.: Designing of customer and employee churn prediction model based on data mining method and neural predictor. In *2017 2nd International Conference on Computer and Communication Systems (ICCCS)* (pp. 74–77). IEEE (2017, July)

Efthymiou, M., Usher, D., O'Connell, J.F., Warnock-Smith, D., Conyngham, G.: The factors influencing entry level airline pilot retention: an empirical study of Ryanair. J. Air Transp. Manag. **91**, 101997 (2021)

Ekawati, A.D.: Predictive analytics in employee churn: a systematic literature review. J. Manage. Inf. Decis. Sci. **22**(4), 387–397 (2019)

Freeman, R.B., Medoff, J.L.: What do unions do. Indus. Lab. Rel. Rev. **38**, 244 (1984)

Finney, S. J., DiStefano, C.: Non-normal and categorical data in structural equation modeling. In Hancock, G. R., Mueller, R. O. (eds.), Structural equation modeling: a second course (2nd ed., pp. 439–492). IAP Information Age Publishing (2013)

Firdaus, M., Eliyana, A., Pratama, A. S., Cahyani, A., & Kamil, N. L. M. (2022). Continuance of organizational commitment among flight attendants as an intervening variable to job performance. *Problems and Perspectives in Management*

Fornell, C., Larcker, D.F.: Evaluating structural equation models with unobservable variables and measurement error. J. Mark. Res. **18**(1), 39–50 (1981). https://doi.org/10.2307/3151312

Griffeth, R.W., Hom, P.W., Gaertner, S.: A meta-analysis of antecedents and correlates of employee turnover: update, moderator tests, and research implications for the next millennium. J. Manag. **26**(3), 463–488 (2000). https://doi.org/10.1177/014920630002600305

Gan, E., Voon, M. L.: The impact of transformational leadership on job satisfaction and employee turnover intentions: a conceptual review. In *SHS Web of Conferences* (Vol. 124, p. 08005). EDP Sciences (2021)

Gu, Q., Wang, L., Sun, J.Y., Xu, Y.: Understanding China's post-80 employees' work attitudes: an explorative study. J. Chin. Hum. Resour. Manag. **1**(2), 74–94 (2010)

Hancock, J.I., Allen, D.G., Bosco, F.A., McDaniel, K.R., Pierce, C.A.: Meta-analytic review of employee turnover as a predictor of firm performance. J. Manag. **39**(3), 573–603 (2013)

Hassouna, M., Tarhini, A., Elyas, T., Abou Trab, M.S.: Customer churn in mobile markets: a comparison of techniques. Int. Bus. Res. **8**(6), 224–237 (2015)

Hayes, A. F.: *Introduction to mediation, moderation, and conditional process analysis: A regression-based approach.* Guilford Press (2013)

Hill, M., Hill, A.: Investigação por Questionário. 2ª Ed. Edições Sílabo (2005)

Hoole, C., Vermeulen, L.: Job satisfaction among South African aircraft pilots. SA J. Ind. Psychol. **29**(1), 52–57 (2003)

Hu, L.T., Bentler, P.M.: Cutoff criteria for fit indexes in covariance structure analysis: conventional criteria versus new alternatives. Struct. Equ. Modeling **6**(1), 1–55 (1999). https://doi.org/10.1080/10705519909540118

Huang, S., Chen, Z., Liu, H., Zhou, L.: Job satisfaction and turnover intention in China: the moderating effects of job alternatives and policy support. Chin. Manag. Stud. **11**(4), 689–706 (2017)

Huang, Y.H., Lee, J., McFadden, A.C., Murphy, L.A., Robertson, M.M., Cheung, J.H., et al.: Beyond safety outcomes: an investigation of the impact of safety climate on job satisfaction, employee engagement and turnover using social exchange theory as the theoretical framework. Appl. Ergon. **55**, 248–257 (2016)

Hussain, K., Abbas, Z., Gulzar, S., Jibril, A. B., Hussain, A.: Examining the impact of abusive supervision on employees' psychological wellbeing and turnover intention: the mediating role of intrinsic motivation. Cogent Bus. Manage., **7**(1) (2020), Article 1818998. https://doi.org/10.1080/23311975.2020.1818998

Iaffaldano, M.T., Muchinsky, P.M.: Job satisfaction and job performance: a meta-analysis. Psychol. Bull. **97**(2), 251–273 (1985). https://doi.org/10.1037/0033-2909.97.2.251

Isnanto, S. H.: Improving employee job satisfaction in aviation industry. Angkasa: Jurnal Ilmiah Bidang Teknologi (2021)

Jain, N., Tomar, A., Jana, P. K.: A novel scheme for employee churn problem using multi-attribute decision making approach and machine learning. J. Intell. Inf. Sys., **56**, 279–302

Joseph, D., Ng, K.Y., Koh, C., Ang, S.: Turnover of information technology professionals: a narrative review, meta-analytic structural equation modeling, and model development. MIS Q. **31**(3), 547–577 (2007)

Jou, R.C., Kuo, C.W., Tang, M.L.: A study of job stress and turnover tendency among air traffic controllers: the mediating effects of job satisfaction. Transp. Res. Part E: Logis. Transp. Rev. **57**, 95–104 (2013)

Judge, T.A., Thoresen, C.J., Bono, J.E., Patton, G.K.: The job satisfaction-job performance relationship: a qualitative and quantitative review. Psychol. Bull. **127**(3), 376–407 (2001). https://doi.org/10.1037/0033-2909.127.3.376

Kalleberg, A.: Work values and job rewards: a theory of job satisfaction. Am. Sociol. Rev. **42**(1), 124–143 (1977)

Karatepe, O.M., Eslamlou, A.: Outcomes of job crafting among flight attendants. J. Air Transp. Manag. **62**(July), 34–43 (2017)

Karatepe, O. M., Kim, T. T.: Investigating the selected consequences of boreout among cabin crew. J. Air Transp. Manage., **82**, 101739. Chicago

Kiernan, K.M.: Calculating the cost of pilot turnover. J. Aviat. Aero. Educ. Res., **27** (1) (2018). https://doi.org/10.15394/jaaer.2018.1737

Lim, T.L., Omar, R., Ho, T.C.F., Tee, P.K.: The roles of work–family conflict and family–work conflict linking job satisfaction and turnover intention of academic staff. Aust. J. Career Dev. **30**(3), 177–188 (2021)

Lin-Jing, W.: Research on the relationship between job satisfaction and job performance of private airlines' staff. Journal of Chengdu Textile College (2010)

Mansour, D., Bhardwaj, A., Chopra, A.: Relating OCEAN (Big Five) to job satisfaction in aviation. *2021 International Conference on Computational Intelligence and Knowledge Economy (ICCIKE)*, 285–289 (2021)

Margherita, A.: Human resources analytics: a systematization of research topics and directions for future research. Hum. Resour. Manag. Rev. **32**(2), 100795 (2022)

Marler, J.H., Boudreau, J.W.: An evidence-based review of HR analytics. Int. J. Hum. Resour. Manage. **28**(1), 3–26 (2016)

Mihelic, K.K., Tekavcic, M.: Work-family conflict: a review of antecedents and outcomes. Int. J. Manage. Inf. Syst. (ijmis) **18**(1), 15–26 (2014)

Mosadeghrad, A.M., Ferlie, E., Rosenberg, D.: A study of the relationship between job satisfaction, organizational commitment and turnover intention among hospital employees. Health Serv. Manage. Res. **21**(4), 211–227 (2008)

Motshegwa, B.: Comparative analysis of teachers' perception of equity, pay satisfaction, affective commitment and intention to turnover in Botswana (2010)

Mowday, R.T., Porter, L.W., Steers, R.M.: Employee–organization linkages – The psychology of commitment, absenteeism and turnover. Academic Press, New York (1982)

Mulki, J.P., Jaramillo, F., Locander, W.B.: Effects of ethical climate and supervisory trust on salesperson's job attitudes and intentions to quit. J Pers Sell Sales Management **26**(1), 19–26 (2006)

Ng, T.W.H., Sorensen, K., Yim, F.: Does the job satisfaction—job performance relationship vary across cultures? J. Cross Cult. Psychol. **40**(5), 761–796 (2009)

Pestana, M. H., Gageiro, J.: Análise de dados para ciências sociais—A complementaridade do SPSS. Edições Sílabo (2003)

Pirrolas, O.A.C., Correia, P.M.A.R., Nascimento, J.L.: The scale of causes of churning: elaboration and validation for Portuguese human resources. Soc. Sci. **12**(1), 1 (2022)

Petty, M.M., McGee, G.W., Cavender, J.W.: A meta-analysis of the relationships between individual job satisfaction and individual performance. Acad. Manag. Rev. **9**(4), 712–721 (1984). https://doi.org/10.5465/AMR.1984.4277608

Rahim, A., Cosby, D.M.: A model of workplace incivility, job burnout, turnover intentions, and job performance. J. Manage. Develop. **35**(10), 1255–1265 (2016)

Robbins, S.P., Judge, T.: Organizational Behavior. Pearson Prentice Hall, New Jersey (2009)

Rombaut, E., Guerry, M.: Predicting voluntary turnover through Human Resources database analysis. Manag. Res. Rev. **41**(1), 96–112 (2018)

Saradhi, V.V., Palshikar, G.K.: Employee churn prediction. Expert Syst. Appl. **38**(3), 1999–2006 (2011)

Sharma, S.: Applied multivariate techniques. Wiley (1996)

Shu, X., Gong, Y.Y., Xiong, J., Hu, X.: Job satisfaction, turnover intention and work performance in Chinese family enterprises. Manag. Int. **22**(2), 84–95 (2018)

Singh, P., Loncar, N.: Pay satisfaction, job satisfaction and turnover intent. Relations industrielles **65**(3), 470–490 (2010)

Sisodia, D. S., Vishwakarma, S., & Pujahari, A. (2017, November). Evaluation of machine learning models for employee churn prediction. In *2017 international conference on inventive computing and informatics (icici)* (pp. 1016–1020). IEEE

Siu, O.L., Cheung, F., Lui, S.: Linking positive emotions to work well-being and turnover intention among Hong Kong police officers: the role of psychological capital. J. Happiness Stud. **16**, 367–380 (2015)

Smith, T.D.: An assessment of safety climate, job satisfaction and turnover intention relationships using a national sample of workers from the USA. Int. J. Occup. Saf. Ergon. **24**(1), 27–34 (2018)

Spector, P.E.: Job satisfaction: Application, assessment, causes and consequences. CA, Sage Publications Inc, Thousand Oaks (1997)

Srivastava, P.R., Eachempati, P.: Intelligent employee retention system for attrition rate analysis and churn prediction: an ensemble machine learning and multi-criteria decision-making approach. J. Global Inf. Manage. (JGIM) **29**(6), 1–29 (2021)

Stock, R.M.: Is boreout a threat to frontline employees' innovative work behavior? J. Prod. Innov. Manag. **32**(4), 574–592 (2015)

Suifan, T.S., Diab, H., Abdallah, A.B.: Does organizational justice affect turnover-intention in a developing country? The mediating role of job satisfaction and organizational commitment. J. Manage. Develop. **36**(9), 1137–1148 (2017)

Supriyanto, S.: Compensation effects on job satisfaction and performance. Hum. Syst. Manag. **37**(3), 281–285 (2018)

Trevor, C.O., Gerhart, B., Boudreau, J.W.: Voluntary turnover and job performance: Curvilinearity and the moderating influences of salary growth and promotions. J. Appl. Psychol. **82**(1), 44 (1997)

Tursunbayeva, A., Di Lauro, S., Pagliari, C.: People analytics—a scoping review of conceptual boundaries and value propositions. Int. J. Inf. Manage. **43**, 224–247 (2018)

Vandenberghe, C., Tremblay, M.: The role of pay satisfaction and organizational commitment in turnover intentions: a two-sample study. J. Bus. Psychol. **22**(3), 275–286 (2008)

Vatankhah, S., Darvishi, M. (2018). An empirical investigation of antecedent and consequences of internal brand equity: evidence from the airline industry. J. Air Transp. Manag. **69** (June), 49–58

Vroom, V.H.: Work and motivation, p. 99. John Wiley and Sons, New York (1964)

Wee, K.H., Bang, W.S., Park, J.: A study on effect relationships of coaching leadership job satisfaction, organizational commitment, turnover intention. Int. J. Soc. Welfare Promot. Manage. **7**(1), 1–8 (2020)

Wells, J.E., Welty Peachey, J.: Turnover intentions: do leadership behaviors and satisfaction with the leader matter? Team Perform. Manage.: An Int. J. **17**(1/2), 23–40 (2011)

Whitman, D.S., Van Rooy, D.L., Viswesvaran, C.: Satisfaction, citizenship behaviors, and performance in work units: a meta-analysis of collective construct relations. Pers. Psychol. **63**(1), 41–81 (2010). https://doi.org/10.1111/j.1744-6570.2009.01162.x

Wyrwa, J., Kaźmierczyk, J.: Conceptualizing job satisfaction and its determinants: a systematic literature review. J. Econ. Soc. **21**(5), 3–18 (2020)

Yahia, N.B., Hlel, J., Colomo-Palacios, R.: From big data to deep data to support people analytics for employee attrition prediction. IEEE Access **9**, 60447–60458 (2021)

Yang, J., Pu, B., Guan, Z.: Entrepreneurial leadership and turnover intention in startups: mediating roles of employees' job embeddedness, job satisfaction and affective commitment. Sustainability **11**(4), 1101 (2019)

Yiğit, İ. O., Shourabizadeh, H.: An approach for predicting employee churn by using data mining. In *2017 International Artificial Intelligence and Data Processing Symposium (IDAP)* (pp. 1–4). IEEE (2017, September)

Zimmerman, R.D., Darnold, T.C.: The impact of job performance on employee turnover intentions and the voluntary turnover process: a meta-analysis and path model. Pers. Rev. **38**(2), 142–158 (2009)

Zhang, F., et al.: The relationships between burnout, general wellbeing, and psychological detachment with turnover intention in Chinese nurses: a cross-sectional study. Front. Public Health **11**, 1216810 (2023)

Employer Branding in Public Organization

Xavier David[1] (✉), Dias Damasceno[2], and Agrochão Miguel[3]

[1] ISCTE, Institute University of Lisbon, Lisbon, Portugal
djvx@iscte-iul.pt
[2] ISCSP, Lisbon University, Lisbon, Portugal
[3] INA, National Institute of Administration, Lisbon, Portugal

Abstract. The study is carried out in the context of the Public Sector, seeking to bring the issue of the Employer to the fore. Branding in Portugal compared to what has been done in some European countries. The literature explains the concept, the elaboration of strategies, the complexities of the involvement of the actors, who create their own meanings around the concept. This study aims to add a new paradigm to organizational image and reputation, seeking to adapt, recreate an employer brand strategy in the public sector through the lens of Employer theory Branding. This is a major challenge for public organizations, since public employment is perceived as being unattractive.

The aim of this study was to explore some employer branding practices within the public sector and see to what extent the idea can be endogenized through interpretation, understandings of the actors involved and interest in the concept. The research context is merely exploratory, and it is hoped that it can contribute to explaining how the employer brand is constructed in practice and how it emerged within public organizations for understanding the contribution of the employer brand.

The restructuring of government functions, privatization and cuts in personnel costs have led, in many countries, to a decrease of public servants and one of the crucial challenges facing the public sector will be how to attract personnel to the functions of the future. The practice will be a complex construction process, which obviously must be adapted to different work contexts to be successful.

Keywords: Employer Branding · Nudge · Public sector · PSM

1 Introduction

Demographic changes in the global labor market tend to transform the labor market the fragility of the current context itself, the anxiety we all live in, the non-linearity of events and permanent uncertainty, increasingly call for the search for more qualified civil servants with new valences in terms of skills and new attitudes. The comparative advantage, of organizations and especially public ones, tend to adapt and demand more and more, to have more qualified employees, and with excellent levels of delivery of a good service to the citizen, because this desideratum is a differentiating and determining factor for the Country.

M. J. Sousa (Ed.): UFHRD 2024, ISEM 52, pp. 230–239, 2025.
https://doi.org/10.1007/978-3-031-92548-1_14

Fierce competition in markets and processes of globalization mean that products and services are easily imitated by competitors, but the intrinsic motivation for public service is difficult, or even impossible, to imitate (Backhaus & Tikoo, 2004). To become more attractive, some of these organizations begin to use the so-called employer brand (EA) strategy to precisely differentiate themselves from their competitors; thus building a unique and strong brand, with the aim of being the employer of choice (Backhaus & Tikoo, 2004; Kucherov & Zavyalova, 2012). Organizations have recognized that human capital is a valuable asset and, therefore, consider it important to employ strategies to attract and retain current and potential talent (Moroko & Uncle, 2008).

Over the next 10 years, many OECD member countries will be forced to hire a notable number of new staff to replace current staff when they reach retirement age. The restructuring of government functions, privatization and cuts in personnel costs have led, in many countries, to a decrease in the number of public employees, a crucial challenge facing the public sector will be precisely how to attract qualified personnel for the functions of the future.

This generically pervaded idea that there are difficulties in attracting qualified human resources to the AP because it is not competitive in the market, could be reversed! The relevance of the contribution of the Public construct service Motivation (PSM) in scientific research, supported by several studies in different cultural contexts, seems in fact significant and complementary, insofar as it points out that there is an intrinsic motivation based on the idea "of an individual's propensity to respond to motives fundamentally or exclusively from institutions public" Perry & Wise, (1990).

Regardless of the changes the public sector has undergone or will undergo recently, governments will continue to perform their fundamental core functions (the judiciary, defense, education, health, taxation…) into the future. These functions depend on the human workforce, although the development of new technologies may reduce this need to some degree. Although the availability of human resources in the public sector depends on the national (and international) economy and the situation in the labor market, there is a general trend towards a decrease in the number of new employees entering the labor market and the accumulation of a lack of specialized knowledge. in certain areas. These problems, shared by many member governments, require strategies for their solution that call for a radical reconfiguration of how OECD countries traditionally approach the management of public servants.

Ambler and Barrow (1996) created the concept of employer branding in the 1990s and since then it has been frequently used as a strategy in the private sector. However, Welan et al. (2010) state that marketing concepts is still a fairly recent field in the public sector. However, recently, public organizations are more interested in publicizing their employer brand due to the difficulties they have in attracting and retaining qualified employees (Lavigna & Hays, 2004). The idea of the emergence of a new public and some management theories explain that the public sector can be improved by imitating the management principles of the private sector (Gromark & Melin, 2013). Although some people tend to criticize the spending of public money on marketing, the interest of this subject in the context of employment is still receiving increasing attention in the public sector. Human resources, marketing and communications professionals are often the ones with the responsibility of promoting an Employer Branding as an organization strategy (Christiaans, 2013; Edwards, 2010; Barrow & Mosely; 2005).

There is a huge literature in the field of employer branding, however, in the scientific arena, most research has been done in the context of private organizations, and little or nothing done within public organizations (Whelan et al., 2010).Please note that the first paragraph of a section or subsection is not indented. The first paragraphs that follows a table, figure, equation etc. does not have an indent, either.

Subsequent paragraphs, however, are indented.

2 Theoretical Framework

Ambler and Barrow (1996) were the first to use the concept of employer branding and concluded that branding is important in the context of employment. They defined the concept as "the package of functional, economic and psychological benefits provided by the job, and identified with the employing company" (Ambler & Barrow, 1996:8). The functional benefit has to do with employee job satisfaction. Economic benefits are about rewards, and psychological benefits explain employees' feelings of belonging or purpose. Backhaus & Tikoo (2004:502) state that previous definitions of the term were about company promotion and define employer branding as "the process of building an identifiable brand and unique identity of the employer and the concept of the company that differentiates it from its competitors".

These definitions explain how an organization can differentiate itself from its competitors by building a unique and strong brand as an employer, to be the employer of choice. The Employer Branding is described as a process that includes planning and implementing a strategy, with the aim of influencing the target employee's perception of an attractive employer. Research has shown that there are many benefits of an employer brand being one of the strategic concerns (Barrow & Mosely, 2005). First, because it helps an organization to attract and retain current and potential employees (Backhaus & Tikoo, 2004). Furthermore, research has also shown that this strategy can reduce costs, increase employee performance and lead to competitive advantage (eg Backhaus & Tikoo, 2004; Barrow & Mosely, 2005). Although the employer brand strategy lies in the area of human resources, communication and marketing (Christiaans, 2013; Edwards, 2010; Barrow & Mosely; 2005), the different functional areas must cooperate intensely in order to create a better branding strategy. Employer brand.

The employer branding process is usually developed in three steps (Backhaus & Tikoo, 2004). The first step in the process is to develop the employer value proposition, which are the values that the employer would like employees to associate with their brand. The second step is to use external marketing to attract potential candidates. The third step is to use internal marketing to retain current employees. Chhabra and Sharma (2014) argue that the internal employer brand (Employ Brand) increases an employee's loyalty, while the external employer brand increases the organization's attractiveness to potential candidates. According to Backhaus and Tikoo (2004) the internal perception of an organization influences the external perception of an organization, increasing its attractiveness. Therefore, the employer brand needs to be built from the inside out of the organization. Moroko and Uncles (2008) state that the employer's brand strategy follows the employee throughout the entire life cycle. Employees within the organization influence the external view of the organization. In other words, satisfied and proud employees act as ambassadors for the organization and inspire others to seek jobs within their organization (Whelan et al., 2010). The better the reputation, the more employees perceive the organization as an attractive employer (Edwards, 2010).

2 - Employer Brand as an inducer to attract candidates for Public Administration. The chronic shortage for public affairs to capture and retain qualified resources is an original sin that runs through the entire public sector and completely derails the course it is intended to serve the citizen with efficiency and with high standards of delivery of the public service to the citizen.

The key question that arises is to find which are the "key" factors that contribute to "employer attractiveness" and which, in Berthon 's perspective, et al. (2005) are identified in the table below (Table 1):

Table 1. Measuring Employer Attractiveness

Measuring employer attractiveness

Berthon et al. (2005) propose five factors for assessing employer attractiveness:

- Factor 1, labelled 'Interest value', assesses the extent lo which an individual is attracted to an employer that provides an exciting work environment. novel work practices and that makes use of its employee's creativity to produce high-quality, innovative products and services,
- Factor 2, labelled 'Social value' assesses the extent to which an individual is attracted to an employer that provides a working environment that is fun, happy, provides good collegial relationships and a team atmosphere,
- Factor 3, labelled 'Economic value', assesses the extent to which an individual is attracted to an employer that provides above-average salary, compensation package, job security and promotional opportunities,
- Factor 4 labelled 'Development value', assesses the extent to which an individual is attracted to an employer that provides recognition, self-worth and confidence, coupled with a career-enhancing experience and a springboard lo future employment,
- Finally, factor 5, labelled 'Application value', assesses the extent to which an individual is attracted to an employer that provides an opportunity for the employee to apply what they have learned and to teach others, in an environment that is both customer orientated and humanitarian.

Source: Berthon, Ewing ad Hah (2005), "Captivating company: dimensions of attractiveness in employer branding" The Quarterly Review of marketing communications, vol. 242/2, pp. 151-172, https://doi.org/10.1080/02650487.2005.11072912

The data in the following Fig. 1 show that although salary is often cited as a barrier, work-life balance, general concern for the public good, and job stability are the 3 most important factors. Note that 22 of the countries highlight learning and development opportunities as very important.

Interestingly, only 8 countries classify the quality of management as a "very important" aspect and only 7 classify a digital environment as "very important". This situation reflects the current state of public services and less an ideal state needed to attract the best talent.

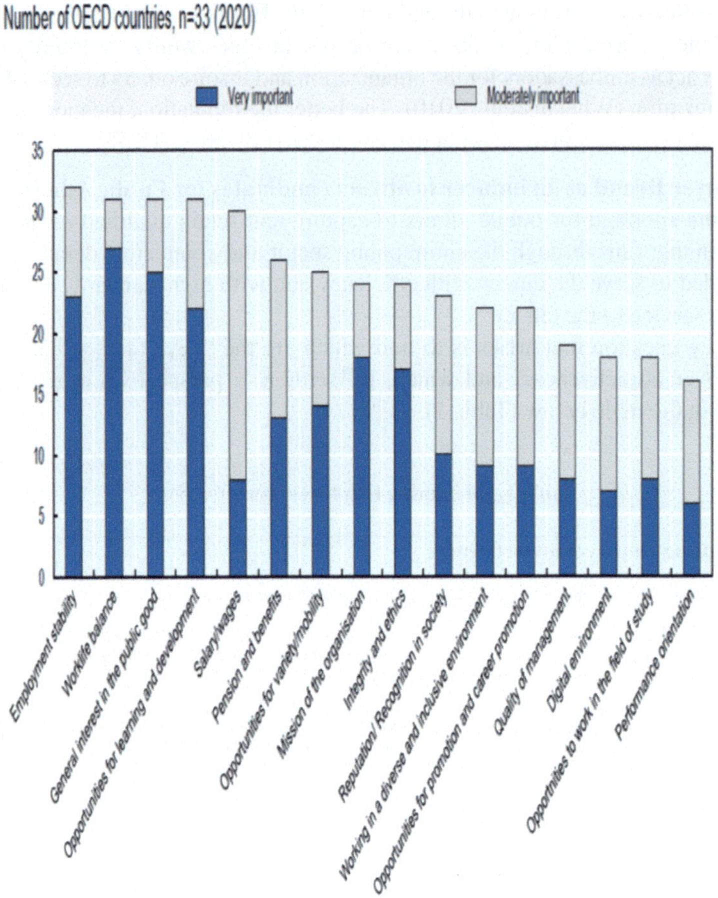

Fig. 1. Measures of Employer Attractiveness

Based on the attractiveness aspects below, there has been a growing recognition of employer branding strategies in public services in OECD countries with differing degrees of importance, however, it should be noted that 14 of the member countries did not have employer branding strategies and only 5 indicated that they had no specific branding strategy (Fig. 2). For the rest, it is either left to individual organizations or mostly part of a management-wide strategy.

As shown below, OECD countries use a variety of channels to attract candidates in order to meet relevant candidates, obtain labor market information and actively promote career opportunities and employer branding (Fig. 3).

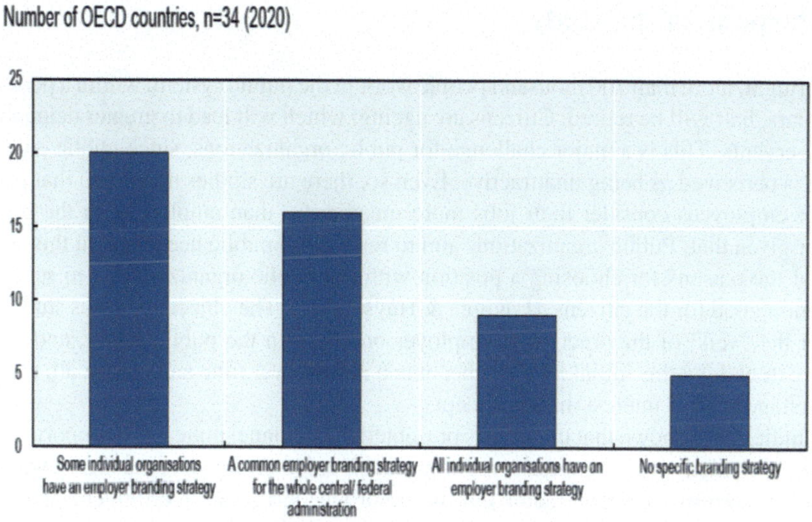

Note: Original survey question: "Which of the following employer branding strategies exist in the central/ federal administration?"
Source: OECD survey on Public Service Leadership and Capability, module 2, 2020.

Fig. 2. Employer Branding Strategies

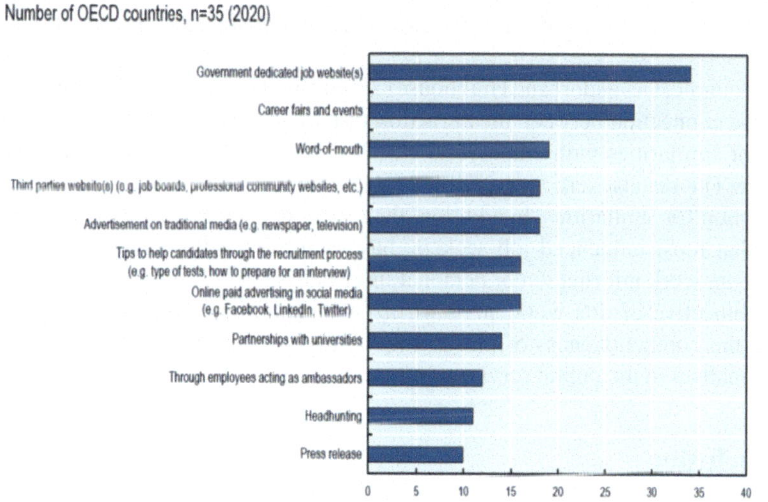

Note: Original survey question: "How does the central/ federal administration communicate about its recruitment campaigns and job openings?"
Source: OECD survey on Public Service Leadership and Capability, module 2, 2020.

Fig. 3. Recruitment campaigns

3 Purposes of the Study

In Portugal, more than 700 thousand people work in the public system, within a period of ten years, half will be retired. Citizens are ageing, which will lead to greater demand for civil servants. This is a major challenge for public organizations, since public employment is perceived as being unattractive. Even so, there are studies that claim that public sector employees consider their jobs more meaningful than employees in the private sector given that, Public organizations aim to respond to public needs and in this sense, one of the reasons for choosing a position within a public organization is to serve the common good for the citizens. (Lavigna & Hays, 2004). The objective of this study was to lift the "veil" of the practice of employer branding in the public sector, and see to what extent, how this initial idea can become a reality over time and, above all, capture understanding and interest in the concept.

Studies have shown that the factors presented in personnel policies are important; for example, challenging assignments, training, development, rewards or managing working hours. In addition, another factor that is important during the recruitment phase are the symbolic benefits (Christiaans, 2013. When a candidate receives a job offer from a known attractive employer, the candidate feels special and can gain a higher level of self-image, because their personality is perceived as compatible with the attractive employer (Christiaans, 2013). The theory of person-organization fit has been used to analyze this phenomenon, which describes the importance of the congruence of values held by the employee and the employer. Employees who identify with an organization tend to be loyal, and generally tend to stay with the organization (Backhaus & Tikoo, 2004).

For example, Devendorf and Highhouse (2008) conducted a study where they demonstrated the connection between the attraction of female university students and the perception of similarities with employees, which influenced their attraction to a specific employer. The results were analyzed with the theory of social identity, which explains the attraction for similarities. In addition, the researchers investigated the type of information that most influenced job seekers, based on three different sources; employer information, work information or personal information.

The objective of this work, in short, is to support the efforts of governments to increase the competitiveness of public employment in the labor market and improve professionalism in the public service.

4 Conclusion

During the EU Presidency, Estonia conducted a survey among EUPAN members on their experience regarding employer brand management and service activities where 26 countries were involved, some of the countries developed a slogan such as (Table 2);

Table 2. Country Slogans

Luxembourg	*"Commit yourself to the service of the citizen"– "Engagez-vous au service du cytoyen";*
In Sweden	**in addition to the employer branding activities carried out by each agency, SAGE (Swedish Agency for Government Employers) had some centrally initiated activities**
Switzerland	*"Switzerland - Our Company"*
Latvia	*"You are Latvia – get involved and do it!"*
Finland	*"Make a difference – jobs that count"*
France	***launched an action plan for 2018 in September***
Netherlands	*"Working for the central government is working for the Netherlands"*
Bulgaria	*"We work for the people"*

In the same survey, when asked about the definition in relation to the employee value proposition, most responses to this question were related to money. If the country's economic situation does not allow spending money in a field like this or the salaries of civil servants are too low, which inhibits the promotion of the government as an employer and the hiring of new employees.

Growing skepticism and the current uncertain and complex social situation, which the public sector has to navigate, make it difficult to sell the PA brand when, in reality, what goes against now more than ever, the need for talented and motivated employees to achieve its citizens and engage with them proactively and positively.

Employer Branding therefore involves developing and communicating the culture of a public organization as an employer. Various communication strategies and methods can be used to promote the government as an attractive employer so that the industry is able to develop an Employer brand of Choice. In this war for talent, public administrations need to actively emphasize and market the benefits they offer. Attracting great skills into the public service will require some strategic thinking around the employer value proposition and market segmentation to attract the right candidates and the summary of this research indicates difference between countries that have done some employer branding work and countries that have not made. With a strong employer reputation, you will be able to attract and retain talented employees.

It also seems that the ESG theme until some time ago, which focused mainly on environmental and community impact, in the current situation, there is, however, a gradual and increasingly evident change, providing greater emphasis on Social issues and Governance. This mutation also opens up a window of opportunity for recruiting and managing talent and, consequently, appeals to the employee's experience.

What seems to be worth mentioning is that in Portugal some measures have even been adopted, such as:

- The ONLINE Program, which aims to promote a better balance between professional, personal and family life, as a condition for effective equality between men and women and for full citizenship, which allows free choices to be made in all spheres of life.

- Flexibility of working hours
- Telework
- The insertion of doctorates in Public Administration
- The recent work agenda

...

Some actions have been developed, but without a strategic guideline. The existence of an instance that is careful to design a " frame" consisting of the 6 pillars of EVP advocated by the Indeed + Glassdoor model, namely; (1) Reconciliation (work + family) – (2) Corporate Culture – (3) Diversity and Inclusion – (4) Growth Opportunities – (5) Quality Management/Leadership – (6) Salary and Benefits and think about which ones Public Administration can truly be competitive. It is also urgent to know what work is being done in each of these pillars and with what visibility and even if there are good examples, will they be known? As the Nobel Prize in Economics Thaler J. (2009) adds, many public policies lack " nudging" and in this case, it is important to endogenize that personal options tend to have greater influence when, in fact, job seekers make the choice of employer based on your EP.

This approximation can, in fact, leverage the Public Administration to stand shoulder to shoulder with the private sector, when, above all, what was its main value proposition, job stability, is no longer valued by the younger generations. Moreover, what is seen as the main tool for attracting young talent, growth opportunities, also negatively compares the public sector with the private sector.

Thus, what the future envisions is, with the growing inability of the public sector to attract the best, it will tend to get worse without a clear and distinctive value proposition, centered on the expectations of candidates.

What is needed is to increase the efficiency of current processes while driving new service and innovation processes requires successful management of contradictions. It therefore requires leadership versatility: that is, leaders will need to trigger and encourage their teams to question daily routines and encourage new behaviors "keeping our core business on track". For this, we need different profiles to promote change and be a deep transformation.

5 Study Limitations

Most of the scientific literature emphasizes the need to reinforce research oriented to specific cases applied to the context of public organizations (Whelan et al., 2010). Therefore, there is a need for additional depth, as the process is rarely rational and cannot be done in the same way in all organizations. Many of the surveys consulted were unable to explain how the employer brand idea can adapt when transported to different cultures, and what adjustment is needed to adapt to local practice.

By trying to focus on the public administration, the study leaves out the comparison with the private sector, emphasizing results that could be in the same environment, within the same parameters and especially in the comparison between pairs of public organizations and not between them and the private sector, which could leave room for future studies.

References

Ambler, T., Barrow, B.: The employer brand. Panágra working paper. No. 96–902 (1996)

Backhaus, K., Tikoo, S.: Conceptualizing and researching employer branding. Career Dev. Int. **9**(5), 501–517 (2004)

Barrow, S., Mosley, R.: The Employer Brand- Bringing the Best of Brand Management to People at Work. Chichester. John Wiley & Sons Ltd, England (2005)

Christiaans, L.: International employer brand management: a multilevel analysis and segmentation of students' preferences. Weisbaden: Springer Gabler (2013)

Czarniawska , B. Joerges, B.: Travel of ideas. In: Translating Organizational Change (1996)

Czarniawska, B. Sevon, G., De Gruyter, B.: 13–48

Devendorf, S.A., Highhouse, S.: Applicant-employee similarity and attraction to anemployer. J. Occupat. Organ. Psychol. **81**(4), 607-617 (2008)

Edwards, M.R.: An integrative review of employer branding and OB theory. Pers. Rev. **39**(1), 5–23 (2010)

Fonseca, J., Nascimento, J., Dias, D.: Preliminary Adaptation Study of Perry's Public Service Motivation Model (1996) for the Portuguese Context. Magazine of the Center for Administration and Public Policy, ISCSP, University of Lisbon, Portugal (2020)

Gromark, J., Melin, F.: From market orientation to brand orientation in the public sector. J. Mark. Manag. **29**(9–10), 1099–1123 (2013)

Lavigna, R., Hays, S.: Recruitment and selection of public workers: an international compendium of modern trends and practices. Pub. Person. Manage. **33**(3), 237–250 (2004)

Moroko, L., Uncles, M.D.: Characteristics of successful employer brands. J. Brand Manag. **16**(3), 160–175 (2009)

Perry, J.L., Wise, L.R.: The motivational bases of public service. Public Adm. Rev. **50**, 367–373 (1990)

Thaler, R.: Sunstein Cass Improving Decisions About Health, Wealth, and Happiness Paperback (2009)

Definitions of Organization Well-Being in the New Normal Era: A Qualitative Research

Dawisa Sritanyarat(✉), Oranuch Pruetipibultham, and Asma Tehmarn

Graduate School of HRD, National Institute of Development Administration, Bangkok 10240, Thailand

{dawisa.s,oranuch.p}@nida.ac.th, asma.tehmarn@gmail.com

Abstract. This research studied the definitions of organization well-being and how it is interpreted by working people in Thailand. Data collection was conducted by interviewing practitioners that represent 20 organizations in Thailand known to have supporting functions and policies for employee well-being. The findings were discussed alongside with the previous study of Sritanyarat and Pruetipibultham (2023) that examined the definitions based on the Nine Cell framework as well as used to fulfilment the concept of Happy 8 from the ThaiHealth organization. This research provides both academic and practitioners with the insights of organization well-being in Thailand in the new normal era and ways to accomplish it. This research, together with its initial stage of literature review, is determined to lead to the development of an organization well-being diagnosis tool that allows the Thai organizations to monitor and assess the current state of well-being and take actions toward fostering well-being more systematically.

Keywords: Organization Well-being · Definition · Nine-cell Framework

1 Background

1.1 Significance of the Study

The digital era and the new normal risen after the COVID-19 pandemic at the national and global levels, impacted perceptions, notions, and ways toward organization well-being. Thailand is not exempted from these rapid changes. The lifestyle has changed through the years of digital transformation and the new normal led by the global pandemic. Likewise, the way working people interpreting happiness at work, quality of work-life, and well-being has changed too.

The pandemic affected work arrangements from on-site to online settings. Employees adjust to work in the new and unintentionally designed environments. Such adjustments affected well-being, both in terms of the interpretation and the literal quality. The Bank of Thailand (2021) stated that working remotely were predicted to continue. The increasing number of organizations have planned or decided to move toward the use of automation in key processes. Online activities could be expected, either within organization or externally. Workforce would shift from being full-timer to freelancers.

M. J. Sousa (Ed.): UFHRD 2024, ISEM 52, pp. 240–254, 2025.
https://doi.org/10.1007/978-3-031-92548-1_15

Rather, they should be taken into consideration as a context for more extensive research to explain the newly exposed social phenomena and guide the practices with data and systematic interpretation. It is important for those who work within the area of HRD and other related areas to aware that the new sets of explanations for organization well-being are needed. This study explores the extended realms of the organization well-being as well as illuminates the new ways of looking at this term particularly, but not limited to, Thailand.

Over the past decades, the Thai Health Promotion Foundation (ThaiHealth) has been a governmental mechanism supporting healthiness for the Thai people. Since 2009, ThaiHealth has been promoting the concept of Happy 8 as the means of organization well-being. Happy 8 was established under the concept of balancing work-life, positive and happy work atmosphere, teamwork, and person's holistic development (Thai Health Promotion Foundation, n.d.). This concept stresses on personal health and happiness and relationship with one's family and society. However, this concept has not been sufficiently stressed on work and work-life aspect. The concept has not taken work results nor organization's goals into consideration. The major limitation of Happy 8 as organization well-being definition seems to be the lack of organization's aspect. Despite this limitation, the Happy 8 has been one of the major efforts in fostering well-being in Thailand.

This paper builds on the work of Sritanyarat and Pruetipibultham (2023) which investigated definitions of organization well-being in the new normal era, using integrative literature review to broaden the concept of organization well-being, from personal happiness to a more holistic organization-related concept. Efforts should span from individual level, process level, to organization level. Changes should be facilitated considering organization strategy, structure, system design, process management, as well as employee well-being and satisfaction.

1.2 Research Purpose

This extended study explored the concept of organization well-being by collecting empirical data using the qualitative method to explore insight about definitions of organization well-being in the context of Thailand especially in the new normal era. How organization well-being is interpreted by working people in Thailand is worth studying, due to the changes and diversity in social context. The research question, what organization well-being would be defined in Thailand in the new normal era? was asked to explore and seek explanations of organizational well-being, hoping to provide both academics and practitioners with insights into organization well-being and ways toward it. This research serves as a starting point for a discussion of the concept of a happy workplace and healthy organization. Also, it can function as the seed of an organization's well-being diagnosis tool.

2 Literature Review

The concept of organization well-being in Thailand has been related to ThaiHealth's happy workplaces or Happy 8 concept, a together with the concept of healthy organization. The literature relating to the two concepts—happy workplace and healthy organization are reviewed. Another part of the literature review is dedicated to the concept of the Nine-cell organization diagnosis, which would be used as a foundation of data interpretation. Accordingly, the important elements of the Nine Cells are briefly summarized to be used as the foundations and understanding of results from the analyzed data.

2.1 Happy Workplace

The concept of happy workplaces has received a great deal of attention internationally for the past years (Poopan & Jiawiwatkul, 2017). Workplaces can refer to various types of organization. It is not limited to company or business organizations, schools can be apparently considered a happy organization in Thailand as well (UNESCO, 2022). The concept of happy workplace has mainly been ties with the ThaiHealth's Happy 8, which includes the following elements.

1. Happy Body, includes both physical and mental well-being of employees and the readiness to deal with difficult situations.
2. Happy Heart, means generosity and helpful behaviors.
3. Happy Relax, means ability to relief stress, and find relaxation.
4. Happy Brain, means continuous learning and self-development that lead to the career advancement and professionalism.
5. Happy Soul, means to have a spiritual anchor and morality leading to consciousness and mindfulness in dealing with challenges in either work or life.
6. Happy Family, means having a loving and warm family leading to family members being good members of society.
7. Happy Money, means being free of debt while having savings and spending on necessities.
8. Happy Society, means social bond that people share with one another either at work or in communities where one belongs to.

The concept of Happy 8 was publicized since 2009 (Health Promotion in Organizations Plan, 2009). It has been an initial development from the public sector aiming to influence the Thai organizations to become happy organizations. Even though this concept is easy to comprehend and convincing for decision makers and employees, it lacks focus on how to make organization's success as a happy entity. Essentially, the concept of Happy 8 mainly informs organization and human resource department about what should be achieved, rather than what should be done.

2.2 Healthy Organization

Healthy organization refers to an organization that can succeed despite encountering changing circumstances (Ho, 2000); operating with a mission to motivate internal members to succeed by providing support so that they can learn, grow, and continuously develop themselves and the organization (Dive, 2004; Verschoor, 2003). Healthy organization operates as a system that processes needs and raw materials to deliver products or services, utilizes of up-to-date data in decision making, takes clear objectives and action plans in accordance with its vision into consideration, just as other types of organizations. Just that healthy organization aspires for nurturing clear communication, balancing the profits for individual employee, society, environment, to the same level as business profit. Employees' learning and development, and work-life balance are important elements in administration. While creativity and innovation are still considered important for healthy organization, social and environmental responsibility also needs attention. It was believed by many that these components allow an organization to continue moving forward despite future changes (Beckhard, 2006; Tassanapak, 2007; Wattanamekhinkul, 2007).

Nevertheless, the holistic nature of healthy organization mentioned above could be considered picture-perfect and demanding for an organization to achieve. It looks like one healthy organization must be perfectly balanced between business profit and social and environmental accountability, also between innovative and excellence work process and employees' wellness and learning. It would not be unfair to say that the concept of healthy organization is supreme, yet surreal. It is nice to envision, but very challenging to implement.

2.3 The Nine Performance Analysis (Nine-Cell) for Organization Diagnosis

Considering the concepts of happy organization and healthy organization, the challenge rises for organization to seek guidelines toward having a standard for well-being. This research follows the framework proposed by Rummler and Brach (2013) about multi-level and multi-facet organization diagnosis to understand the concept deeper. This framework was used to examine the process and organization-related components, which happy organization in general does not seem to cover. Furthermore, it provides guideline for implementation that administrators could simply manage.

Essentially, Rummler and Brach (2013) defined an effective organization as one with systematic management. The overall effectiveness of the organization comprises nine performance variables. Organizational Performance is separated into 3 categories: 1) organization level; 2) process level; and 3) job/performer level. This concept extends the boundary from the individual level proposed by the Happy8 concept to process and organization level.

As for the organization effectiveness, there are three desirable factors to be considered for fulfilling organization's performance needs; these include goals, design, and management. This concept elaborates how performance needs can possibly be achieved. More explicit explanation of 'how' to achieve is added from 'what' to achieve as suggested by the concept of healthy organization.

The Nine Cells proposed by Rummler and Brach (2013) is shown in Fig. 1 below.

		The Three Performance Needs		
		Goals	Design	Management
The Three Levels of Performance	Organization Level	Organization Goals	Organization Design	Organization Management
	Process Level	Process Goals	Process Design	Process Management
	Job/Performer Level	Job Goals	Job Design	Job Management

Fig. 1. The Nine Performance Variables

The following are definitions of the nine variables according to Rummler and Brach (2013).

Organization goals relates to visions, missions, vision indicators and organizational strategies.

Organization design refers to policies, work systems, organization structures, management alignment, workforce capacity and business models, all of which to ensure that an organization can succeed according to its goals.

Organization management covers actions that drive strategies; assess work outcomes, cultures, and values, leadership either as influence or as role models.

Process goals is strategic goals of work processes set to ensure that work processes move in alignment and to the same direction as the organizational goals.

Process design relates to crafting of operational plans and designing of workspaces, systems, and work processes to achieve outcomes effectively.

Process management covers the process of creating ownership of activities to push plans toward targets or strategic objectives.

Job goals are considered as performer's work goals that move in the same direction with processes. Such goals could be implemented using different management tools, including job specifications, performance indicators, self-development plans, individual work outcome evaluation, and teamwork effectiveness assessment.

Job design covers designing of roles and duties, identifying necessary competency model for jobs, and training and development for performance.

Job management relates to the verification of operational outcomes, provision of feedback, work outcomes, provision of advice and support, interpersonal relationship, communication, and cooperation.

The horizontal-axis variables help organization plan, monitor, or diagnose its strategy in a cascading manner. The three variables guide the process of transferring goals to planning and implementation. Alignment of purpose, plan, and action is nurtured respectively.

Sritanyarat and Pruetipibultham (2023) found that for organization to achieve its well-being, it needs to design and manage process and jobs to meet the desirable goals in every level. Based on the Nine Cells concept, Sritanyarat and Pruetipibultham (2023) concluded each cell according to integrative literature review as follow.

1. Organization Goals Should Cover Cost Management Goals, Profit Goals, Management Goals, and Sustainability Goals.

2. Organization design should cover organization's structure and system to enable readiness to change, healthy job design, organization's policies that concern employees' happiness and well-being.
3. Organization Management Comprised Good Relationship with Community and Society, Focus the Operation on Organization's Vision and Mission.
4. Process goal comprised setting clear goals of employees' health promotion mechanism, maintaining employees' satisfaction and engagement, maintaining employees' satisfaction toward corporate social responsibility (CSR) policies, ensuring clear and healthy work processes.
5. Process design should consider having lean and effective work system, opportunity to grow, work environment is taken into consideration, reward and recognition system is motivating employees, factors that contribute to employees' happiness are incorporated in work system design.
6. Process management comprised elements of equity and unity at work, communication about organization's goals as common goals, having activities about physical and mental health.
7. Job goals should cover employees' positive emotion life, work, work environment, employees' awareness about well-being, employees' behavior such as engagement, personal health concern, teamwork, and morality.
8. Job design comprised having mechanism to encourage employees' work accountability and self-value, promoting active participation and free decision of employees, performance development in alignment with core value, creating work ecosystem to reduce stress.
9. Job management should take the following into account: creating good relationship among employees, positive communication, rewarding participation, rewarding behaviors that are good for physical and mental health.

According to the findings from Sritanyarat and Pruetipibultham (2023), not only definition of organization well-being is revealed, but also elements of organization well-being too. However, it is suggested that integrative literature review may lack the contextual element. This research is therefore designed to fulfill such gap by collecting the qualitative empirical data to further explore the concept in the context of Thailand in the new normal era.

3 Research Method

This research based on the qualitative approach. Data were collected through in-depth interviews with participants from organizations reported having practices relating to happy workplace, healthy workplace, or well-being in organization. Participants were selected by the snowball technique (Glesne, 2016). Those who were referred to and consented to participate in the research were approached. There were 20 participants from 20 different organizations from different industries as shown in Table 1.

Table 1. Organizations from which participants were selected.

No	Industries in which the companies fit	Numbers of company	Code for audit trail
1	Startup	2	A1 and A2
2	Multinational	2	B1 and B2
3	State enterprise	2	C1 and C2
4	Government agency	1	D
5	Social enterprise	1	E
6	Consumer product	2	F1 and F2
7	Financial	2	G1 and G2
8	Manufacturing	2	H1 and H2
9	Hospitality	1	I
10	Energy	1	J
11	Healthcare	2	K1 and K2
12	Food	1	L
13	Chemical	1	M
	Total	20	

The semi-structure interviews were operated to collect data from participants. The main interview questions were as follows:

What is the definition of organization well-being in your perspective?

What does your organization do to achieve organization well-being? for what purpose?

What policies or activities have been done at the organizational, departmental, group, and individual levels? How is the success measured?

Are there any changes in organization--in terms of input, process, outcomes, or environment, that occur in relation with the effort of creating organization well-being?

Probing questions were used when there were issues other than those from the main questions, in exploring the issues that the participants had mentioned for both breadth and depth (Glesne, 2016). Ethical standards were applied by respecting the participants and avoiding sensitive probing questions. The participants were free to skip responding to the questions they were not comfortable with.

Data analysis was conducted using the theory-based coding method (Ruona, 2005). The data derived from the definitions of organization well-being from the participants was used as the starting point. It was also coded based on the framework of Rummler and Brache's (2013) Nine Cells.

4 Findings

The results from the empirical data collection provided wider and deeper elaboration of Sritanyarat and Pruetipibultham's (2023) elements of organization well-being. The findings can be used for policy makers to drive their organizations toward well-being organizations. The research findings are as follows:

4.1 Organization Level

Organization Goals

Cost management goals should consider managing the overall cost that include effective human resource management. Considering human resource management cost could involve managing absenteeism and medical expenses ["…I think a well-being organization also means the one that is profitable and taking good care of their staff even when they have physical or stress issues (A1)"].

Management goals could include the following issue: ensuring healthiness of employees both while working in the office and work from anywhere.

Employee engagement and sense of ownership should be included in organization goals. ["One of the goals of organization is to create employees' ownership (D)"].

Organization Design

There were organizational restructures to correspond to the changing situations especially for the new normal era ["Our structure has just been redesigned to cope with the situation, but it does not seem to have much impact to the overall employees (G1)].

Work design, work structure and workforce allocations needed to be reconsidered for future situations. There was also a choice of outsourcing for both skilled and low-skilled cases to assist employees' work-life balance and for those who required special assistance [We hired deft and blind employees and provided them knowledge and tools to be good baristas. We had set up ergonomic counter and chairs as well as the braille-written menu (E)].

Organizations were expected to issue policies and act by adhering to happiness and well-being of their employees and even former employees. These policies were to be communicated to all members in the organizations to ensure understanding in the same direction ["One of the policies in my organization is called the revolving-door policy that enable former employees who have good relationship with us to return home. We like to think of our workplace as a home (G1)"].

Most of the participants mentioned the need to have clear employees' development planning, good relationships between employees and the top, middle, as well as the supervisory management ["Here, we emphasize working together with empathy. Leaders need to be trained to listen with empathy (L)"].

Organization Management

It was important to foster linkage between organizations and the society and community by focusing on direct relationships with customers and creating a sense of community between the organization and various agencies ["It is essential that our clinic focuses on building cordial relationships with communities around us. (A2)"].

Leaders were accountable to carry out the organizational goals by having aligned vision and missions such as focusing on innovation, and communicating the vision to employee, being responsible and be able to supervise subordinates ["Our leaders show that they care to talk and listen to us when it comes to goal setting (A1)"].

It was vital to diagnose and design the implementation and development phases by having executives to push forward well-being policies and serve as role models. There needed to be evaluation creating well-being sustainably. Data-driven method such as engagement survey, pulse survey and Happiness-related survey were important ["We conduct engagement survey here every now and then. It is important to consistently check and react on how our staff engage with us (C1)"].

There was organizational culture that supported the policies on work-life balance that were suitable for the organizational context. This included support for employee to adjust in terms of using technology as part of their work for achievement and well-being ["We highly support the work-from-anywhere policy by investing in technological equipment and software. Our tech dev(elopers) can choose to work where they like (A1)"].

Leaders have clear roles to push organization well-being. They created an environment and communicated to employees that all are equal. They should perform as role models for employee with open-mind and empathy [My boss truly supports diversity and equality among the subordinates. I appreciate his kindness (I)"].

4.2 Process Level

Process Goal

Goal should be set to support both physical and mental health management process as well as other dimensions relating to health and well-being.

There should be a goal for employee engagement by maintaining good relationships between the organizations and employees ["…and we are assigned to be partners with our employees to serve the well-being goals…they are serious about us being healthy (L)"].

Clear and specific goals to designing work systems that adhere to employee well-being would enable smooth implementation ["I like that this organization has clear work flows to serve the goals. It makes me work very comfortably (H1)"].

Process Design

Work systems should be designed to be simple and lenient. Work systems should not consume too much time and efforts but rather utilize resources appropriately. The system should facilitate performance and minimize stress. Technology-assisted design was mentioned here as well ["Compared to my ex-workplace, the work system here is designed to assist work quite efficiently. I am pretty happy (B1)"].

Work systems should be designed for employees' self-development through working in new roles and various processes. Work process should encourage employees' continuous self-development and work improvement ["Career development is clearly established in our firm. I know how and why I would register for a training program (H2)"].

Working environments should be designed to be suitable for contexts related to employees, including work methods, equipment, and workspace. Flexibility should be considered in designing the work process too ["I like to come to work because the workspace here is very well designed. I can freely design my clock-in/clock out time as well (F1)"].

The total compensation and benefits package must be appropriately considered in the process of designing. The appropriate package that reflects operations needed careful consideration. For example, project-based pay or rewards might be utilized when needed to facilitate both performance and satisfaction [I am satisfied with the total package here. The salary is not that much, but the benefit is great (G2)"].

The designing and setting of work systems should take physical environments, such as work facilities and options for working off/online, into consideration.

Process Management

It is essential to establish principles of fairness and equity are significant for the organization's well-being ["Our activities take diversity and inclusion into consideration. We just have a training about understanding generation gaps (H2)"].

The communication about organization goals and culture cannot be omitted ["Our leaders pay a lot attention on culture. We have many channels to promote culture to our staff (G1)"].

Activities that promote physical and mental health need to be executed. Channels to collect employees' feedback and happiness pulse checks are important. In addition to activities that are directly work-related, life-focused activities are needed as well ["Pulse checking is absolutely important. We also need to take employee feedback into account, so we provide them channels to do so (F1)"].

Having specified team or people who are assigned to be responsible for activities or projects related to employee well-being on a regular basis is recommended ["Managers of each department are responsible for employee engagement and wellbeing. HR team hired professional people to help us with personal-debt issue—just awesome (B2)"].

4.3 Job Level

Job Goals

Job goals should also incorporate the challenges and values of the job. People should be proud of their jobs, the outcomes of jobs, and their performances ["I'm proud to know that I contribute to the organization and that they appreciate what I have done. (F2)"].

Members should be able to select their teammates. A preferable working society should be considered as a goal of a job as well.

The awareness for the importance of personal health and attention to work-life balance should be included in the performer's goal. Job performer should not focus only on work performance but also their health ["Our CEO pays attention to our mental and physical health. It is clear that our personal goal is to keep fit and healthy. This goal reflects our mission as a wellness provider (K2)"].

Engaging in personal health, work, and organization is to be considered an overall job goal.

Job Design

Clear job design that provides guidance to employees about their roles and responsibilities, the significance of this job, as well as their relationship with others is important.

Job should be designed to allow employees to have a work-life balance. They could be able to manage their personal time and find leisure that make the employees content with their lives ["I love that I can discuss with my bosses about work and family balance. They help me with time management, and ways to relax (K1)"].

Employees should be encouraged to be involved in designing how the job should be done. They should be able to participate in decision-making about their own work ["Employees can initiate new project here, which is great (G1)"].

Employees' competencies development could be considered during the process of job design to ensure that performers have a suitable competency set to be able to work effectively and aligning their work with the organizational values [...provide support such as self-development courses or opportunities for working abroad as expats (J)"].

Physical environments should be considered in job design to ensure a desirable working atmosphere and stress minimization. The needs of employees from different generations or occupational groups could be reflected in this mechanism ["As a website developer, my boss allows me to work from everywhere. I ended up working in Chiang Mai in a workspace paid by the company. It has been a totally blissful working experience (A1)"].

Job Management

Good relationships should be nurtured and encouraged among employees in organizations. The working atmosphere of collectivism and family-liked was found to be favorable for organizational well-being in Thailand. Leaders tend to be expected to work closely with employees ["With this small amount of salary, I can still work here because of the working atmosphere that makes me feel like I am surrounded by family and friends (C2)"].

Constructive communication and regular feedback system are desirable. Employees should be provided with opportunities to express opinions and receive feedback from others. Mentoring and coaching are taken as positive forces for problem-solving and motivation ["My colleagues and I like that we can express our opinion freely and our supervisor is not just our boss but mentor who always be there for us (M)"].

Mutual understanding of shared goals at either the organization or department level is important for the well-being of the organization. Freedom to speak and encouragement to present ideas about work are equally vital ["Our management makes sure we understand the department goals and we agree to it. If we cannot achieve such goal we cannot make it because of any reason we can discuss with our boss (F2)"].

Both organizations and individuals are accountable for keeping the members' health physically and mentally well. Health-promoting activities could be arranged for employees. Leaders should lead or be involved in those activities to set shared determination ["Leaders create good vibe and co-create good atmosphere with us. It feels great (F2)"].

5 Discussion and Recommendation

Based on this qualitative study, it was found that organization wellbeing in the new normal era extend to both physical and mental health of employees. This includes employees' work-life balance and employee engagement. To enhance wellbeing, organizations need to make an effort to redesign system/policy, restructure organizations, and allocate workforces to better serve the changing operational situations. After all, employees themselves need to be aware of 'why' and 'how' they are required to keep themselves fit and healthy.

The organization's policies and regulations should be revisited to allow the redesign of either goals, processes, or management approaches that nurture the well-being of employees. Clear processes should be put in place by adhering to principles relating to considering the well-being of health and work designed for them to have freedom in decision-making and be able to self-develop. In addition, appropriate physical working environments should be prepared for multi-platforms. Positive relationships at work across hierarchical levels should be ensured.

Apparently, the results show that the definitions of organization well-being that emerged from this qualitative research resembled the one based on the integrative literature review presented by Sritanyarat and Pruetipibultham (2023). In addition, this research discovered that employee engagement plays an important part in an organization's well-being.

It is also worth noting that employee engagement in this case extends to the ex-employees too. Specifically, the policies that were open for former employees to return to work with the organization were highlighted. HR policies in accordance to enable such engagement were mentioned significantly as well. In addition to work-based competencies, recruitment and selection should be redesigned and established to seek for people who have personal goals that aligned with those of the organization. The redesigned policy needed to be inclusive and embrace diversity issues; these would lead to the design work process that supports employees with diverse needs.

Most of the participants significantly mentioned the freedom to choose whether to work remotely or in a hybrid manner as the reason of their work-based well-being. Job design that can ensure transparency and fairness at work was highly stated too. The collective work environment was considered a part of the organization's well-being.

Following up with the comparative analysis with the research results in the systematic literature review by Sritanyarat and Pruetipibultham (2023), it was found that the concept of Rummler and Brache (2013) could cover the definitions of an organization's well-being.

Definitions of organization well-being that emerged from this qualitative research resembled the integrative literature review presented by Sritanyarat and Pruetipibultham (2023). Back then there were thirty-six performance variables emerged from the analysis of a good number of peer-reviewed sources. After qualitative data collection and interpretation, a total of thirty-nine performance variables were found, along with an elaboration of the existing performance variables.

To summarize the thirty-nine performance variables, and an elaboration of the existing performance variables, Table. 2 here presents the revised concept of organization well-being based on an integrative literature review and qualitative data. The add-up from this qualitative research is presented in italic. It can be noticed that there are variables being added, clarified, and elaborated.

Table 2. Organizations from which participants were selected.

	Goals	Design	Management
Organization Level	1. Cost management goals: *Ensure that processes and mechanisms are effective and efficient* 2. Profit goals 3. *Operational* management goals 4. HR Management goals: *Ensure long-term engagement* 5. Sustainability goals	6. Organization's structure and system that are ready to change 7. Healthy job design 8. Organization's policies that concern about employees' happiness and well-being: *Focus on long-term relationship, including returning of ex-employees*	9. Good relationship with community and society 10. Focus the operations on organization's vision and mission: *HR policies and process highlights the common and shared values* 11. Diagnosis before making decision 12. Promoting participative management 13. Clear management's roles in supporting organization's well-being 14. *Inclusive and diversity embraced organization's policies: Enforce the policies to* support differently abled people to work with equal opportunities
Process Level	1. Setting clear goals of employees' health promotion mechanism: *Including different dimensions of life, such as financial well-being, mental well-being, social well-being, and* so on 2. Maintaining employees' satisfaction and engagement 3. Maintaining employees' satisfaction toward CSR policies: *Including external CSR and internal CSR* 4. Ensuring clear and healthy work processes	5. Lean and effecting work system 6. Opportunity to grow: *Consider both vertical and horizontal career advancement* 7. Work environment is taken into consideration: *Ensure suitable work process, platform, and workspace for employees* 8. Reward and recognition system is motivating: *Including total compensation package* 9. Factors that contribute to employees' happiness are incorporated in work system design: *Consider the total work atmosphere, such as onsite-online-hybrid working policy, investment in workspace, convenience at work, etc.*	10. Equity and unity at work: Distribution justice and process justice 11. Communication about organization's goals as common goals: *Communicate clear and specific directions of organization* 12. Having activities about physical and mental health: *Such as stress reduction, club activities, and mingling activities.* 13. Having representatives of well-being for employees
Job Level	1. Employees' positive emotion, life, work, and work environment 2. Employees' awareness about well-being 3. Employees' behaviors: engagement, personal health concern, teamwork, and morality 4. Employees' success in job performance, emotion stability, and life aspects: *Consider challenging tasks in balance with work-life*	5. Having mechanism to encourage employees' work accountability and self-value 6. Promoting active participation and free decision of employees 7. Performance development in alignment with core values: *Transparent and fair performance appraisal needs to be implemented* 8. Creating work ecosystem to reduce stress	9. Creating good relationship among employees: *Collectivism and family-like vibe* 10. Positive communication 11. Rewarding participation: *Ensure participation of employees by rewards and recognition* 12. Rewarding behaviors that are good for physical and mental health

6 Conclusion

This study elaborates how the Thai employees contemplate 'wellbeing' at a multi-organizational level. Using the Nine-cell framework, the paper reveals how leaders can rethink and redesign the organizational system, policies, and structure to facilitate the transition of well-being in a new normal era. The prior concept regarding happy workplace (Happy 8) that had been predominantly established in Thailand was found to be rather short of the job-based dimension. It is worth nothing that the results of this qualitative data interpretation demonstrated the alignment with the results of the research based on an integrative literature review previously studied by Sritanyarat & Pruetipibultham (2023). The combined study is determined to use for the future tool development to enhance the understanding of organization well-being and happy workplace in Thailand. Based on the present study, individuals, academia, and practitioners who are interested in driving organization well-being may study these preliminary results and use them as a guideline or lesson to apply for their further work designs, future research questions or escalating the issue for a better and concrete development.

Key take-away from the previously mentioned findings for HR practitioners is that organizational well-being is born out of nurturing, not by nature. It is a part of the results of HR and OD practices, with clear objectives to nurture organizational well-being. Moreover, the findings here can be interpreted that to nurture organizational well-being, practitioners need to work on HR and OD processes, not to only create activities. Having employee satisfaction-related activities is nice, yet not enough. The systematic processes are vital.

Acknowledgement. This research project would not have been possible without the financial support from the Thai Health Promotion Foundation. We also would like to thank our interviewees who shared their insightful experiences and precious time.

References

Bank of Thailand, https://www.bot.or.th/th/research-and publications/articles-and-publications/bot-magazine/Phrasiam-64-2/cover-story-64-2.html Accessed 17 Mar 2024

Beckhard, R.: The healthy organization. In: Gallos, J.V. (ed.) Organization development, pp. 950–952. Jossey-Bass, San Francisco, CA (2006)

Dive, B.: The healthy organization: A revolutionary approach to people & mangement. Kogan Page, United Kingdom (2004)

Glesne, C.: Becoming qualitative researchers: An introduction. 5th edn. Pearson, United Kingdom (2016)

Health Promotion in Organizations Plan, https://dol.thaihealth.or.th/resourcecenter/sites/default/files/documents/khuumuuemaasraangngkhkraehngkhwaamsukhkanetha.pdf Accessed 17 Mar 2024

Ho, J.T.S.: Managing organizational health and performance in junior colleges. Int. J. Educ. Manag. **14**(2), 62–73 (2000)

Poopan, S., Jiawiwatkul, A.: The development of indicators of good practice happy workplace organizations in Thailand. In: New Trends and Issues Proceedings on Humanities and Social Sciences, pp. 221–227 (2017)

Rouna, W.E.A.: Analyzing qualitative data. In: Swanson, R.A., Holton, E.F. (eds.) Foundations Research in organizations and methods in inquiry, pp. 233–263. Berrett-Koehler, San Francisco, CA (2005)

Rummler, G.A., Brache, A.P.: Improving performance: How to manage the white space on the organization chart, 3rd edn. Jossey-Bass, San Francisco, CA (2013)

Sritanyarat, D., Pruetipibultham, O.: Definitions of organization's well-being in the new normal era: an integrative literature review. In: McCabe, T. J. (ed.), The 23rd University Forum for Human Resource Development, National College of Ireland. https://www.ncirl.ie/University-Forum-for-Human-Resource-Development-Conference Dublin (2023)

Tassanapak, K., Healthy Organization [Master's thesis, National Institute of Development Administration]. http://library1.nida.ac.th/termpaper5/hrd/2550/19238ab_co.pdf Accessed 17 Mar 2024

Thai Health Promotion Foundation, https://happy8workplace.thaihealth.or.th/happy-8 Accessed 17 Mar 2024

UNESCO, https://www.unesco.org/en/articles/what-you-need-know-about-unescos-happy-schools-project Accessed 17 Mar 2024

Verschoor, C.C.: https://link.gale.com/apps/doc/A118951795/AONE?u=anon~834c986f&sid=googleScholar&xid=e3a93522 Accessed 17 Mar 2024

Wattanamekhinkul, B.: The study of healthy organizational conceptual model : Case studies of 100 year-old Thai-Owned private organizations in Thailand [Master's thesis, National Institute of Development Administration]. https://repository.nida.ac.th/items/dcf6e847-3051-4c33-a8f6-aaee2c785caa Accessed 17 Mar 2024

Exploring Organizational Practices for Gender Equality: A Narrative Inquiry Approach

Marica Liotino[1(✉)], Monica Fedeli[1], and Burcu Simsek[2]

[1] University of Padova, Via Beato Pellegrino 28, 35137 Padova, Italy
marica.liotino@studenti.unipd.it
[2] Hacettepe University, 06800 Beytepe Yerleşkesi, Ankara, Turkey

Abstract. The contribution focuses on Diversity Management (DM) and the implications of its use for gender equality in Italian companies, where its development has mainly centered on issues regarding women in the workforce. In Italy, women face lower employment rates despite being more educated than men, and this further narrows with an increase in the number of children. They also tend to have lower incomes because of the precariousness of their employment contracts, gender imbalance in economic sectors, and limited opportunities for advancement.

This paper builds on previous studies that highlight the need for a contextual understanding of DM approaches using the lens of critical feminism and social constructionism.

Critical feminism focuses on the dismantling of patriarchal organizational structures, emphasizing reflection on privilege, structural reform for emancipation, increased gender awareness, and the reconstruction of solidarity. Simultaneously, the contextual and interactive nature of issues, rooted in the formulation of claims and narrative analysis is underscored by social constructionism.

Narrative inquiry is used to understand perceptions of gender equality in organizations in Vicenza (in the North-East part of Italy). Fifteen semi-structured interviews with human resource managers will explore policies and organizational practices that promote gender equality. Subsequently, three digital storytelling workshops will be organized with voluntary employees, allowing them to share their experiences. This effort aims to examine and disseminate organizational models for improving gender equality practices, contribute to the development of greater awareness and understanding of gender equality in Vicenza's organizations, and provide insights and tools for promoting positive organizational change.

Keywords: Social Constructionism · Critical Feminism · Organizational Processes · Gender Equality · Narrative Inquiry

1 Introduction

Research on diversity management (DM) represents a growing and increasingly necessary field in today's world (Olsen and Martins 2012), yet developments are particularly concentrated in the United States of America (Jonsen et al. 2011).

M. J. Sousa (Ed.): UFHRD 2024, ISEM 52, pp. 255–267, 2025.
https://doi.org/10.1007/978-3-031-92548-1_16

Although a clear and detailed definition has not yet been reached, DM describes methods of human resource management to solve the critical issues that arise from the simultaneous presence of heterogeneity of workers in the workplace, which comprises demographic, gender, ethnic, and disability factors (Ferrara 2019). DM proposes human resource management aimed at creating inclusive workplaces, where the enhancement of the specificity of each individual also promotes better achievement of organizational goals (Buemi et al. 2015; Ravazzani et al. 2021).

In Italy, research on DM is limited, its adoption in organizations is low compared to other European countries, and when adopted, it tends to focus on the issue of women's work (Colella and Di Lorenzo 2023). Previous studies suggest a different conception of DM in Italy, where its observance is embedded in an interplay between coercion and voluntarism, privileging reputation, and social expectations to cater to legal or market demands (Ravazzani 2016). For instance, reconciliation measures, including smart working, calibrated to the lives, work situations, and paths of diverse employees, are the most important practices from a DM perspective (Santoni and Crespi 2022). At the same level, and from a gender perspective, collective agreements mainly cover maternity protection, childcare leave, part-time work, working time organization and flexibility, remote and smart working, as well as combating sexual violence and harassment (Ferrara 2019, p. 107). This is not enough to make workplace organizations value gender diversity, and it seems that there is still a long way to go before concrete DM initiatives that consider the needs of female workers and their potential can be established.

Research has shown that the Italian quota system (Law 120/2011) has stimulated an increase in gender diversity in large companies, but it also reveals that women continue to be managed mainly as non-executive and independent members (De Vita and Magliocco 2018). Moreover, despite important legislative measures to prevent the occurrence of unfair treatment in the workplace, discrimination in the workplace still recurs (Palumbo and Manna 2019). Despite significant legislative efforts to prevent unfair treatment in the workplace, discrimination persists (Palumbo and Manna 2019). The Italian National Recovery and Resilience Plan introduced the Italian Gender Equality Certification System, which was subsequently regulated by specific national laws (Law 162/2021 and Law 234/2021). However, there is still little evidence of the outcomes of the initiative.

Thus, these aspects thus highlight the central role of contextual conditions in shaping organizations' approach to DM and how the specificity of national systems are neglected when these issues are studied (Jonsen et al. 2011). Hence, the perception that "a contextual understanding of diversity management approaches should therefore be favored over assumptions about time or stage progression and their uncritical transposition across national contexts and organizational realities" (Ravazzani 2016, p.165).

According to ISTAT (National Statistic Institution) 2022 data, women in Italy are better educated than men, where 65.3% have at least a high school diploma (60.1% of men) and 23.1% have a university degree (16.8% of men)—differences that are much more noticeable than those observed for the EU27 average. However, the female advantage in education does not translate into an advantage in employment (Guarascio et al. 2023); in fact, the female employment rate is much lower than the male rate (55.7% versus 75.8%). In general, the Italian employment rate is about ten points lower than the European average (74.7%), and this gap is mainly determined by the female employment rate, which is more than 14 points lower than the European average (ISTAT 2023). The full-time equivalent (FTE) employment rate has barely increased for women (31%) and has decreased for men since 2010. The gender gap is far wider in couples with children than in couples without children (*Gender Equality Index 2020: Italy*, 2020; see Fig. 1a), confirming what was previously reported in the 2019 ISTAT data: the more children, the greater the gap in female and male employment rates.

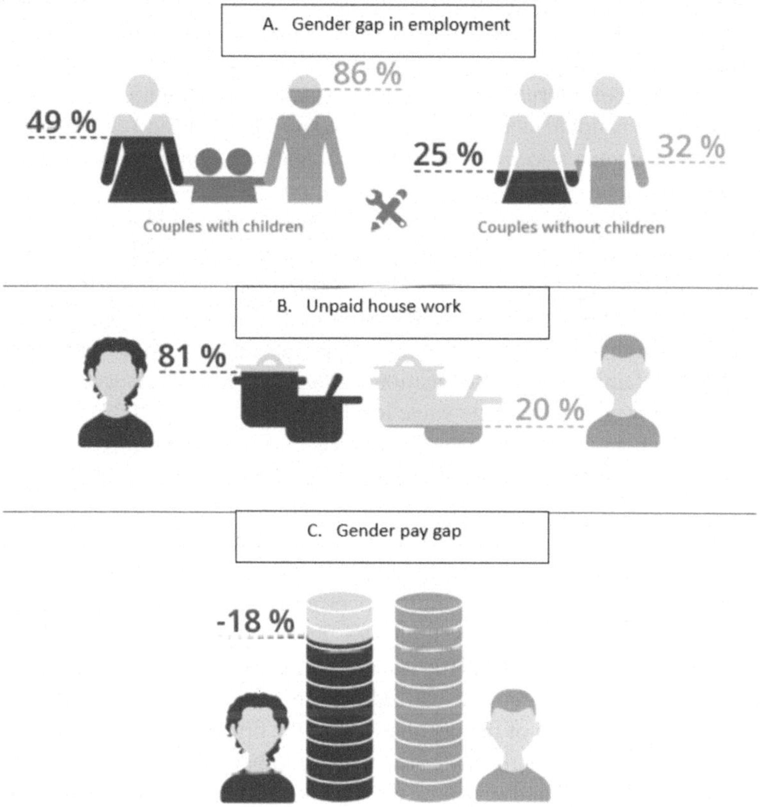

Fig. 1. Gender inequalities: A. Gender gap in employment; B. Unpaid housework; C. Gender pay gap (Gender Equality Index 2020: Italy, 2020, p.4).

In line with what has been previously mentioned, the ISTAT report of 2022 testifies that within couples, the rise in equity of distribution of housework slows down compared to previous years. This affirms the prevalence of women in this type of task, where women were already four times more likely than men to spend time cooking and doing housework every day for at least one hour in 2020 (Gender Equality Index 2020: Italy, 2020; see Fig. 1b).

Moreover, in 2019, nearly a third of employed women worked part-time (30% of women versus 8% of men). Even in 2022, many women are employed because they are unable to find full-time work. Furthermore, the incidence of involuntary part-time employed women out of the total number of part-time employed women is well above 50%. Gender inequalities in earnings persist. Women's mean monthly earnings amount to almost a fifth less than men's (Gender Equality Index 2020: Italy, 2020; see Fig. 1c). Men occupy higher positions than women, as evidenced by the fact that one-third (33%) of managers in the EU in 2019 were women (ISTAT 2020), in Italy in 2021, only 20.5% of managers were female (Manageritalia 2023). Nevertheless, greater gender diversity in the boardroom and the achievement of a critical mass of women have been shown to have a positive impact on corporate creditworthiness, especially in small and medium-sized Italian companies (Manello et al. 2023, p. 1398).

What's more, a high presence of women in leading positions seems to have a positive impact on the degree of internationalization of organizations (Bannò et al. 2019).

In summary, the national landscape, as reported by Guarascio and colleagues (2023), clearly shows that women's access to the paid labor market is consistently lower than that of men, even when equal characteristics are considered. Furthermore, women work fewer hours on average than men, which negatively impacts both current income and future pensions (Guarascio et al. 2023). Women are more likely to be employed in precarious jobs such as temporary contracts, collaborations, or fictitious self-employment positions. The distribution of employment by economic sector also shows a significant gender imbalance, with women concentrated in a few sectors, such as trade, accommodation and food services, family services, education, health care, and other services, which are characterized by lower-than-average wage levels (Guarascio et al. 2023).

There is overwhelming evidence that Italian organizations fail to provide women with equal access and opportunities for professional development, which affects women's career progression (Poggio et al. 2010), and the picture becomes less clear when on a regional or more local level. From a DM perspective, it is essential to start with an accurate analysis of local contexts to activate effective innovation processes, which should include specific connotations, dominant practices and cultural models, the needs of the organization and people, possible asymmetries, and potential resources and solutions to overcome them (Poggio et al. 2010; Ravazzani 2016). This research contributes to the findings of particular and contextualized data in Vicenza's province. Its goal is to promote a greater knowledge of the peculiarities of the organizations in the area, which is one of Italy's most successful in terms of the number of companies, productivity diversification, commercial and structural dynamism, and export propensity (Camera Di Commercio Vicenza 2024).

2 Theoretical Framework

The lens of critical feminism and constructionism is used throughout the research. Constructionism is based on constructivism theory, which asserts that social reality is constructed and contextual. Pioneers in the field, Spector and Kitsuse (2017) posit that social problems are not objective entities but rather social constructions arising from the interactions and interpretations of social actors. A key feature is a focal point on claims-making, or the process by which certain issues are identified and brought to public attention as social problems. Miller and Holstein (1989) clarify that this perspective looks at how ethical and ameliorative discourse embedded in the broader culture helps to create or transform the institutional spaces in which people involved in this process are considered.

In addition to the constructivist view of learning as a reconstruction and not a mere transmission of knowledge, Papert (1986) argues that learning is most effective when it is not only mental but also accompanied by actual construction. Therefore, concrete thinking is central to this conception of syntonic learning and is based on three principles: continuity with prior experience and knowledge; power of the personal learning process, imbued with meaning; and cultural resonance of the knowledge to be learned (Varisco 2002; Papert 1980). Active learning is privileged. Teaching is limited to providing more space for solving and finding problems in a process of continuous adjustment. This recursive process of learning generates useful, shared, practical, and intentional knowledge based on the prerogatives of the learner and the learning context (Harel and Papert 1991). The public's crucial role has been well established in the theoretical discourse on social constructionism. Nichols (2003) emphasizes that the public is not a passive recipient but an active and constitutive part of the process. This aligns with the idea that social problem setting involves a "joint production" similar to everyday conversations. This paper connects the constructivist view of learning and critical feminism, focusing on the narratives of working women in the Vicenza area.

Critical feminism adds a crucial perspective to this view of social spaces. It challenges the gender narrative in organizational development, offering a critical understanding of patriarchal organizations and how they perpetuate implicit biases that hinder women's access to opportunities for professional development and advancement (Bierema 2017). A deep examination of the concept of privilege characterizes critical feminism. It rejects the notion of simply correcting women to make them more self-confident, instead promoting a critical awareness of power dynamics and the asymmetries that arise in relation to these dynamics. It also pays attention to reforming the structural inequities existing in neoliberal postfeminist societies. It aims to emancipate women from structural domination, rather than to adhere to postfeminist narratives (Bierema et al. 2022). This perspective prompts an increased gender consciousness through critical action and questioning of organizations to promote real change toward gender equality (Bierema et al. 2022). Solidarity is a cornerstone of critical feminism, which underscores the value of collective awareness of differences and collaboration between women and allies to challenge postfeminist practices that perpetuate oppression (Bierema et al. 2022). Therefore, critical feminism aims for social transformation through critical awareness, collective action, and challenging existing power structures (Bierema et al. 2022). This perspective takes an intersectional approach, highlighting the interconnectedness between different dimensions of identity, such as gender, race, class, and sexuality (Crenshaw 1989; Hooks 1984). Through a critical analysis of social, political, and economic institutions, critical feminism identifies and addresses entrenched gender inequalities (Collins 1990). It also highlights the need to recognize and challenge power inequalities within feminist movements. Greater inclusion and representation of diverse groups of women must be promoted (Butler 1990; Mohanty 1984). Agency and resistance are at the heart of feminist theory, which emphasizes the need to empower those on the margins to challenge oppressive structures. This calls for concrete steps toward social and political change, such as defending women's rights and opposing gender-based violence (Davis 1981; Ahmed 2017; Crenshaw 1991; Hooks 2000).

3 Methodology

This study adopts Narrative Inquiry (NI), a form of qualitative research that has emerged in the social and psychological sciences (Barkhuizen et al. 2014) and has since been developed into other fields. These include cognitive sciences, organizational studies, theory of knowledge, occupational sciences, and educational sciences (Snowden 2010). It focuses on the analysis of personal narratives and stories, aiming to understand human experiences in social contexts (Chase 2018). This approach examines stories, autobiographies, diaries, field notes, conversations, interviews, and other narrative texts to understand human experiences and the processes of social meaning (Clandinin and Connelly 2004).

Furthermore, NI is an excellent vehicle for professional development (Goodson and Gill 2011; Wells 2011) and is designed to drive social change (Andrews et al. 2013). This involves presenting marginalized stories to change perceptions and analyzing how institutions regulate narrative practices, that contribute to either oppression or resistance. Its theoretical maturity, focus on the social life of stories, and commitment to social change make it an invaluable asset in the in-depth analysis of social contexts and their change.

A constructionist approach to NI is based on the concept that narratives are socially constructed through the interplay of interpersonal, social, and cultural relationships (Flick 2013). Thus, narratives are interpreted as products of the dynamic interaction between individuals and their social context (Andrews 2007; Riessman 2002). The focus on the story itself is a distinctive feature, as well as the way it is told and its meaning for both the teller and the listener or reader, including researchers and research audiences (Esin et al. 2014). Moreover, NI's constructionist approach is renowned for its critical perspective on the nature of narratives. It acknowledges that narratives are not merely objective reflections of reality; they are shaped by specific historical, social, cultural, political, and economic contexts (Tamboukou 2008; Esin et al. 2014). In this context, it is essential to examine the power relations involved in the construction of narratives. It is crucial to highlight how some narratives emerge as dominant while others are marginalized (Tamboukou 2008). Finally, this type of research highlights the importance of considering the role of the audience in the constitution and understanding of narratives. It is recognized that the meanings of narratives are ascribed not only by the audience at the time and in the social context in which the story is first told but also by audiences positioned in different ways; thus, audience construction becomes an integral part of the approach itself (Bakhtin 2010). In this case, NI with a constructionist approach is distinguished by its attention to the socially constructed nature of narratives, consideration of the interactive process of meaning-making, and critical awareness of the dynamics of power and cultural context that influence the production and interpretation of narratives.

3.1 Research Design

This research aims to identify organizational models, policies, and practices that promote gender equality in enterprises of different sizes, with an emphasis on small and medium-sized enterprises. Manufacturing and service enterprises, trade, tourism and service enterprises, craft enterprises, and cooperatives in Vicenza (Veneto region) are involved.

The research design involves conducting 15 semi-structured interviews (Trinchero 2002) with resource managers and labor union representatives to identify good practices, policies, and models for promoting gender equality contextualized in the Vicenza area. The dimensions explored in the interviews mainly deal with information on the characteristics of the organization and its people, DM strategies, and practices to promote gender equality. Table 1 provides a more detailed description of the dimensions; in particular, it is worth noting that the interviews also serve as an opportunity to request documentary sources and databases, such as those related to employee information, which will be analyzed together with the interview transcripts. In addition, the collection of data related to corporate welfare and employment processes will be facilitated by supporting tools, such as checklists, developed by the researchers based on existing scientific literature to ensure the accuracy of the information collected.

Table 1. Dimensions, descriptors, and indicators investigated through interviews and their references.

Dimensions	Descriptor/indicator
Respondent demographic information	Age Gender Qualification level Role Duties Years of experience
Organization information	Foundation year Sector Mission Vision Values Number of employees Organizational model (Mintzberg 1981) and decentralization of power (Filippi et al. 2023) Competitive climate (Murayama and Elliot 2012) Work-life interference and fit (De Simone et al. 2018)

(*continued*)

Table 1. (*continued*)

Dimensions	Descriptor/indicator
Employee variety	Feminization rate (Poggio et al. 2010) Gender wage differential (Poggio et al. 2010) Gender segregation index (Poggio et al. 2010) Turnover rate
Employee information	Biographical data (Poggio et al. 2010) Data related to professional and contractual status (Poggio et al. 2010) Data related to the relationship between work and personal life (Poggio et al. 2010) Data related to career paths (Poggio et al. 2010) Data related to participation in training activities (Poggio et al. 2010)
Diversity management	Organizational openness to change (Di Fabio 2016; Harris et al. 2007) Symptoms of diversity-related problems (Di Fabio 2016; Harris et al. 2007) Diffusion of inclusive language (Di Fabio 2016; Harris et al. 2007; Parks and Roberton 2000) DM's current position (Di Fabio 2016; Harris et al. 2007)
Policies and practices to foster gender equality	Policies and processes of the organization (Gay 2013; Poggio et al. 2010) Corporate welfare processes (Razetti, and Santoni 2019; Income Tax Consolidation Act) Employment welfare processes (Razetti and Santoni 2019; Law 81/2017; Consolidated Maternity/Paternity Act) Other initiatives of the organization

Based on the results of the interviews, three organizations that stand out for good DM practices and gender equality initiatives will be identified and involved in three separate digital storytelling workshops. Employees will also voluntarily participate to offer their perspectives on the organization's gender dynamics and initiatives. In fact, DM research suggests giving employees a voice to share their perspectives, as "employee perceptions are a more telling indicator of the organization's actual support for diversity" (Herdam and McMillan-Capehart 2010, p. 40). The next section clarifies the second part of the qualitative research.

Qualitative content analysis of the collected data will be supported by AtlasTi.23 software. In addition, statistical analysis will be conducted using Excel sheets if the respondents do not provide database processing of employee information.

3.2 Using Digital Storytelling to Share Organizational Best Practices

Digital Storytelling (DS) can enrich professional development by improving, disseminating, and continuing to create good practices already implemented in Vicenza organizations.

DS is a process that uses multimedia tools and digital resources to tell stories and share information. This approach has been adopted in diverse educational settings, including formal, informal, and non-formal. Among its benefits, DS is considered an effective tool for enhancing the pedagogical experience by providing space for creative thinking, reflection, listening, and sharing, as well as for developing skills and competencies related to information, data literacy, and communication (Mittiga 2018). Furthermore, DS promotes active learning by fostering collaborative knowledge exchange, enhancing critical thinking, and developing the ability to seek additional interpretations when faced with new issues and conceptual frameworks (Mittiga 2018). It is a powerful tool for constructing narratives that facilitate learning and long-term retention of information (Scarinci 2011). It also promotes inclusivity, as it supports the creation of personalized narratives, giving voice to different perspectives, and contributing to more diverse and inclusive representation (Scarinci 2011). DS has already been used to analyze and narrate practices for gender equality in organizational settings and has proven to be a powerful tool to empower women and girls and reduce gender stereotypes (Simsek 2016).

Digital Storytelling Workshops (DSWs) have been used in multiple contexts and for multiple purposes as a co-creative community practice. It has been adopted for awareness-raising and community-building around social and educational issues such as gender, poverty, health conditions, and inclusion to change learning ecologies, health contexts, and society at large, thanks to the work of Dana Atchley and later Joe Lambert (2013). It has been readily associated with qualitative research because of its participatory nature. NGOs, educational institutions (from primary to higher education), and private organizations have already employed DSWs for human resources and corporate culture initiatives, and marketing purposes. A DSW is a collaborative and co-creative learning process facilitated by professionals. They engage people in telling, listening to, and sharing personal stories in a multimedia format. A digital story is created through the creative blending of a voice-recorded story with personal images. These are fragments of personal life narratives that reflect the reality of the author on social issues such as domestic violence, major health conditions, women's political participation (Simsek 2012), migration, and education. Hartley and McWilliam (2009) referred to the dialogic nature of DSWs as follows:

> Atchley's innovation was to develop an exportable workshop-based approach to teach "ordinary" people- from school students to the elderly, with or usually without knowledge of computers or media production- how to produce their own personal videos.But despite the term "digital" in the digital storytelling, the emphasis is on the *story* and the *telling* (p.3).

There are six steps in conducting a DSW: 1. Story circle, 2. Text writing, 3. Voice recording, 4. Creating/finding images, 5. Assembling the digital story by combining voice recording and images, and 6. Group projection of the created digital stories. The first and the last steps (story circle and group projection respectively) are dialogic stages in which participants tell and listen to each other's stories. In addition, DSWs can support qualitative data collection during the story circle and group projection phases, following dynamics like the one used for focus groups. Digital stories can also be analyzed as textual content. DSWs in qualitative research projects pursue the narrative shift in the social sciences, which relies on the power of first-person narratives to gain insight into social issues through life stories.

4 Conclusions

In conclusion, adopting the NI methodology will allow an in-depth analysis of the current processes related to women's employment in the Vicenza area, promoting reflection and awareness of the current situation. The research will help identify gender equality policies and practices in the organizational context from a DM perspective, informing and promoting professional development initiatives for companies' cadres. In addition, using DSW as a tool for sharing good practices will encourage the dissemination of organizational experiences and promote participatory dialog between employees and organizations, supporting empowerment processes and prompting a more gender-sensitive corporate culture.

Thus, this study provides important insights for training employees and improving corporate culture and offers valuable insights for promoting gender equality and fostering an inclusive work environment. Narrative approaches can stimulate organizational learning, fostering change and innovative processes (Poggio et al. 2010). The situated nature of the development and co-construction of knowledge at the service of local entities to foster greater awareness of their practices is a major strength of this research. The knowledge developed and the practices promoted will also go a long way toward informing the deficient existing literature at regional and national levels.

Acknowledgements. Conceptualization M.L., and M.F.; Introduction M.L.; Theoretical framework M.L.; Methodology M.L.; Using digital storytelling to share organizational best practices B.S. and M.L.; Conclusions M.L. All authors have read and agreed to the published version of the manuscript.

References

Ahmed, S.: Living a Feminist Life. Duke University Press (2017)

Andrews, M.: Shaping History: Narratives of Political Change. Cambridge University Press, Cambridge (2007)

Andrews, M., Squire, C., Tamboukou, M. (eds): Doing Narrative Research, 2nd edn. Sage, Thousand Oaks (2013)

Bannò, M., D'Allura, G.M., Dawson, A. (2019). Le traiettorie di crescita internazionale e la presenza delle donne nei ruoli decisionali: un'indagine empirica sui CdA delle imprese italiane. In: Sinergie-SIMA 2019 Conference. Management and Sustainability: Creating Shared Value in the Digital Era, 20–21 June 2019, Sapienza University of Rome (2019)

Bakhtin, M.M.: The dialogic Imagination: Four Essays. University of texas Press (2010)

Barkhuizen, G., Benson, P., Chik, A.: Narrative Inquiry in Language Teaching and Learning Research. Routledge, New York (2014). ISBN 978-0-415-50933-6.

Bierema, L.L., Sim, E., He, W., Cox, A.B.: Double jeopardy: the paradox and promise of coaching women leaders from a critical feminist perspective. Gend. Manag. Int. J. (2022)

Bierema, L.L.: No woman left behind: critical leadership development to build gender consciousness and transform organizations. In: Madsen, S.R. (ed.) Handbook of Research on Gender and Leadership, pp. 145–162. Edward Elgar (2017)

Buemi, M., Guazzo, G., Conte, M.: Il Diversity Management per una crescita inclusiva: strategie e strumenti. Franco Angeli (2015)

Butler, J.: Gender Trouble: Feminism and the Subversion of Identity. Routledge, Abingdon (1990)

Camera Di Commercio Vicenza. Conoscere Vicenza 2023: guida statistica della provincia di Vicenza (2024). https://www.vi.camcom.it/it/servizi/studi-statistica-e-ricerca-economica/conoscere-vicenza.html

Chase S.E.: Narrative Inquiry: towards theoretical and Methodological maturity. In; Lincoln, Y. S., Denzin, N.K. (eds.) The Sage Handbook of Qualitative Research, 5. edn. Sage (2018)

Clandinin, D.J., Connelly, F.M.: Narrative Inquiry: Experience and Story in Qualitative Research. John Wiley & Sons, Hoboken (2004)

Colella, F., Di Lorenzo, S.: Diversità, inclusione e innovazione sociale nei sistemi organizzativi complessi: Una riflessione critica sulla Gender Diversity intersezionale in Italia. MONDI MIGRANTI **2**, 21–37 (2023)

Collins, P.H.: Black Feminist Thought: Knowledge, Consciousness, and the Politics of Empowerment. Routledge, Abingdon (1990)

Crenshaw, K.: Demarginalizing the intersection of race and sex: a black feminist critique of antidiscrimination doctrine, feminist theory and antiracist politics. Univ. Chic. Leg. Forum **1989**(1), 139–167 (1989)

Crenshaw, K.: Mapping the margins: intersectionality, identity politics, and violence against women of color. Stanford Law Rev. **43**(6), 1241–1299 (1991)

Davis, A.Y.: Women, Race, & Class. Vintage Books, New York (1981)

De Simone, S., Agus, M., Lasio, D., Serri, F.: Development and validation of a measure of work-family interface. Revista de Psicología del Trabajo y de las Organizaciones **34**(3), 169–179 (2018)

De Vita, L., Magliocco, A.: Effects of gender quotas in Italy: a first impact assessment in the Italian banking sector. Int. J. Sociol. Soc. Policy **38**(7–8), 673–694 (2018). https://doi.org/10.1108/IJSSP-11-2017-0150

Di Fabio, A. (2016). Diversity Management Questionnaire: Primo contributo alla versione italiana. *COUNSELING*, 9, 0–0

Esin, C., Fathi, M., Squire, C.: Narrative analysis: the constructionist approach. In: Flick, U. (ed.) Handbook of Qualitative Data Analysis, pp. 203–216. Sage, Thousands Oaks (2014)

Ferrara, M.D.: Lavoro femminile e diversity management tra mediazione sindacale e nuove opportunità del lavoro agile. POLITICHE DEL LAVORO103–128 (2019)

Filippi, S., Peters, K., Suitner, C.: Power to the people: a social identity perspective on organizational decentralization and employee well-being. J. Commun. Appl. Social Psychol. (2023)

Flick, U.: The SAGE Handbook of Qualitative Data Analysis. SAGE Publications Limited, Thousands Oaks (2013)

Gay, V., Santacreu-Vasut, E., Shoham, A.: The Grammatical Origins of Gender Roles, Berkeley Economic History Laboratory. hal-04046900 (2013)

Gender Equality Index 2020: Italy. European Institute for Gender Equality (2020). https://eige.europa.eu/publications-resources/publications/gender-equality-index-2020-italy#eige-files

Goodson, I.F., Gill, S.R.: The Narrative turn in social research. Counterpoints **386**, 17–33 (2011). http://www.jstor.org/stable/42981362

Guarascio C., Musumeci R., Villa P.: Il Contesto Macro: Cambiamenti Legislativi e Disparità di Genere. In: Naldini, M., Poggio, B. (eds.) Genere e accademia. Carriere, culture e politiche, pp. 35–62. Il mulino (2023)

Harel, I., Papert, S. (eds.): Constructionism. Ablex (1991)

Harris, C., Rousseau, G.G., Venter, D.L.J.: Employee perceptions of diversity management at a tertiary institution. South Afr. J. Econ. Manag. Sci. **10**(1), 51–71 (2007)

Hartley, J., McWilliam, K.: Computational power meets human contact. In: Story Circle: Digital Storytelling Around the World, pp. 1–15 (2009)

Herdam, A.O., McMillan-Capehart, A.: Establishing a diversity program is not enough: exploring the determinants of diversity climate. J. Bus. Psychol. **25**(1), 39–53 (2010)

Hooks, B. (1984). *Feminist Theory: From Margin to Center*. South End Press

Hooks, B.: Feminism is for Everybody: Passionate Politics. South End Press (2000)

Istat. The life of women and men in Europe (2020).https://www.istat.it/donne-uomini/index.html?lang=it

Istat. Rapporto Bes 2022: il benessere equo e sostenibile in Italia (2023). Accesse 3 Dec 2023. https://www.istat.it/it/archivio/282920

Jonsen, K., Maznevski, M.L., Schneider, S.C.: Special review article: diversity and its not so diverse literature: an international perspective. Int. J. Cross Cult. Manag. **11**(1), 35–62 (2011). https://doi.org/10.1177/1470595811398798

Lambert, J.: Digital Storytelling: Capturing Lives, Creating Community. Routledge (2013)

Manageritalia. Donne manager: prosegue la marcia verso la parità. Manageritalia (2023).https://www.manageritalia.it/societa/donne-manager-prosegue-la-marcia-verso-la-parita/

Manello, A., Falavigna, G., Isaia, E., Rossi, M.C.: Female top managers and credit risk: evidence from Italian firms. Corp. Gover. Int. J. Bus. Soc. **23**(6), 1379–1403 (2023). https://doi.org/10.1108/CG-03-2022-0092

Miller, G., Holstein, J.: On the sociology of social problems. In: Miller, G., Holstein, J. (a cura di), Perspectives on Social Problems, pp. 1–16. JAI Press (1989)

Mintzberg, H.: Organization design: fashion or fit?. Harvard Bisiness Review, Innuary-February (1981). https://hbr.org/1981/01/organization-design-fashion-or-fit

Mittiga, S.: Il valore educativo del digital storytelling. Media Educ. **9**(2), 308–328 (2018)

Mohanty, C.T.: Under western eyes: feminist scholarship and colonial discourses. Fem. Rev. **30**, 61–88 (1984)

Murayama, K., Elliot, A.J.: The competition–performance relation: a meta-analytic review and test of the opposing processes model of competition and performance. Psychol. Bull. **138**(6), 1035 (2012)

Nichols, L.T.: Voices of social problems: a dialogical constructionist model. In: Studies in Symbolic Interaction, pp. 93–123. Emerald Group Publishing Limited (2003)

Olsen, J.E., Martins, L.L.: Understanding organizational diversity management programs: a theoretical framework and directions for future research. J. Organ. Behav. **33**(8), 1168–1187 (2012)

Palumbo, R., Manna, R.: A portrait of workplace discrimination in Italy: empirical evidence from a nationwide survey. Evid.-Based HRM Glob. Forum Empir. Scholarship **8**(1), 92–112 (2019). https://doi.org/10.1108/EBHRM-03-2019-0029

Papert, S.: Mindstorms: Children, Computers, and Powerful Ideas. Basic Books. Trad. It. Mindstorms. Bambini computer e creatività. Emme, 1984 (1980)

Papert, S.: Constructionism: a new opportunity for elementary science education. NSF Grant Application (1986)

Parks, J.B., Roberton, M.A.: Development and validation of an instrument to measure attitudes toward sexist/nonsexist language. Sex Roles **42**, 415–438 (2000)

Poggio, B., Murgia, A., De Bon, M.: Interventi organizzativi e politiche di genere. Carocci (2010)

Ravazzani, S.: Understanding approaches to managing diversity in the workplace: an empirical investigation in Italy. Equal. Diver. Inclus. Int. J. **35**(2), 154–168 (2016). https://doi.org/10.1108/EDI-08-2014-0062

Ravazzani, S., Mazzei, A., Fisichella, C., Butera, A.: Diversity and inclusion management: an analysis of practice developments in Italy. Sinergie Italian J. Manag. **39**(3), 213–229 (2021)

Razetti F., Santoni V.: Il mercato del welfare aziendale: l'intermediazione e il ruolo dei provider. In: Maino, F., Ferrera, M. (a cura di) (2019), Nuove Alleanze per un welfare che cambia. Quarto Rapporto sul secondo welfare in Italia. Giappichelli, pp. 119–152 (2019)

Riessman, C.K.: Analysis of personal narratives. In: Gubrium, J.F., Holstein, J.A. (eds.) Handbook of Interview Research, pp. 695–710. Sage, Thousands Oaks (2002)

Santoni, C., Crespi, I.: Conciliazione famiglia e lavoro tra smart-working e diversity management. Una riflessione su pratiche e nuove semantiche. Autonomie locali e servizi sociali **1**, 45–66 (2022). https://doi.org/10.1447/104028

Scarinci, A.: Digital storytelling: un'applicazione didattica per ripensare ai media attraverso i media (2011)

Simsek, B.: Enhancing women's participation in Turkey through digital storytelling. Cult. Sci. **5**(2), 28–44 (2012)

Simsek, B.: Neoliberal burdens for women in academy: "How Dare We Not Work." Kadın/Women **17**(1), 51–72 (2016)

Snowden, D.: Naturalizing sensemaking. In: Mosier, K.L., Fischer, U.M. (eds.). Informed by Knowledge: Expert Performance in Complex Situations, pp. 223–234. Psychology Press (2010)

Spector, M., Kitsuse, J.I.: Constructing Social Problems. Routledge, Abingdon (2017)

Tamboukou, M.: A Foucauldian approach to narratives. In: Andrews, M., et al. (eds.) Doing Narrative Research, pp. 102–120. Sage, Thousands Oaks (2008)

Trinchero, R.: Manuale di ricerca educativa. Franco Angeli (2002)

Varisco, B.M.: Costruttivismo socio culturale. Carocci, Roma (2002)

Wells, K.: Narrative inquiry. In: Pocket Guides to Social Work Research Methods. Oxford University Press, Cambridge (2011)

Organizational Capability Building in Transformation with University-Company Collaboration

Muhammad Shaheer and Jyri Vilko(✉)

LUT University, Kouvola Unit, Kauppalankatu 13, 45100 Kouvola, Finland
jyri.vilko@lut.fi

Abstract. Transformation in the modern developed societies challenges the companies in them, where especially the smaller companies acting in traditional and rigid fields are experiencing difficulties. There are companies that do not have the experience of drastic changes or have rather aimed their efforts in resisting or shielding themselves against change are now back against the wall in face of the megatrends of digital and green transformation. In order to cope with the change, organizations have to expand and heterogenize their value production, which requires developing their capabilities but also developing a change-positive organizational culture. The capability development can be done by educating the current personnel or hiring new skilled professionals. In a highly competitive labour market, where most of the companies are looking for new people, the recruitment is challenging, especially when field- and company-specific tailored skillset are sought after. University-company collaboration with tailored learning and internship can provide new focused capabilities for the organizations, but the culture for such collaboration needs development and planning. This paper studies the opportunities and challenges in a university-collaboration in a tailored internship-course development for organizations challenged by changes in the field. The study focuses on a case of an international master's programme on a regional context.

Keywords: Organizational Capability Building · Collaboration · Digital Transformation · Green Transformation · Tailored Learning Programmes · Professional Association · Generational Differences · Sustainable Development Goals · Industry Partnerships · Practical Learning · Curriculum Development · Internship Programmes

1 Introduction

In the contemporary business environment, evolving digital and green initiatives are rapidly disrupting industries and these megatrends pose significant challenges to small companies that are associated with traditional and rigid sectors (Linthorst and de Waal 2020). Firms that have in the past either avoided change or did not know how to innovate and manage drastic shifts are now experiencing the necessity of rapid change to survive

M. J. Sousa (Ed.): UFHRD 2024, ISEM 52, pp. 268–281, 2025.
https://doi.org/10.1007/978-3-031-92548-1_17

in the markets. To manage these transformations, companies need to expand and diversify value generating processes, which means that they not only have to develop new capabilities but also provide a culture that supports change (Stouten et al. 2018).

Organizational capability dictates an organization to integrate and adapt competencies and skills to deliver organization's goals and meet emerging needs (Andrews et al. 2016). This capability can be developed by training and developing existing staff or by employing competent staff with proper skills and knowledge. However, recruitment can be very difficult in a competitive labour market especially when candidates with specialized skills are required in certain fields and organizations (Reman 2022). Tailored learning facilitated through academia-company collaboration can become a feasible solution for developing these specific learning needs (Turunen 2024).

University-company partnerships have a rich background and theoretical rationale and serve as a bridge between the academic research and the practical industry applications. These partnerships not only generate value for universities and companies but also provide invaluable real-world experience to the students (Fraser and Mancl 2021). The inclusion of professional associations as intermediaries strengthens these partnerships through curriculum mapping that aligns academic programmes with industrial requirements and facilitate transfer of knowledge. Through conventions, workshops, and internships, professional associations can create opportunities for the students connect directly with professionals and apply their academic learning in practical settings (Research 2005).

This paper seeks to identify the opportunities and challenges of collaboration involving academia, industry, and professional associations in building organizational capabilities. To achieve that aim, we will investigate "How can firms, academic institutions, and professional associations effectively collaborate to develop tailored training programmes that improve organizational capacities, cater to industry-specific requirements, and benefit all stakeholders involved?". It focuses on a case study involving an International Master's programme, exploring how tailored curriculum involving courses of case studies and projects-based curriculum can help organisations facing specific industry transformations and how it can improve students' skillset. Therefore, the study will offer insights by providing some recommendations suitable for future cooperation between involved stakeholders.

The literature review examines prior findings about challenges and benefits of collaborations between academia, industries, and professional organizations and the impact of generational gaps on these collaborations. Subsequently, the research design section describes the steps involved in the research process, methodology and strategies involving stakeholders' engagement. The empirical case and data collection section outlines the case study and qualitative data collection techniques. The analysis and results section then analyses the data collected to identify key findings. Lastly, the discussion and conclusion section sum up the conclusion of the study to highlight the theoretical and practical implications of the identified findings and formulates recommendations for future research and collaborations.

2 Literature Review

2.1 Organizing the Collaboration Between Academia, Industry and Professional Associations

Collaborations between universities, organizations and companies have an important place in society as facilitators of technological spillovers and as a source of addressing innovation market failures and these interactions range from patenting and spin-offs to joint research, consultancy, and training (Kauppila et al. 2015). Developing large-scale research infrastructures requires robust partnerships, where universities should ensure strategic alignment and provide leadership in relation to both organizations and smaller companies (Kohn Rådberg and Löfsten 2023). The collaborations between the academic and the professional associations are essential in the research and practical applications in entrepreneurship (Song et al. 2019). Such partnerships facilitate the sharing of knowledge and innovation, not only enhancing research quality but also translating the academic research directly into economic gains in the fastest way possible to the benefit of both academia and industry. (Song et al. 2019). In such collaborations the research findings must be effectively collected, documented, and disseminated to make sure effective knowledge sharing and use of knowledge for learning and innovation across the sectors (Kettunen et al. 2022). Collaboration is an essential component in sustainable development research, and highly contributes to the achievement of the United Nations Sustainable Development Goals (SDGs) (Payumo et al. 2021).

2.2 Challenges in Building the Collaboration

Academia-industry collaboration channels face certain challenges in their applicability and effectives and one of the reasons can be that academia lacks entrepreneurship aspects such as absence of technology transfer offices and career centres which encourage innovation and partnerships (Mkhitaryan 2018). Another constraint in the establishment of co-operative relationship between small enterprises and universities is that small enterprises have comparatively little awareness about the university programmes that are available and how to access them, thus need communication channels that can help improve the sharing of opportunities for co-operation (Pereira and Franco 2022). Also, there are a variety of interaction channels in university-industry collaborations with traditional and formal two-way channels being dominant over those involving commercial transactions which were perceived to be less significant by both academics and industry professionals (Nsanzumuhire and Groot 2020).

One key challenge that might occur while developing collaboration between professional associations and academia can be a risk of biasness in scientific communication and so can cause impartiality of educational content leading to concerns about the influence of commercial interests (Komajda et al. 2012). While developing collaboration between academia and professional associations communication gap can occur which can then cause difficulties in task distributions with in large and diverse teams, so efficient communication is very important to prevent misunderstandings and ensure cohesive progress (Leogrande and Nicassio 2021). Professional organizations can face difficulties when collaborating with academia and companies' terms of potential misalignment of objectives and expectations of all stakeholders involved which can lead to misaligned goals (Arinaitwe 2021. Students mentioned that during project-based learning facilitated by organizations, guidance from professionals should be more consistent and structured otherwise it poses difficulties (Ashruf et al. 2021).

2.3 Opportunities and Benefits of Collaboration for Organizational Capability

One of the collaborative barriers between universities' research and development and organizations can be reduced by beginning with smaller initiatives that can be expanded and greater complexity in the future (Rossoni et al. 2023). There are seven key factors that enable successful collaborative innovation between universities and industries: resources, organizational structure, boundary spanning roles, intergroup relations, cultural diversity, status differentiation, and environmental factors (Sjöö and Hellström 2019). Professional associations can play vital role in promoting and maintaining ethical standards and developing industry collaborations, which ensures scientific research, educational programmes to be solely geared towards promoting knowledge and professional development (Komajda et al. 2012). Structured collaboration between associations and academia in scientific research requires proper distributed responsibilities, efficient communication systems, and teamwork to produce informed progress (Leogrande and Nicassio 2021). Professional societies increase cross-disciplinary interactions through initiatives such as publishing journals, organizing workshops, and symposiums, thus addressing complex problems, and enhancing innovation across disciplines (Research 2005). These bodies also provide awards and recognitions which also encourages these interdisciplinary efforts (Research 2005). The benefits from synergy focus on the efficiency in leveraging joint resources and knowledge in coordinating and executing large scale research projects (Leogrande and Nicassio 2021). At the institutional level, a growth trend has been observed in context of collaborative efforts over time across different fields which suggests that that core institutions have the ability to sustain research collaborations in the long run (Payumo et al. 2021).

Professional associations can help connecting the academia and firms by organizing conventions, events, conferences, and workshops to share experiences and develop networks, creating opportunities for project courses and internships for students, and setting up collaborations between academics and companies (Arinaitwe 2021). They can promote cooperation between the academic community and industry through implementing project-based educational activities that involve students solving business tasks in cooperation with employers under the supervision of industry professionals (Ashruf et al. 2021).They can perform middleman roles and can facilitate matching of educational programmes with industry requirements to the extent that both parties gain and maintain long-term partnership (Arinaitwe 2021). Students can get practical experience through this structured approach, such as "Professional Practice" programme; firms can obtain innovative solutions; and universities can also ensure that they incorporate the right programmes to meet the industry needs (Ashruf et al. 2021). Academia-Industry Collaboration Plan is a design model that aligns academic and industrial needs to accomplish shared objectives while facilitating internships and other cooperative projects (Ahmed et al. 2022).

2.4 Collaborations and Generational Differences

Understanding the viewpoint and expectations of gen z is very important while onboarding them in any collaborative setting. And indirectly they are the major stakeholders in university collaborations as young researchers and students (Chillakuri 2020). Understanding the differences in motivation between Gen Y and Gen Z is valuable as people from these generations have varying traits due to generational difference and greatly influence their behaviors, work values and attitudes. Thus, in all such collaboration students' preferences and requirements should be considered and this information can be taken from academic research and data from industries where gen z is mostly employed (Kuzhabekova and Temerbayeva 2018). One of the barriers in these collaborations can be generational differences and these influence the dynamics of research and development environments in work values and motivations and thus necessitates the tailored approaches to engage new generation and integrate them with older generation (Güngörl 2019). To better understand it, "kaleidoscope career model" provides a lens through which impact of generational differences on individuals' career trajectories can be explored. This model can help understand that how varying perspectives on authenticity, balance and challenge can influence collaborative dynamics between universities, organizations, and companies (Bulgur and Esen 2022).

3 Research Design

3.1 Research Process

In this research context, a mixed method of reflexive thematic analysis and coding reliability thematic analysis was used as we took an inductive approach to produce codes and themes (Walsh et al. 2019). Since major source of data was primary so Teams platform was chosen to conduct the interviews. Questionnaires were made beforehand and communicated with interviewees before the interviews for the appropriate preparation. All interviews were recorded and transcript, so the nature of data was unstructured. Average duration of each interview was 35–40 min. In the following picture we have given visual representation of our research process (Fig. 1).

3.2 Methodology and Data Collection

In the methodology section we discussed the research paradigm which we used for this study. The suggested research onion model created by Saunders provides a clear illustration of the various processes required in our research process. Research technique will follow a methodical, logical, step-by-step strategy that proceeds from the outer to the inner layer (Saunders et al. 2019).

In our case Inductive technique was most appropriate, since it is qualitative research conducted under the Interpretivism research philosophy (Thomas 2006). For the second layer, this research utilized the Inductive technique and for the third layer 'Mono method Qualitative' approach was utilized. This approach involved solely relying on interviews conducted with different stakeholders as the primary source of data for analysis. It is important to highlight that in the third layer, the interview questionnaire was

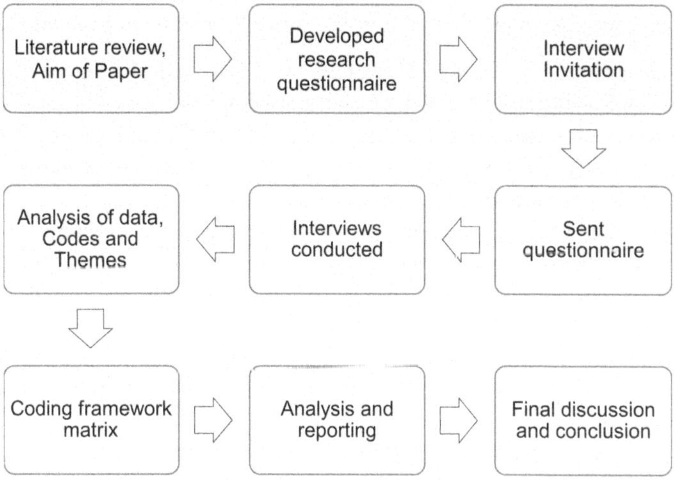

Fig. 1. Research Process

created by the authors of this study using the empirical data collected during the literature review. Subsequently, the interviews were done independently, and the modified interview transcript served as the major data source for this research.

The fourth layer discussed the employed research approach, which is a 'Case Study'. This strategy was chosen due to the tiny sample size, consisting of a single case. Utilizing a single case allowed for a comprehensive and detailed investigation of a specific event, as opposed to making broader generalizations. The fifth layer discussed that the duration of time, and the considerations are cross-sectional, meaning that the interview information is provided as of the date of the interview. In the final layer, the topic of data analysis was addressed, specifically focusing on the application of Thematic Analysis for this qualitative investigation.

3.3 Empirical Case Selection

In this study, we used information-oriented selection strategy that allowed to obtain data from a limited number of carefully selected cases (Flyvbjerg 2006). This approach was applied to give qualitative and valuable insights to analyze the nature of university-company partnership in capability development. Based on the objectives of the current study, we found critical case approach to be the most appropriate (Flyvbjerg 2006). The selected stakeholders were diverse and covered individuals from various sectors in academia, industry, and other professional organizations. They presented different perspectives of the cooperative processes involved in organizational capability development. Our goals were to achieve comprehensive insights that allowed making logical conclusions about the generalizability of the findings in related settings. In doing so, this approach was well-suited to addressing our objective of tailored educational curriculum to better address industry-specific needs and support long-term partnerships.

For this research, we had selected four key stakeholders from Finland (Table 1):

- Master's Programme at Finland University X: This master's degree offer specialization in "Innovation and Logistics" and is a center point of our case study. This programme is being offered at Leading University in Finland which is the world´s top 15th university for sustainable consumption and production and is among the world's top 11 small universities.
- Company A: A worldwide company specialized in offering technologies, automation, and services for the pulp, paper, and energy sectors. Its presence brought a viewpoint of a major industrial corporations that are involved in partnerships with educational institutions.
- Company B: An organization of regional development and innovation. Its participation gave geographic perspective and elucidated the position of the middleman organizations in the university-industry cooperation.
- Association C: The Finnish Association of Purchasing and Logistics, which is actively involved in bridging academia and industry through conventions; workshops; and networking. It has over 4000 individual members and over 300 corporate members. Its angle of view was valuable for revealing the overall professional and organizational contexts in which these partnerships occur.

Table 1. Informants

Informants	Organizational role	Collaboration network role	Experience
Informant 1	Post Doctoral Researcher	Course Lecturer	10 years
Informant 2	Communication and Marketing Manager	Main Contact / Coordinator in the Organization	18 years
Informant 3	Executive Director	Main contact person in Association	25 years
Informant 4	Project Management, Funding Applications, University Collaboration, Teaching & Development	Programme Coordinator	3 Years
Informant 5	Sourcing Engineer	Main contact member in company and board member in association	7 years

4 Analysis and Results

After collecting primary qualitative data from respondents, we derived themes and codes as shown below (Table 2):

Table 2. Themes and codes applied

No	Themes	Codes
1	Nature of current collaborations	1. Limited Effectiveness in terms of current course 2. Current successes 3. Improvement areas 4. Lacking big success
2	Role of Professional Associations	1. Facilitation 2. Support Functions 3. Resources Limitation 4. Long Term Engagement
3	Student Engagement and Benefits	1. Practical Experience 2. Skill Development 3. Gen Z preferences
4	Challenges in Collaboration	1. Communication Gaps 2. Resources Constraints 3. Misalignment of Goals 4. Consistency in guidance
5	Opportunities for Improvement	1. Tailored Programmes 2. Long term sustainability 3. Curriculum Integration
6	Contribution to SDGs	1. Alignment with SDGs 2. Support from Collaboration 3. Strategic Importance
7	Addressing Generational Gaps	1. Diverse work preferences 2. Integration Strategies 3. Organizational Adaptation

4.1 Nature of Current Collaborations

Current university-industry partnerships and particularly under studied International Master's programme have demonstrated a moderate success. Respondents also stated that these collaborations have offered practical learning opportunities for their students and valuable industry connections. Focusing on the case study and project courses one participant pointed out that these courses help in solving real world challenges of industry. Despite this, there are many possibilities for expansion and the improvement of future cooperative initiatives, which are already evidenced by numerous ongoing cooperative projects, including those funded by the EU. Given these first experiences, further development and enhancement of such partnerships offer great promise for increasing organizational capabilities and for meeting sector-specific needs more adequately.

4.2 Role of Professional Associations

Professional associations can play a very important role in form of bridging the gap between academia and industry, however there are certain challenges they face. Some of them we found are the resource limitations in terms of staff, time, and money and prioritization issues. Another issue is the extensive list of activities in their agendas which hinder the effectiveness of these associations for such kinds of collaborations. These associations can play their role in networking, serve support functions, and ensure that the academic programmes offered match the job market needs. One of the respondents had the vision about the professional associations to become the service or platform models, giving the continuous support to the academic – industry collaborations. This highlights the important but limited role that these professional associations have in facilitating sustainable cooperation that can strengthen organizational capacity and address issues unique to different sectors.

4.3 Student Engagement and Benefits

Strategic collaborations between Universities and Industries provide numerous advantages aimed at the enhancement of students' effective learning within academic and professional context to meet evolving needs of modern industries. Many of related initiatives, especially the case studies and project courses, are beneficial to the students as they offer a chance to solve industry specific challenges but there is a further need of improvement. During the interview, some of the participants addressed the needs of the students of Generation Z, whose work attitudes and requirements are different and should be incorporated into the programmes. Thus, involving students into collaborative projects and asking their opinions, organizations can better align their goals and initiatives with the preferences of younger students. This engagement is critical in availing students for the future roles and ensuring that they have gained adequate skills and experience, which will incrementally develop the organizational capabilities especially in the light of digital and green transformations.

4.4 Challenges in Collaboration

This paper identifies some key challenges that affect the university-industry collaboration and hampers their capacity to improve organizational capabilities. These challenges include lack of effective communication, as it results in misunderstandings and misaligned expectations and resource constraints to support collaborative initiatives. The problem of misalignment of goals and objectives of academic institutions and industries is another major challenge, which results in ineffective partnerships. Survey respondents pointed out that students enrolled in project-based learning require more consistent and structured guidance from industry professionals and lack of this approach causes a time loss of all stakeholders which ultimately affects these partnerships. To sum up, these challenges should be solved with the help of better communication, optimal distribution of resources, and the desired objectives alignment for creating the effective and lasting partnerships that will enable organizational capabilities building.

4.5 Opportunities for Improvement

Opportunities for enhancing university-association-industry collaborations lie in strategies such as customizing the programmes, achieving appropriate longevity of their sustainability, and incorporating these collaboration efforts into curricula. All stakeholders should be adaptable to achieve long term sustainability in their partnerships. Programmes designed to meet specific industry requirements could enhance the effectiveness of such collaborative efforts. Through a process of curriculum mapping and by developing specific interdisciplinary internship courses that incorporate the partnerships with the professional associations, benefits for the students as well as the industry partners can be optimized. These opportunities demonstrate some of the possible ways of enhancing the development and sustainability of the university-industry relations, which would improve the organizational competencies to address the requirements of the technology and environmental shifts.

4.6 Contribution to SDGs

The respondents stressed that integrating business strategies with SDGs is one of the most important strategic priorities of organizations. An industry university partnership can contribute to the achievement of specific SDGs through specific intervention programmes or projects. One respondent mentioned that currently through this master's programme they are achieving Goal 4, Goal 7, and Goal 9 namely as Quality Education, Affordable and Clean Energy, and Industry, Innovation and Infrastructure. All interviewees had a unanimous view that sustainable development goals should be considered as central to the goals of cooperation, increasing its significance and applicability for the participants. This alignment also benefits global sustainability initiatives and enhances organizational capacities to address the dynamics of digital and green transitions.

4.7 Addressing Generational Gaps

One respondent stressed that addressing the expectations of different generations and diverse work preferences is essential for the success of such collaborative projects. The differences in work values of Generation Y and Generation Z require certain strategies to involve students in the development process and gather their feedback. Companies need to address new trends in the work environment, develop diverse and motivating work environments for the younger generations. Flexible workplaces that provide options for environments beyond the traditional corporate office can appeal to and foster Generation Z employees. According to one other view, some of the organizations are now getting adaptable and already started accommodating these differences. These strategies are important because they help to create a solid and effective collaborative atmosphere among the representatives of different generations and meet organizational capabilities needed in accordance with emerging requirements required for digital and green evolution.

5 Discussion and Conclusions

This study explored the opportunities and challenges in university-industry collaborations with professional associations as a major stakeholder between them in the context of developing organizational capability to address digital and green transformations. The findings of the research show the importance of these partnerships in developing practical and up-to-date skills among students, aligning curriculum development with industrial needs, and addressing generational differences in the workplace. However, there are numerous challenges that reduce the effectiveness of these collaborative measures which includes communication gaps, resource constraints, and misalignment of goals. Addressing these challenges through strategic communication, resource allocation, and alignment of goals can enhance the effectiveness of these partnerships, benefiting all stakeholders involved.

5.1 Scientific Implications

This research contributes to the extant literature by showing that academic programmes can be made more relevant to the requirements of specific industries: by modifying curriculum and introducing practical value-added internships. This synergy is crucial for promoting creativity and creating the specific competencies that are needed for the industry today (Song et al. 2019; Kauppila et al. 2015). Based on the emphasis of the practical usage of academic findings, this research aligns with the concept of the significant contribution of universities to the formation of workforce competencies (Turunen 2024).

This research is useful in providing insights on the prospects and difficulties of cooperation between universities and industries. It added some contribution to solve previously raised challenges including the lack of effective communication, inadequate resources, and mismatched goals (Nsanzumuhire and Groot 2020; Pereira and Franco 2022). The study also presents recommendations that can be implemented to enhance the outcomes of future collaborations between education and business through strategies aimed at enhancing the alignment of the educational expectations with industry requirements and expectations. This aligns with the previous studies on the need for well-structured coordination in order to facilitate the sharing of knowledge and innovation (Leogrande and Nicassio 2021; Komajda et al. 2012). To the current body of knowledge, the study brings additional perspective that reveals how generational differences affect collaboration. Evaluating motivational dissimilarities between Generation Y and Generation Z is essential for engaging young students in collaborative settings (Alp et al. 2020; Chillakuri 2020). The results and suggestions by this study can be used as a starting point for future research that aims to build on these partnership approaches.

5.2 Managerial Implications

From a managerial point of view, the study offers insights for both academic institutions and the industrial partners interested in collaboration. To improve effectiveness of collaborations, robust communication channels and alignment of goals between all stakeholders should be prioritized. This can be achieved by feedback mechanism and frequent communication to align educational outcomes with industry expectations.

In this ecosystem, professional associations are critically important. They can offer connections, materials, and services that help reduce the divide between academia and industrial practice. Managers should capitalize on these associations in order to ensure that academic programmes remain current and useful in the ever-evolving market place. Additionally, the study highlights the importance of considering intergenerational relations in the context of partnerships. The managers should ensure that he or she formulates policies and plans of working that accommodate the members of the Generation Z, making them to work under conditions that suit them. Not only it will help in attracting and retaining young talents, but it will also promote culture of innovation and learning in organizations.

5.3 Limitations and Suggestion for Further Research

There are some limitations that need to be acknowledged in this study. First, the study was confined to a few companies, one association and was based on a specific master's programme. Thus, the results cannot be necessarily generalized to all other settings or educational fields. Future studies should preferably involve a larger number of participants across diverse geographical locations and incorporate several academic courses to improve the study's external validity. Secondly, the study had qualitative data from the interviews conducted on a few stakeholders only. Although this approach brought about richness to the study, further research could employ a comparative methodology that incorporates quantitative data to complement the qualitative results.

Lastly, the research uncovered various challenges and opportunities in University-Industry partnership but did not investigate the in-depth approaches to tackle these challenges. Rather the findings were limited to the data obtained from interviews. Future research should focus detailed practical strategies for bridging communication breakdowns, limited availability of resources, and lack of purpose congruity. It should also explore the long-term impact of these collaborations on organizational capabilities and their contribution to sustainable development goals.

References

Ahmed, F., Tahir, M., Ali, S., Enam, R.: Strengthening the bridge between academic and the industry through the academia-industry collaboration plan design model. Front. Psychol. **13**, 875940 (2022). https://doi.org/10.3389/fpsyg.2022.875940

Alp, G., Karacaoğlu, K., Gungor, A.: Motivational differences between generations Y and Z in working life. In: Chapter-11, pp. 135–146 (2020)

Andrews, R., Beynon, M.J., McDermott, A.M.: Organizational capability in the public sector: a configurational approach. J. Public Adm. Res. Theory **26**(2), 239–258 (2016). https://doi.org/10.1093/jopart/muv005

Arinaitwe, D.: Practices and strategies for enhancing learning through collaboration between vocational teacher training institutions and workplaces. Empir. Res. Vocat. Educ. Train. **13**(1), 13 (2021). https://doi.org/10.1186/s40461-021-00117-z

Ashruf, S., Plaats, D., Malinowska, A., Alers, H., Peng, X.: Professional practice: a case study of collaborations between education and industry. J. Educ. Pract. **12**, 84–95 (2021). https://doi.org/10.7176/JEP/12-6-10

Bulgur, N.E., Esen, E.: Generational differences in career perspective: applying kaleidoscope career model. Middle East J. Manag. **1**(1), 1 (2022). https://doi.org/10.1504/MEJM.2022.10047340

Chillakuri, B.: Understanding generation Z expectations for effective onboarding. J. Organ. Change Manag. **33** (2020). https://doi.org/10.1108/JOCM-02-2020-0058

Flyvbjerg, B.: Five misunderstandings about case-study research. Qual. Inq. **12**(2), 219–245 (2006). https://doi.org/10.1177/1077800405284363

Fraser, S., Mancl, D.: Exploring the dimensions of university-company collaborations: research, talent, and beyond. In: 2021 IEEE/ACM 8th International Workshop on Software Engineering Research and Industrial Practice (SER&IP), pp. 57–64 (2021). https://doi.org/10.1109/SER-IP52554.2021.00017

Kauppila, O., Mursula, A., Harkonen, J., Kujala, J.: Evaluating university-industry collaboration: the European foundation of quality management excellence model-based evaluation of university-industry collaboration. Tert. Educ. Manag. **21**(2), 229–244 (2015). https://doi.org/10.1080/13583883.2015.1045550

Kettunen, P., Järvinen, J., Mikkonen, T., Männistö, T.: Energizing collaborative industry-academia learning: a present case and future visions. Eur. J. Futures Res. **10**(1), 8 (2022). https://doi.org/10.1186/s40309-022-00196-5

Komajda, M., et al.: Relations between professional medical associations and the healthcare industry concerning scientific communication and continuing medical education: a policy statement from the European society of cardiology. Eur. Heart J. **33**, 666–674 (2012). https://doi.org/10.1093/eurheartj/ehr480

Kuzhabekova, A., Temerbayeva, A.: The role of conferences in doctoral student socialization. Stud. Grad. Postdoct. Educ. **9** (2018). https://doi.org/10.1108/SGPE-D-18-00012

Leogrande, E., Nicassio, R.: Collaborative processes in science and literature: an in-depth look at the cases of CERN and SIC. Front. Res. Metrics Anal. **5** (2021). https://doi.org/10.3389/frma.2020.592819

Linthorst, J., de Waal, A.: Megatrends and disruptors and their postulated impact on organizations. Sustainability **12**(20) (2020). https://doi.org/10.3390/su12208740

Mkhitaryan, D.: University-Industry Collaboration in Armenia, 2nd edn. Yerevan University Press, Yerevan (2018)

Nsanzumuhire, S.U., Groot, W.: Context perspective on university-industry collaboration processes: a systematic review of literature. J. Clean. Prod. **258**, 120861 (2020). https://doi.org/10.1016/j.jclepro.2020.120861

Payumo, J., He, G., Manjunatha, A.C., Higgins, D., Calvert, S.: mapping collaborations and partnerships in SDG research. Front. Res. Metrics Anal. **5** (2021). https://doi.org/10.3389/frma.2020.612442

Pereira, R., Franco, M.: Cooperation between universities and SMEs: a systematic literature review. Ind. High. Educ. **36**(1), 37–50 (2022). https://doi.org/10.1177/0950422221995114

Rådberg, K.K., Löfsten, H.: The entrepreneurial university and development of large-scale research infrastructure: exploring the emerging university function of collaboration and leadership. J. Technol. Transf. **49** (2023). https://doi.org/10.1007/s10961-023-10033-x

Reman, P.: Developments in HR: The Future of Recruitment, Current Trends, and Challenges (2022)

Research, C.: Facilitating Interdisciplinary Research (2005). https://doi.org/10.17226/11153

Rossoni, A.L., de Vasconcellos, E.P.G., de Castilho Rossoni, R.L.: Barriers and facilitators of university-industry collaboration for research, development, and innovation: a systematic review. Manag. Rev. Q. (2023). https://doi.org/10.1007/s11301-023-00349-1

Saunders, M., Lewis, P., Thornhill, A., Bristow, A.: Research Methods for Business Students. Chapter 4: Understanding Research Philosophy and Approaches to Theory Development. 2nd edn. Pearson, Harlow (2019)

Sjöö, K., Hellström, T.: University-industry collaboration: a literature review and synthesis. Ind. High. Educ. **33**(4), 275–285 (2019). https://doi.org/10.1177/0950422219829697

Song, R., Xu, H., Cai, L.: Academic collaboration in entrepreneurship research from 2009 to 2018: a multilevel collaboration network analysis. Sustainability **11**(19) (2019). https://doi.org/10.3390/su11195172

Stouten, J., Rousseau, D., Cremer, D.: Successful organizational change: integrating the management practice and scholarly literatures. Acad. Manag. Ann. **12** (2018). https://doi.org/10.5465/annals.2016.0095

Thomas, D.: A General inductive approach for analyzing qualitative evaluation data. Am. J. Eval. **27**(2), 237–246 (2006). https://doi.org/10.1177/1098214005283748

Turunen, J.: Exploring the Dynamics of University-Industry Collaboration (2024). https://doi.org/10.13140/RG.2.2.12327.32169

Utami, H., Alamanos, E.: Resource-based theory: a review. Water Act **2016**, 1–26 (2022)

Walsh, R.S., et al.: A thematic analysis investigating the impact of positive behavioral support training on the lives of service providers: 'it makes you think differently.' Front. Psychol. **10**, 2408 (2019). https://doi.org/10.3389/fpsyg.2019.02408

Gamifying Knowledge Management Systems in Small and Medium-Sized Enterprises: Development and Validation of a Data Collection Tool

Marta Sampaio[1,2](✉), Maria José Sousa[3], and Andreia Dionísio[1]

[1] Management Department, Universidade de Évora, Cefage, Évora, Portugal
martamotasampaio@gmail.com
[2] School of Management and Technology, Santarém Polytechnic University, Santarém, Portugal
[3] ISCTE Instituto Universitário de Lisboa, Lisbon, Portugal

Abstract. The efficient utilization of human capital within organizations to foster innovation and problem-solving is crucial for long-term organizational performance. Knowledge management practices serve as internal mechanisms to facilitate learning and employee engagement, thereby promoting the generation of new ideas, processes, and solutions. This study aims to develop a data collection tool that acknowledges the significance of gamification in knowledge management processes, particularly in addressing the challenges faced by small and medium-sized enterprises (SMEs) in motivating knowledge sharing among employees. Through the Delphi method, a panel of experts in gamification and knowledge management systems contributed to the development of the tool over three rounds.

The study considered two major dimensions: knowledge management system enablers, encompassing leadership, organizational culture and structure, and information technologies, and gamification elements that can enhance knowledge management systems, such as game dynamics and mechanics. The primary contribution of this study lies in the refinement of a Conceptual Model of Analysis of the Contribution of Gamification to Knowledge Management Systems in SMEs and the development and validation of a new tool to assess the role of gamification in fostering a knowledge-oriented culture and structure, promoting trust, collaboration, learning, and employee motivation and involvement. However, limitations include challenges in recruiting experts and ensuring their participation in each round, as well as the pioneering nature of the tool, which necessitates further testing in diverse gamified knowledge management system contexts.

Keywords: Gamification · Knowledge management · SME · Delphi method · Questionnaire development and validation

1 Introduction

In recent years, gamification has aroused greater interest among researchers, a fact demonstrated by the growing number of academic and non-academic articles (Hamari et al. 2014; Patricio et al. 2022).

M. J. Sousa (Ed.): UFHRD 2024, ISEM 52, pp. 282–320, 2025.
https://doi.org/10.1007/978-3-031-92548-1_18

The term Gamification refers to the use of game design elements in non-game contexts (Deterding et al. 2011). It is often used to create applications or systems in organizations with the aim of encouraging specific behaviors or changing behavioral patterns. Gamification can be found in various fields, including health, marketing, education, crowdsourcing and customer loyalty, and is also used to stimulate team motivation and productivity (Raftopoulos et al. 2015; Seaborn and Fels 2015).

In recent times, Micro, Small, and Medium Enterprises (MSMEs) have begun exploring gamification as a solution. This approach is increasingly recognized as an effective means for organizations to fully engage their workforce in business processes and tasks, as highlighted by Cardador et al. (2017).

With business competitiveness essentially based on knowledge (Kianto et al, 2016; Abubakar et al. 2019), organizations that achieve an intensive environment in creating and sharing knowledge gain a competitive advantage over their competitors (Hajir et al. 2015; Li et al. 2020).

Consequently, this study aims to develop a data collection tool on the potential importance of Gamification in Knowledge Management processes, namely in creating and maintaining an organizational environment conducive to learning, skills development and knowledge generation, relying on the analysis and contribution of a panel of experts in the field of Gamification and Knowledge.

Initially, a framework of the concepts of Knowledge Management and Gamification is presented, followed by an analysis model that can help identify at what levels Gamification can facilitate Knowledge Management dynamics. Subsequently, the data collection tool and the analysis process of this questionnaire carried out by a panel of experts are presented, using the Delphi method, where the group of experts makes a careful analysis of the data collection tool developed, attesting to its quality.

Providing organizations with plausible and grounded information about the tools at their disposal to raise knowledge sharing is the main relevance of this work, and the data collection instrument presented here is very relevant for future research in Gamification and Knowledge Management.

2 Theoretical Framework

According to Gloet and Terziovski (2004), knowledge management is the formalisation of experience and how to access that knowledge and proficiency, which produces new skills. Knowledge has been established as fundamental to organizations (Polanyi 1966; Nonaka and Takeuchi 1995; Davenport and Prusak 1998; Strauhs 2003). Jorge and Faléco (2016: p. 69) state: "one of the most valuable tools that organizations have today comes from the intellectual capital of each employee". Therefore, organizational strategies should increasingly focus on human capital, its qualifications, skills and knowledge, since effective management of this resource will create value and achieve superior performance (Serrano and Fialho 2005).

In an increasingly precarious global landscape, the adept management of information and knowledge emerges as a paramount undertaking for SMEs (Costa and Monteiro 2016; Bell and Cooper 2018; Durst et al. 2022). Such endeavors hold the promise of bolstering their readiness to navigate current and future adversities, encompassing both internal organizational hurdles and external environmental exigencies. (Durst et al. 2022).

Knowledge Management is relevant for all SMEs regardless of the category and thus can be successfully implemented even in very small companies.; the SME owners/managers to should consider the existing Knowledge Management practices and strategies, and they should guide their decisions regarding where and how to channel financial and non-financial resources to maximize the benefits of KM (Bell and Cooper 2018; Petrov et al. 2020; Durst et al. 2022).

The concept of Gamification refers to an emerging organizational practice, defined as using game mechanics and design to measure, influence and reward the behaviors of target users. Gamification uses the essence of game characteristics - goals, rules, fun elements, feedback, rewards and progression - to solve day-to-day business challenges. These game mechanics, in a non-game context, act as a trigger that makes technology more engaging, influencing the behavior of those who use it and the methods of social interaction in organizations (Deterding et al. 2011; Zichermann and Cunningham 2011; Maan, 2013, Rauch 2013; Kumar and Raghavendran 2015; Petelczyc et al. 2018).

In a business context, according to Rauch (2013), the use of Gamification increases employee and customer engagement and motivation in companies and can help organizations to respond to various business needs, such as innovation (Patricio et al. 2022), training (Attali and Arieli-Attali 2015) marketing and consumer behavior (Rodrigues et al. 2016) or employee engagement (Shpakova et al. 2017; Suh et al. 2018).

According to several authors, the use of Gamification can have a positive impact on Knowledge Management Processes (Mettler and Pinto 2015; Shpakova et al. 2017; Weretecki et al. 2021). There are indicators that the dynamics and mechanics applied while creating the serious game to be implemented in the organization can enhance knowledge sharing (Jorge and Sutton 2016; Suh and Wagner 2017; Hamari and Keronen 2017; Li 2018). For Weretecki et al (2021), gamification can be used to promote the exchange of information and thus enhance the co-creation of knowledge among employees without necessarily having to change attitudes first.

Based on a systematic literature review, Sampaio et al. (2019) developed a Conceptual Model for Analysis of the Contribution of Gamification to Knowledge Management Systems, as presented below (Fig. 1).

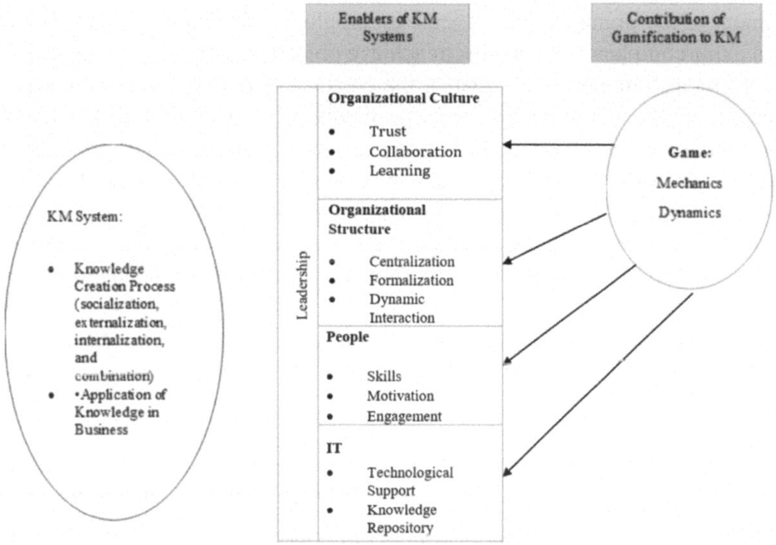

Fig. 1. Conceptual Model to Analyze the Contribution of Gamification to Knowledge Management Systems (Source: Sampaio et al. 2019)

According to this model, the facilitating elements for implementation of a Knowledge Management system are: (1) Leadership, as this is the driving force behind the system's implementation, development and operation; (2) the organizational culture, as it is in organizational environments with collaborative characteristics, including trust and a focus on organizational learning, that knowledge management systems are more successful; (3) the organizational structure, finding that it is in more formalized, centralized structures that knowledge circulates more effectively, and the organizational structure must allow for the dynamic interaction between employees from different areas or sectors of activity; (4) the organization's people, since for a successful knowledge management system, people must be skilled in specific technical areas and at the same time aware of the impact of their specific tasks, while being motivated and involved with the organization, their function and the knowledge management system itself, since this is the only way to share knowledge, disseminate it throughout the organization and enhance it with a view to new knowledge creation; (5) information technology, technological support being essential for knowledge management to take place, as it means easier creation of knowledge repositories in the organization that are available for consultation and use by the employees, and simultaneously, allow the maintenance of an organizational memory over time.

In the dimension of Gamification, this model argues that Gamification, as a management practice, can be an external facilitator for the implementation of knowledge management systems. From the dynamics and mechanics applied in creating the serious game to be implemented in the organization, collaboration, motivation, employee involvement and learning, knowledge sharing, increased competition going towards achievement of the organization's objectives, employee development, and establishing intra and extra-company partnerships to achieve objectives can all be enhanced.

According to Sampaio et al. (2019), the creation of performance indicators (KPI) associated with the Gamification process, involving cooperation and collaboration between employees and their contributions in terms of knowledge, may also be an effective way to collect knowledge that can be directed from the game directly to the knowledge repositories the organization wishes to create and maintain.

Taking as a starting point the Model for Analysis of the Contribution of Gamification to Knowledge Management Systems (Fig. 1), we collected information in order to characterize and contextualize the theoretical constructs of each dimension of the Model (Facilitators of Knowledge Management Systems and Contribution of Gamification to Knowledge Management) as well as their component parts.

2.1 Leadership

In the realm of small and medium enterprises (SMEs), leadership stands as a cornerstone in establishing and implementing effective knowledge management systems, echoing the sentiments of Ologbo and Nor (2015) and Stankosky & Baldanza (cited by Mohajan 2017). It falls upon organizational leaders to delineate and execute measures conducive to the successful integration of such systems, ensuring all requisite conditions are met for their efficacy. Three fundamental pillars of a knowledge management system surface: organizational culture and structure, personnel, and information technology infrastructure. As Sorin-George Toma et al. (2020) elucidate, leadership embodies a multifaceted and context-dependent concept without a universally accepted definition. Historical research predominantly focuses on attributes like charisma and transformational competence in leaders' interactions with employees since the early 1980s (Santos et al. 2008). Bruns (1978) introduced the distinction between transactional and transformational leadership, where the former involves an exchange of interests to achieve the leader's objectives, and the latter entails motivating and inspiring followers by aligning their values and goals with the leader's. Bass (1985) further elaborated on these leadership behaviors, forming the foundation of the Multifactor Leadership Questionnaire (MLQ). Gaviria-Rivera and López-Zapata (2019) have evidenced that transformational leadership significantly influences organizational climate, fostering teamwork and motivating employees, underscoring the significance of understanding and prioritizing transactional and transformational leadership styles in organizational leadership practices. Organizational Culture.

Culture can be defined as a single system within which values are shared and employee behavior is built accordingly (Jeng and Dunk 2013). For successful implementation of knowledge management, organizational culture is one of the most important foundations (Demarest 1997; Davenport and Prusak, 1998; Gold et al. 2015).

With regard to organizational culture, the existence of a collaborative environment where there is trust and a focus on organizational learning is essential for a successful knowledge management system (Choo 1996; Krogh 1998; Eppler and Sukoswski 2000; Nejatian et al. 2013; Ologbo and Nor 2015; Stankosky and Baldanza cited by Mohajan, 2017).

Collaboration

Collaboration is defined by Gupta and Govindarajan (2000) as the extent to which people in a group actively support and help each other in their work.

A collaborative culture is at the basis of effective knowledge management (Dor and Segev 1982; Hansen and Järvelin 2005). Collaborative interactions such as social interaction and joint activity can help create organizational knowledge (Hedlund 1994).

For Peng et al. (2012), people involved in game processes feel more involved with other players, especially when games provide social interaction. (Pruzbulski et al. 2010).

Pe-Than et al. (2014) demonstrate that people have a sense of belonging to the group when playing, since those involved who share common interests can develop a strong sense of connection. The same author goes on to say that the gaming context allows players to create and maintain social relationships that are highly enjoyable for those involved, making them willing to share more information than they would share with close friends.

Trust

Trust has a profound impact within the context of small and medium enterprises (SMEs). Trust, inherently tied to human beliefs, feelings, and intentionality, entails maintaining mutual faith in others regarding their intentions and behaviors (Hurley and Hult 1998; Aryee et al. 2002). Numerous studies have delved into the role of trust in knowledge sharing within SMEs (Iansiti 1993; Hansen and Järvelin 2005; Ozlati 2015). Politis (2013) asserts that trust within organizations underpins collaboration and knowledge sharing, fostering an environment where individuals are more inclined to exchange knowledge and engage in social interactions (Hedlund 1994). Employees, guided by trust, seek advice from trusted colleagues to enhance their problem-solving capabilities, demonstrating a willingness to share knowledge with select individuals deemed trustworthy (Olander et al. 2015). The institutionalization of trust among employees, as proposed by Iansiti (1993), serves as a cornerstone for facilitating knowledge transfer within organizations.

Furthermore, studies on gamification (Vallat et al. 2016; Lithoxoidou et al. 2019) underscore its impact on fostering collaboration and interaction within SMEs. Employees engaged in gamified processes perceive them as providing an attractive environment for collaboration and interaction, characterized by transparency and privacy considerations. Additionally, they acknowledge that gamification enhances the appeal of the work environment and strengthens the collaborative spirit among colleagues.

Learning

In the context of small and medium enterprises (SMEs), the process of learning within knowledge management involves individuals acquiring new knowledge that they are both capable of and willing to apply in decision-making or influencing others (von krogh, 1998). SMEs can promote active learning and the discovery of innovative problem-solving approaches by prioritizing learning and development initiatives, as highlighted by Kanevsky and Housel (1998). Encouraging individuals to challenge assumptions is fundamental for successful knowledge creation (Eppler and Sukowski 2000). Gamification, a method that stimulates motivation, engagement, and positive attitudes in activities, has emerged as a valuable tool for enhancing learning (Clark and Rossiter 2008; Gibson et al. 2013). Empirical research indicates that gamification not only enhances learning but also aids in the comprehension of complex subjects (Moccozet et al. 2013; Simões et al. 2013). factors such as the level of challenge, demonstration of skills, and interest generated by gamification are positively associated with perceived learning, which is a critical precursor to effective learning outcomes (Procci et al. 2012; Hamari and Koivisto 2015; Chen et al. 2015; Reese 2016). Moreover, knowledge sharing in smes involves individuals disseminating their expertise (know-what, know-how, and know-why) to other members within the organization (Van Den et al .2012).

Organizational Structure

The implementation of knowledge management systems within small and medium enterprises (SMEs) can be significantly influenced by the organizational structure (Hopper 1990; Caruana et al. 1998). According to Ichijo et al. (1998), maintaining coherence between the organizational structure and the practical utilization of knowledge is paramount for SMEs. It falls upon the organization to ensure that its structure is conducive to fostering a knowledge management-oriented environment. This entails having a formalized and centralized structure that facilitates the collection, ownership, and accessibility of knowledge, while also promoting dynamic interaction among members across different areas or sectors of activity (Nonaka and Takeuchi 1995; Krogh and Roos 1995; Stacey 1996; Wenger and Snyder 2000; Kakabadse et al. 2003; Ologbo and Nor 2015; Stankosky and Baldanza as cited by Mohajan 2017).

Centralization and Formalization

In small and medium organizations, centralization pertains to how power is distributed, with decision-making authority typically concentrated within a single entity as outlined by Andrews and Kacmar (2001) and Mintzberg (1989). Conversely, decentralization occurs when decision-making authority is spread among various elements of the organization. Overly centralized decision-making in small and medium organizations can inhibit creativity, while decentralized decision-making encourages spontaneity, expertise, and the expression of ideas. Scholars argue that a highly centralized organizational structure can hinder knowledge creation (Nonaka and Takeuchi 1995; DiBella et al. 1996; O'Dell and Grayson 1999), whereas less centralized structures promote interdepartmental communication and idea sharing (Nonaka and Takeuchi 1995; Hadad 2017; Bernstein et al. 2022).

Formalization, on the other hand, refers to the degree to which decisions and working relationships are governed by formal rules, procedures, and standardized policies (Hopper 1990; Zucker et al. 1996). Extensive formalization may impede the emergence and sharing of ideas necessary for knowledge creation in small and medium organizations, as it limits communication and interaction (Gold et al. 2015). Flexibility, in contrast, encourages interaction and communication among organizational members, thereby facilitating the flow of knowledge (Nelson and Cooprider 1996; Johanenssen et al. 1999).

Dynamic Interaction

Hidi (2001) suggests that in SMEs, dynamic interaction among individuals and various environmental elements cultivates a psychological state of interest, enhancing cognitive and affective functioning, leading to sustained motivation and engagement with specific topics. Suorsa and Huotari (2014) highlight the importance of interaction in SMEs' knowledge creation process, where individuals share experiences, collaboratively constructing new knowledge based on diverse perspectives and previous encounters. These interactions foster innovation by generating novel insights and understandings (Cook and Brown 1999), facilitated by safe, open, trusting, and self-aware processes. Suorsa and Huotari (2014) further demonstrate that contextual and relational factors, including diversity, familiarity, equality, and flexibility, significantly influence knowledge utilization for innovation, underscoring the importance of fostering environments conducive to knowledge exchange and collaboration.

2.2 People

People are at the center of organizational knowledge creation ((Scott 1998; Eppler and Sukowski 2000; Gottschalk 2000; Terra 2003)). For Scott (1998), an important indicator is that Knowledge Management depends 10% on technology and 90% on people.

Thus, the success of any knowledge management system is associated with the degree to which people are skilled in specific technical areas but are simultaneously aware of the potential impact of their specific tasks, while at the same time being motivated to contribute and engage with their work and the knowledge management system. Only then will it be possible to share knowledge, disseminate it throughout the organization and enhance it with a view to creating more, and more updated, knowledge that may be applied in day-to-day business. (Nonaka and Takeuchi 1995; Wenger and Synder 2000; Nejatian et al. 2013; Ologbo and Nor 2015).

Knowledge sharing depends on employees' voluntary decision to share their knowledge. Given this voluntary nature, knowledge sharing means people must be strongly self-motivated (Davenport and Prusak 1998).

Skills

In the context of small and medium enterprises (SMEs), Cheong and Tsui (2011) highlight that the competence and proficiency of employees' individual knowledge serve as foundational elements for successful knowledge management systems. According to Ahmed et al. (2002), knowledge management in SMEs involves the combination and sharing of experiences, skills, intuition, ideas, judgments, context, motivations, and interpretations among individuals.

The effectiveness of a knowledge management system in SMEs hinges not only on the technical skills of its members but also on their understanding of how their tasks contribute to the broader organizational system (Quinn et al. 2005; Ologbo and Nor 2015). Moreover, skills related to cooperation and collaboration within teams are crucial (Madhavan and Grover 1998; Lee and Choi 2003).

Li et al. (2022) propose that three sets of skills are essential for facilitating knowledge sharing within SMEs: social cognitive skills, which involve perceiving knowledge differences among colleagues to ensure timely sharing of relevant knowledge; interpersonal skills, enabling individuals to understand and communicate with other members, resolve conflicts, and achieve common objectives; and strategic orientation skills, where individuals demonstrate goal orientation and the ability to collaborate with multiple stakeholders toward shared goals.

Motivation

The process of knowledge sharing, according to Lin and Lo (2015), is significantly influenced by motivational factors that drive knowledge owners to express and communicate their insights actively. However, motivational barriers can impede knowledge sharing behavior, posing challenges for effective Knowledge Management (Hong et al. 2011; Riege 2005; Singh and Kant 2007). Drawing from the principles of Deci and Ryan's Self-Determination Theory (SDT) (1985, 2000, and 2006), some scholars, such as Werbach and Hunter (2012) and Tondello et al. (2016), highlight the theory's profound impact on understanding behavioral motivation. SDT asserts that individuals possess inherent proactive and autonomous tendencies, capable of engaging intrinsically with tasks (Ryan and Deci 2000). It delineates intrinsic motivation, stemming from internal enjoyment of tasks, and extrinsic motivation, driven by external factors like anticipated rewards. Intrinsic motivation thrives on three fundamental components: competence, autonomy, and relatedness, all of which are essential for fostering enduring positive attitudes towards knowledge sharing (Decy and Ryan 2000; Werbach and Hunter 2012).

In the organizational context, gamification, leveraging game mechanics and elements, can either positively or negatively impact the fulfilment of these fundamental needs (Peng et al. 2012; Ryan et al. 2006). While providing flexibility and choice positively influence autonomy, constraints on user actions diminish this sense of control (Ryan et al. 2006). Game and gamification researchers emphasize the importance of positive emotions in gaming experiences (Francisco-Aparicio et al. 2013; McGonigal, 2011; Harwood and Garry 2015; Robson et al. 2015; Rodrigues et al. 2016). Eppman et al. (2018) reveal that gamification engenders high levels of positive emotions while minimizing negative ones, emphasizing the coexistence of enjoyment and absence of negative emotions as crucial for motivating engagement. Chou (2015) suggests that gamified systems should initially rely on extrinsic rewards to attract users, gradually transitioning towards intrinsic rewards to sustain engagement over time.

Engagement

Engagement in small and medium enterprises (SMEs) is crucial for fostering employee commitment and enthusiasm towards organizational objectives, as defined by Mubarak and Samantha (2021). Butt et al. (2019) emphasizes its pivotal role in assessing the performance of knowledge management systems, highlighting its significance in ensuring system efficacy. Individual involvement in knowledge management activities, crucial for team performance and viability, is essential for SME success (Aubé and Rousseau 2005). To establish effective knowledge management systems, SMEs must cultivate environments that motivate active participation in knowledge creation, sharing, and utilization

(Butt et al. 2019). Gamification offers potential in enhancing long-term employee motivation and engagement, emphasizing the importance of positive emotions and immersive experiences (Chou 2015; Hammedi et al., 2017; Harwood and Garry 2015).

2.3 Technology

In the realm of small and medium enterprises (SMEs), technology infrastructure plays a vital role, encompassing information technology (IT) and its capabilities (Scott 1998; Jarvenpaa and Staples, 2000; Gold et al. 2015). IT serves as a conduit for connecting individuals with reusable codified knowledge and facilitating conversations among employees, naturally facilitating the flow of data (Ein-Dor and Segev 1982; Leonard-Barton 1998). Investment in technological infrastructure is inevitable for bolstering knowledge management systems within SMEs (Davenport and Prusak 1998; Zack 1999; Kakabadse et al. 2003; Nejatian et al. 2013; Ologbo and Nor 2015; Stankosky and Baldanza as cited by Mohajan 2017). Such support simplifies the establishment of knowledge repositories within the organization, serving as organizational memory and accessible resources for employees within the dynamic interaction and collaboration environment.

In this context, gamification emerges as a supportive tool for implementing knowledge management systems, acting as an external facilitator. The dynamics and mechanics embedded in serious games can enhance collaboration, motivation, involvement, and learning among employees (Jorge and Sutton 2016; Suh and Wagner 2017; Li 2018). Moreover, gamification promotes knowledge sharing, fosters healthy competition to achieve organizational goals, facilitates employee development, and encourages intra-company partnerships using game mechanics and dynamics (Werbach and Hunter 2012; Maan 2013; Robson et al. 2015; Salman et al. 2016; Vallat et al. 2016). It serves as an effective means to collect and store knowledge, directing it from the game directly into the knowledge repositories that the organization seeks to establish and maintain.

3 Methodology

3.1 Delphi Method

Since this is an exploratory study, we considered it essential to our research objective to consider the perceptions of experts in Gamification, particularly regarding Knowledge Management, to support us in creating the data collection instrument. To this end, we used the Delphi Method, which consists of applying a questionnaire to a group of experts on a given topic, seeking to achieve convergence of opinion on certain specific issues (von der Gracht 2012; Santos and Almeida, 2022). According to Keeney et al. (2001), the Delphi method is the appropriate technique for consensus building, considering that "several people are less likely to reach a wrong decision than a single individual". It seems the most appropriate technique to validate a study of this nature as it explores participants' views through successive rounds (Donohoe nd Needham, 2009).

As per Velez-Pareja (2005), the Delphi Method application process implies several phases, starting with elaboration of the questionnaire and moving on to its appreciation by a group of experts. After this initial analysis, it is necessary to verify whether a consensus was reached, and if so, the results are compiled into a final summary. If consensus is not reached, the initial questionnaire is re-adjusted based on the responses received, and the questionnaire is re-applied to the group of experts until consensus is reached and the final format of the document is agreed. (Fig. 2).

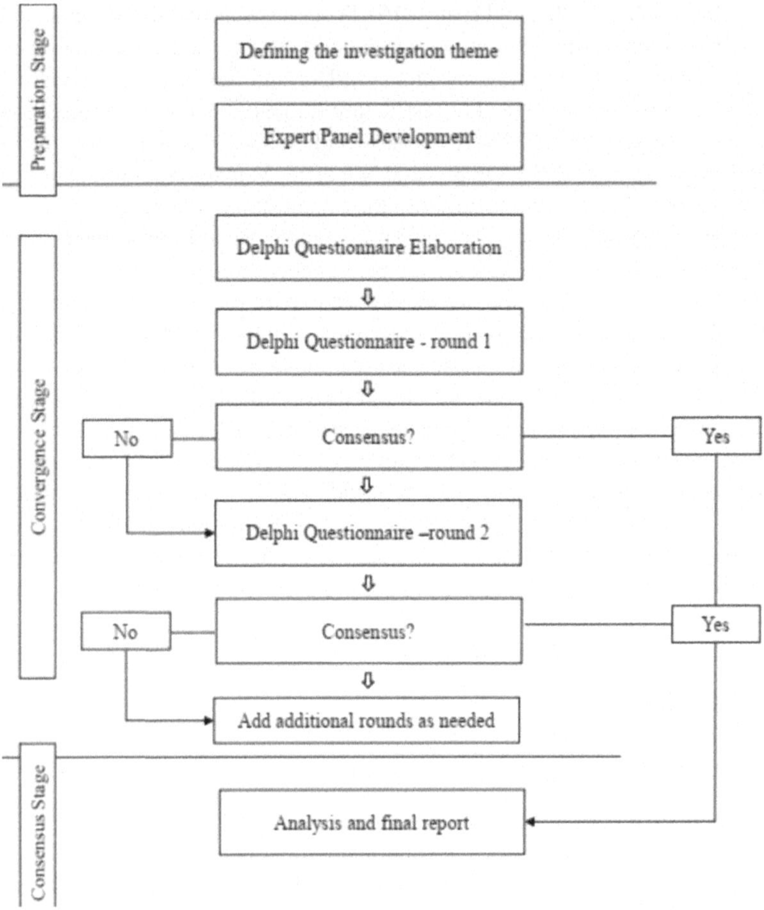

Fig. 2. Application of the Delphi technique process. Source: Adapted from Velez-Pareja (2005) and Donohoe and Needham (2009)

3.2 Preparation Phase

The participants in this study were chosen according to their knowledge, skills, and competencies in the topic under analysis. Based on these assumptions, six experts participated, which is in line with the sample definition parameters advocated by several authors (Mullen 2003; Grisham 2009), who state that a Delphi panel can be as small as three members and as large as 80, it being important to select people with knowledge in the study area and willing to engage in multiple rounds of questions or interactions on the same topic. The participants' profile considers 5 or more years of experience in Organizational Gamification, namely in Knowledge Management and/or Knowledge Sharing processes.

Two rounds of questionnaire application were conducted with the participants, through an online platform, with the researchers explaining the ethical conditions of the study, as well as its purpose and the expected outcome.

Table 1 presents the participants' positions and years of experience:

Table 1. Participants' information

Participants	Position	Experience
1	Consultant	6 years
2	CEO Gamification Company	10 years
3	Training Company Partner	12 years
4	E-Learning Content Manager	9 years
5	Consultant	10 years
6	Senior Researcher and Consultant	13 years

4 Results

4.1 Convergence Phase: Delphi Rounds

In the first round of Delphi, the Conceptual Model for Analysis of the Contribution of Gamification to Knowledge Management Systems by Sampaio et al. (2019) was shared with the panel of experts - Fig. 1 They were asked for their appreciation of the dimensions defined in the model and their suitability for our study. The answers were analyzed and the model items that did not meet consensus were no longer considered, the initial model being adapted (Fig. 3).

In this adaptation, the dimensions of Culture and Organizational Structure were converted into a single dimension "Culture and organizational Structure", including the sub-dimensions of Trust, Collaboration and Dynamic Interaction and Learning and in the dimension of "People" the sub-dimension of "Competencies" was no longer considered. With this transformation, this Conceptual Model of Analysis became more robust, since the Panel of Experts' analysis led to detecting and eliminating sub-dimensions in which Gamification has little, or no, capacity to generate impact, which could bias the results of our future research.

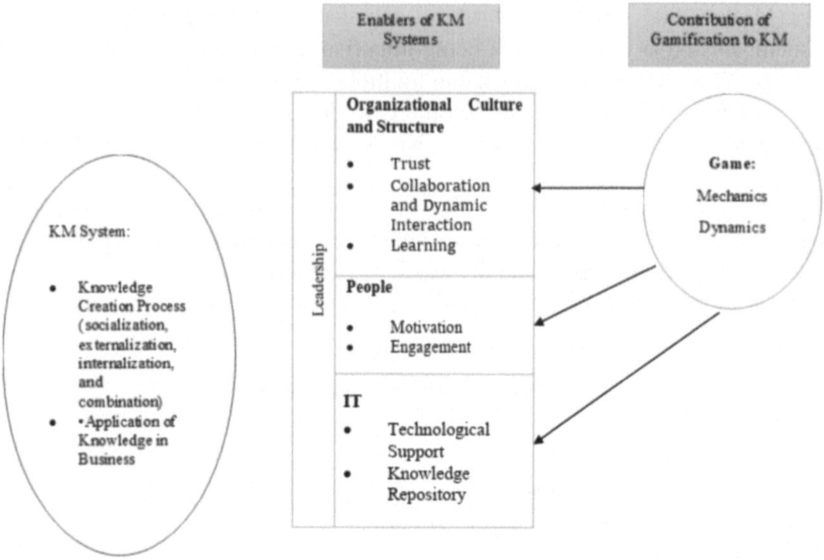

Fig. 3. First round results (Adaptation of the Model by Sampaio et al. (2019).

Thus, in this phase we adopted a new version of the Conceptual Model for Analysis of the Contribution of Gamification to Knowledge Management Systems, considering as facilitators of the Knowledge Management System, (1) Leadership; (2) Culture and Organizational Structure, focusing on trust, collaboration and dynamic interaction and learning; (3) People, focusing on Motivation and Involvement and (4) Information Technology, concerning Technological Support and the existence of Knowledge Repositories.

In the dimensions associated with Gamification's contribution to Knowledge Management, no change was made, keeping the game dynamics and mechanics as the dimensions to be considered.

Based on the results of the first round, it was possible to form statements related to the two major dimensions under analysis: Knowledge Management System Facilitators and Gamification Indicators, to include in the questionnaire to be developed.

Within these dimensions, sub-groups of questions aligned with the areas identified in the model were formed: in the first dimension, questions related to Leadership, Culture and Organizational Structure, People, and Information Technology. In the second dimension, questions were elaborated to gauge the use of Game Mechanics and Dynamics in Gamified processes (Table 2).

Table 2. Statements related to Knowledge Management System Facilitators and Gamification Indicators to include in the questionnaire to be developed.

1 Enablers of Knowledge Management Systems		
Dimension	Item	Affirmation
Leadership		Communicates a clear and positive vision of the future
		Treats employees on an individual basis
		Encourages and recognises employees
		Promotes trust, involvement and cooperation between team members
		Encourages members to think in new ways about problems and questions the ideas that have been proposed
		Is clear about the values practiced and upheld as a leader
		Instigates pride and respect in others and inspires me to be highly competent
		Expresses satisfaction when employees meet expectations
		Focuses on irregularities, mistakes, exceptions and deviations from expected standards
		Only intervenes when problems become serious
		Encourages group decision-making
		Discusses who is responsible for achieving specific performance targets
		Focuses attention on failures to achieve expected standards
		Demonstrates a belief that "you can't change what's right"
		Avoids getting involved when important issues arise
Culture	*Collaboration*	There was more interaction between me and my colleagues
		We now do more activities together
		We start to discuss our company's issues more
		It became easier to have contact with colleagues from other sectors/departments
		I interact more with colleagues from other areas/departments of the company
		Communicating with my colleagues has become easier

(continued)

Table 2. (*continued*)

1 Enablers of Knowledge Management Systems		
Dimension	Item	Affirmation
		I feel that I am part of a group that shares the same knowledge
		I feel that there is a connection with my team
		I feel comfortable with other team members
		There is now an open channel to communicate with other colleagues who share the same interests as me
		I feel that I have learned more about my role
		I know more about my company
	Learning	I feel I have learnt more about my job
		I feel I know more about my company
		I feel more comfortable asking questions about the business
		I feel that when I play I can learn more about the organisation
		I feel that I have learnt something new and complex from the game
		I feel I have recalled things I already knew and couldn't remember from the game
		The game allows me to learn by doing
	Trust	I feel that my team members can be trusted
		I feel that my team members do not mislead me
		I see no reason to doubt the skills of my team members
		After playing together we trust each other more
People	*Engagement*	When I play, I feel committed to helping the team pursue our goal
		When I play, I feel it is important to achieve the team's goal
		I really care about achieving the team's goal
		I feel loyal to the team
		I see myself as part of the team
		I am proud to be a member of the team
		I am proud to think of myself as a member of the team
		I am proud to say that I am part of the team

(*continued*)

Table 2. (*continued*)

1 Enablers of Knowledge Management Systems		
Dimension	Item	Affirmation
		I am willing to share my knowledge to help the team reach its goal
		The game has brought productive competitiveness to the team
		I like it that the game allows me to play in a team
		I feel like playing every day
	Motivation	I enjoy the game very much
		I find the game a boring activity
		The game does not hold my attention
		I would describe the game as a very interesting activity
		When I play, I think about how much I enjoy it
		My experience of the game was enjoyable
		I enjoy playing
		Playing is fun
		When I play, I share information with my teammates
		I make progress in my knowledge when I play
		I do the tasks of the game very well
		When I play I am busy, but I don't feel overloaded
		I feel competent at my game
		When I play I am motivated to share knowledge with other members of my team
IT	*Technological support*	To play you need to use a specific application and tools
		The game is networked with the existing information systems in the organisation
	Knowledge Repository	The company has a knowledge repository where I can obtain relevant information for my performance in the game
Contributions of Gamification to Knowledge Management Systems		
Dimension	Item	Affirmation
Gamification	*Mechanics*	Score
		Levels
		Feedback

(*continued*)

Table 2. (*continued*)

Contributions of Gamification to Knowledge Management Systems		
Dimension	Item	Affirmation
		Clear objectives
		Leader boards
		Challenges
		Badges
		Progress
		Rewards
		Avatar
		Cooperation
		Competition
		Being the winner
		Setting a tactic
	Most motivating mechanics	Score
		Levels
		Feedback
		Clear objectives
		Leader boards
		Challenges
		Badges
		Progress
		Rewards
		Avatar
		Cooperation
		Competition
		Being the winner
		Setting a tactic
	Dynamics	When I play I feel emotion
		The game has a thread that helps me understand it better as a player
		When I play I feel motivated
		The game allows me to relate to other players
		In the game there are rules that I have to follow in order to play and progress in the game

(*continued*)

Table 2. (*continued*)

Contributions of Gamification to Knowledge Management Systems		
Dimension	Item	Affirmation
		The game makes me think about the content of my job or company
		To progress in the game and pass levels I need to apply my knowledge of company products/services
		Having to share information with my teammates to progress in the game makes the game more exciting
		The game encourages collaboration between teammates
		The game brought productive competitiveness to the team
		When I play, I feel excitement

In the second round, the statements were sent to the experts, considering the two areas under analysis - Facilitators of Knowledge Management systems and Gamification - and the dimensions of each area. For each dimension analyzed, the experts were asked to share their opinion on the degree of importance of the statements to the research in question on a Likert-type scale from 1 to 5, where 1 corresponds to "Not important" and 5 to "Very important". The results obtained are shown in Table 3.

Table 3. First round results

Dimension	Affirmation	Not important I Not very important	Moderately Important	Important I Very Important
(1) *Leadership*	**1**			100%
	2	17%		83%
	3			100%
	4		17%	83%
	5		17%	83%
	6		33%	67%
	7		50%	50%
	8	17%		83%
	9	33%	33%	33%
	10	17%	50%	33%

(*continued*)

Table 3. (*continued*)

Dimension	Affirmation	Not important I Not very important	Moderately Important	Important I Very Important
	11	17%	17%	67%
	12		17%	83%
	13		33%	67%
	14		33%	67%
	15	33%		67%
(2) Collaboration	**1**			100%
	2			100%
	3		17%	83,00%
	4		33%	67%
	5		33%	67%
	6		50%	50%
	7		33%	67%
	8			100%
	9		50%	50%
	10		50%	50%
	11			100%
	12		17%	83%
(3) Learning	**1**			100%
	2		17%	83%
	3		17%	83%
	4			100%
	5		17%	83%
	6			100%
	7		17%	83%
(4) Trust	**1**		17%	83%
	2		17%	83%
	3		17%	83%
	4		33%	67%
(5) Engagement	**1**			100%
	2			100%
	3		17%	83%
	4			100%
	5			100%
	6		33%	67%

(*continued*)

Table 3. (*continued*)

Dimension	Affirmation	Not important I Not very important	Moderately Important	Important I Very Important
	7		33%	67%
	8		33%	67%
	9			100%
	10			100%
	11			100%
	12		33%	67%
(6) Motivation	**1**			100%
	2	67%		33%
	3	67%		33%
	4			100%
	5		33%	67%
	6			100%
	7			100%
	8			100%
	9			100%
	10			100%
	11		33%	67%
	12		17%	83%
	13		17%	83%
	14			100%
(7) IT	**1**			100%
	2			100%
	3			100%
(8) Mechanics	**1**		17%	83%
	2			100%
	3			100%
	4			100%
	5			100%
	6			100%
	7		17%	83%
	8			100%
	9		17%	83%
	10			100%
	11		17%	83%

(*continued*)

Table 3. (*continued*)

Dimension	Affirmation	Not important I Not very important	Moderately Important	Important I Very Important
	12		50%	50%
	13			100%
	14		17%	83%
(9) Most motivating mechanics	**1**		17%	83%
	2			100%
	3			100%
	4			100%
	5			100%
	6			100%
	7		17%	83%
	8			100%
	9			100%
	10	17%	50%	33%
	11			100%
	12			100%
	13		17%	67%
	14		17%	83%
(10) Dynamics	**1**			100%
	2			100%
	3			100%
	4		17%	83%
	5		17%	83%
	6			100%
	7			100%
	8		17%	83%
	9		17%	83%
	10			100%
	11		17%	83%

After analyzing the results of the second round, where there was no consensus, we decided to re-adjust the statements of each dimension by removing the ones which were negatively assessed by the experts, except in the characterization of Leadership, since it is a validated scale from which it is not possible to remove items due to the risk of losing the reliability of the results. As no suggestion was made to improve the questionnaire, we kept the initial form.

In the third round, we again presented the statements to the experts and requested their opinion on the assessment scales to be used.

This turned out to be the final round since the answers were consensual with our questionnaire proposal and we received a positive assessment of the proposed statements (between moderately important and very important) from the panel of experts and of the proposed assessment scale.

Consensus Phase: Analysis and Final Report

This last phase consists of the final consensus by the panel of experts in Gamification who were involved in the previous rounds, to validate the Questionnaire on the Contribution of Gamification to Knowledge Management Systems, which will result in a useful and innovative tool to allow data collection on this topic.

In developing the questionnaire, special care was taken to structure this instrument so that it would be aligned with the Conceptual Model of Analysis. We chose to use closed-ended questions which proposed a range of pre-established answers, leaving it up to the respondent to choose one or several of the proposed options, depending on whether they were single or multiple answers. According to Gonçalves (2004), when well-designed, closed-ended questions are easy to understand, fill in, treat, interpret, and quantify, and do not require content analysis. In the present case, the choice of this type of structure also aims to create a relatively easy-to-follow design, to be an objective questionnaire and not too long.

Special care was also given to the language and sequence of questions, the first to try to form questions as close as possible to the linguistic and mental categories of the population to be surveyed and the second considering the importance of a sequence of themes and questions, so that those who answer take it naturally. It is also important to start the questionnaire with questions directly related to the research topic. We also took some care with the sequence of questions to avoid monotony and maintain the interviewees' interest.

In this case, the questionnaire was divided into sections, aligned with the dimensions of the Conceptual Model of Analysis developed - Leadership, Culture and Organizational Structure, People, IT and Gamification - and subsections including the items studied. The section on Culture and Organizational Structure contemplated the subsections of Collaboration, Learning and Trust. The People section included the subsections of Involvement and Motivation, and the Gamification section was segmented into Mechanics, More Motivating Mechanics, and Dynamics.

Several measurement scales were used in this instrument, according to the questions. Most questions were answered using Likert-type scales, considering that, according to Bryman (2012), this type of scale is an appropriate measure for multiple indicators or multiple items of a set of attitudes related to a specific area. This is especially applicable when intending to measure the intensity of feelings about the area in question. In all cases, the choice of scale took into consideration the statistical measures to be used later for data analysis. In the case of nominal scales, these allow non-parametric tests or data analysis through descriptive statistics - modes, percentages, or cross-tabulations. The use of Likert-type scales allows us to perform factor analysis, correlations, and comparison of means, among other statistical tests to allow cross-checking of the data obtained.

We now consider the validity and reliability of the questionnaire. As Bryman (2012) states, it is essential for a researcher to ensure the consistency of the measurement scales used to measure a given concept and check whether the indicators chosen to measure the concepts effectively measure these concepts. Considering that for most questions we will use scales that have not yet been tested, as argued by Hill and Hill (2012), it is necessary to check their internal consistency, consistency in terms of temporal stability, and consistency in terms of equivalence of measures of the latent variable obtained by alternative versions.

As regards content, concept and predictive validity, the Delphi method (Linstone and Turoff, 1975) allowed for analysis and validation by experts to detect inaccuracies and adjust before application.

The result is shown in Table 4:

Table 4. Questionnaire for data collection on the Contribution of Gamification to Knowledge Management Systems

Leadership						
The following are 14 descriptive items. Indicate which of the statements fits the actions of the leader of your organisation, using the following scale: 1 = Strongly disagree 2 = Partially disagree 3 = Neither agree nor disagree 4 = Partially agree 5 = Strongly agree		1	2	3	4	5
1	**My leader:**					
1.1	Communicates a clear and positive vision of the future					
1.2	Treats employees on an individual basis					
1.3	Encourages and recognises employees					
1.4	Promotes trust, involvement and cooperation between team members					
1.5	Encourages members to think in new ways about problems and questions the ideas that have been proposed					
1.6	Is clear about the values they practice and uphold					
1.7	Instigates pride and respect in others and inspires me to be highly competent					

(*continued*)

Table 4. (*continued*)

1.8	Expresses satisfaction when employees live up to expectations					
1.9	Focuses on irregularities, mistakes, exceptions and deviations from expected standards					
1.10	Only interferes when problems become serious					
1.11	Encourages decision-making by the group					
1.12	Discusses who is responsible for achieving specific performance targets					
1.13	Channels their attention to failures in order to achieve expected standards					
1.14	Demonstrates a belief that "you can't change what's right"					
Organisational Culture and Structure						
2	**Collaboration**					
Considering the introduction of gaming in your organisation, please indicate your opinion on the 12 descriptive items below, using the following scale: 1 = Strongly disagree 2 = Partially disagree 3 = Neither agree nor disagree 4 = Partially agree 5 = Strongly agree		1	2	3	4	5
	With the introduction of gaming in the company:					
2.1	There was more interaction between me and my colleagues					
2.2	We now do more activities together					
2.3	We start to discuss our company's issues more					
2.4	It became easier to have contact with colleagues from other sectors/departments					
2.5	I interact more with colleagues from other areas/departments of the company					

(*continued*)

Table 4. (*continued*)

2.6	Communicating with my colleagues has become easier					
2.7	I feel that I am part of a group that shares the same knowledge					
2.8	I feel that there is a connection with my team					
2.9	I feel comfortable with other team members					
2.10	There is now an open channel to communicate with other colleagues who share the same interests as me					
2.11	I feel that I have learned more about my role					
2.12	I know more about my company					
3	**Learning**					
Considering your use of the game, please answer the following statements using the following scale: 1 = Strongly disagree 2 = Partially disagree 3 = Neither agree nor disagree 4 = Partially agree 5 = Strongly agree		1	2	3	4	5
	When I play the game:					
3.1	I feel I have learnt more about my job					
3.2	I feel I know more about my company					
3.3	I feel more comfortable asking questions about the business					
3.4	I feel that when I play, I can learn more about the organisation					
3.5	I feel that I have learnt something new and complex from the game					
3.6	I feel I have recalled things I already knew and couldn't remember from the game					
3.7	The game allows me to learn by doing					
4	**Trust**					

(*continued*)

Table 4. (*continued*)

Considering your use of the game, please answer the following statements using the following scale: 1 = Strongly disagree 2 = Partially disagree 3 = Neither agree nor disagree 4 = Partially agree 5 = Strongly agree		1	2	3	4	5
4.1	When I play, I feel motivated to share knowledge with other members of my team					
4.2	I feel that my team members can be trusted					
4.3	I feel that my team members don't cheat me					
4.4	I see no reason to doubt the skills of my team members					
People						
5	**Engagement**					
Think about yourself as a player. Here below are 12 descriptive items, please indicate which of the statements fit your situation as a player, according to the following scale: 1 = Strongly disagree 2 = Partially disagree 3 = Neither agree nor disagree 4 = Partially agree 5 = Strongly agree		1	2	3	4	5
	As a Player:					
5.1	When I play, I feel committed to helping the team pursue our goal					
5.2	When I play, I feel it is important to achieve the team's goal					
5.3	I really care about achieving the team's goal					
5.4	I feel loyal to the team					
5.5	I see myself as part of the team					
5.6	I am proud to be a member of the team					
5.7	I am proud to think of myself as a member of the team					
5.8	I am proud to say that I am part of the team					

(*continued*)

Table 4. (*continued*)

5.9	I am willing to share my knowledge to help the team reach its goal					
5.10	The game has brought productive competitiveness to the team					
5.11	I like that the game allows me to play in a team					
5.12	I feel like playing every day					
6	**Motivation**					
The following are descriptive items. Please indicate which of the statements fits your situation before the game, considering the following scale: 1 = Strongly disagree 2 = Partially disagree 3 = Neither agree nor disagree 4 = Partially agree 5 = Strongly agree		1	2	3	4	5
	As a player:					
6.1	I enjoy the game very much					
6.2	I find the game a boring activity					
6.3	The game does not hold my attention					
6.4	I would describe the game as a very interesting activity					
6.5	When I play, I think about how much I enjoy it					
6.6	My experience of the game was enjoyable					
6.7	I enjoy playing					
6.8	Playing is fun					
6.9	When I play, I share information with my teammates					
6.10	I make progress in my knowledge when I play					
6.11	I do the tasks of the game very well					
6.12	When I play, I am busy, but I don't feel overloaded					
6.13	I feel competent in the game					

(*continued*)

Table 4. (*continued*)

6.14	When I play, I feel motivated to share knowledge with other members of my team					
6.15	I feel that my team members can be trusted					
6.16	I feel that my team members do not cheat on me					
6.17	I see no reason to doubt my team members' skills					
IT						
	From the situations below indicate which ones exist in your company regarding the game you used:	Yes	No			
7.1	To play you need to use a specific application and tools					
7.2	The game is networked with the existing information systems in the organisation					
7.3	The company has a knowledge repository where I can obtain relevant information for my performance in the game					
Gamification						
8	**Mechanics**					
Of the components below, indicate those that you recognise as existing in your company's game		Yes	No			
8.1	Score					
8.2	Levels					
8.3	Feedback					
8.4	Clear objectives					
8.5	Leader boards					
8.6	Challenges					
8.7	Badges					
8.8	Progress					
8.9	Rewards					
8.10	Avatar					

(*continued*)

Table 4. (*continued*)

8.11	Challenges					
8.12	Cooperation					
8.13	Competition					
8.14	Being the winner					
8.15	Setting a tactic					
9	**Most Motivating Mechanics**					
Consider the components listed and choose the five that you think most encourage you to collaborate with other colleagues in the game		Yes	No			
9.1	Score					
9.2	Levels					
9.3	Feedback					
9.4	Clear objectives					
9.5	Leader boards					
9.6	Challenges					
9.7	Badges					
9.8	Progress					
9.9	Rewards					
9.10	Avatar					
9.11	Challenges					
9.12	Cooperation					
9.13	Competition					
9.14	Being the winner					
9.15	Setting a tactic					
10	**Dynamics**					
The following are 11 descriptive items. Please indicate which of the statements match your own situation before the game, according to the scale: 1 = Strongly disagree 2 = Partially disagree 3 = Neither agree nor disagree 4 = Partially agree 5 = Strongly agree		1	2	3	4	5
10.1	When I play, I feel emotion					
10.2	The game has a thread that helps me understand it better as a player					
10.3	When I play, I feel motivated					

(*continued*)

Table 4. *(continued)*

10.4	The game allows me to relate to other players					
10.5	In the game there are rules I have to follow to be able to play and progress in the game					
10.6	The game makes me think about the content of my job or company					
10.7	To progress in the game and pass levels I need to apply my knowledge of company products/services					
10.8	Having to share information with my teammates to progress in the game makes the game more exciting					
10.9	The game encourages collaboration between teammates					
10.10	The game brought productive competitiveness to the team					
10.11	When I play, I feel excitement					

The data collection tool presented corresponds to the objective initially outlined for this study, i.e., to develop a questionnaire that can be a starting point for data collection in an organizational context, allowing us to verify the applicability of the Gamification Contribution Model as a facilitator of Knowledge Management Systems.

5 Final Considerations

In recent years, with the rise of globalization and the rapid development of information and communication technologies, coupled with the emergence of new forms of work and more flexible and horizontal organizational structures, economies, particularly those of small and medium enterprises (SMEs), have witnessed a significant shift. Knowledge has become a pivotal asset for SMEs, with its role in value creation surpassing that of tangible resources (António 2015). Consequently, SMEs are increasingly focusing on crafting robust knowledge management strategies aimed at capturing, documenting, assimilating, sharing, and applying organizational knowledge effectively. This process is vital for maximizing organizational objectives and ensuring efficient and effective operations (Hadad 2017). To incentivize employees within SMEs to engage in knowledge sharing and acquisition, one promising approach is Gamification—a strategy that employs elements of game design in non-game contexts.

This study is framed within a broader objective of demonstrating that organizational Knowledge Management can be leveraged by resorting to knowledge management systems that integrate elements of game design. After a research phase of an exploratory nature, focused on facilitators of Knowledge Management and elements of Gamification, we were able to gather important knowledge that allowed us to move on to creating a data collection tool. In the future, this will allow us to validate what the theory indicates: that Gamification can make an important contribution to Organizational Knowledge Management.

The starting point of this study was the Conceptual Model for Analyzing the Contribution of Gamification to Knowledge Management Systems by Sampaio et al, (2019), exploring and contextualizing the elements indicated as Facilitators of Knowledge Management systems at SMEs and the elements that, according to the same model, through Gamification can contribute to Knowledge Management.

The Delphi Technique was applied as a qualitative research method, with six experts in the field of Gamification of Knowledge Management Systems being involved in three rounds of analysis.

The two major dimensions and the elements comprising them were analyzed: Facilitators of Knowledge Management Systems and Contributions of Gamification to Knowledge Management. Regarding the Facilitators of Knowledge Management Systems, the following elements were considered (1) leadership, (2) culture and organizational structure (trust, collaboration and interaction and learning); (3) people (motivation and involvement) and (4) IT (technological support and knowledge repositories). The Contributions of Gamification to Knowledge Management considered (1) the dynamics and (2) game mechanics.

This study made significant methodological advancements tailored to the context of small and medium enterprises (SMEs), notably by enhancing the Conceptual Model for the Analysis of the Contribution of Gamification to Knowledge Management Systems by Sampaio et al. (2019). Additionally, it entailed the development and validation of a new tool aimed at assessing the role of Gamification in fostering a culture and structure conducive to knowledge enhancement within SMEs. This tool not only stimulates trust, collaboration, and learning among colleagues but also enhances employees' motivation and involvement in knowledge management systems. Furthermore, the study sheds light on prevalent leadership styles and IT practices within Gamified Knowledge Management systems in SMEs. While the study encountered limitations, such as challenges in accessing specialists and the pioneering nature of the tool requiring further testing in diverse Gamified Knowledge Management contexts, it provides valuable insights for future investigations. Subsequent research will involve data collection on the Gamification of Knowledge Management Systems to contribute to the development of an explanatory model elucidating how Gamification influences employees' willingness to share and acquire organizational knowledge within SMEs.

References

Abubakar, A.M., Elrehail, H., Alatailat, M.A., Elçi, A.: Knowledge management, decision-making style and organizational performance. J. Innov. Knowl. **4**(2), 104–114 (2019). https://doi.org/10.1016/J.JIK.2017.07.003

Ahmed, P.K., et al.: Chapter 20 – The Post Office. In Learning Through Knowledge Management (2002). https://www.routledge.com/Learning-Through-Knowledge-Management/Ahmed-Lim-Loh/p/book/9780750647106

Andrews, M.C., Kacmar, K. M.: Discriminating among organizational politics, justice, and support. J. Organ. Behav. **22**(4), 347–366 (2001). https://doi.org/10.1002/job.92

Antonio, N.: Estratégia Organizacional: do posicionamento ao movimento (2a Edição). Edições Sílabo (2015)

Aryee, S., Budhwar, P.S., Chen, Z.X.: Trust as a mediator of the relationship between organizational justice and work outcomes: test of a social exchange model. J. Organ. Behav. **23**(3), 267–285 (2002). https://doi.org/10.1002/JOB.138

Attali, Y., Arieli-Attali, M.: Gamification in assessment: do points affect test performance? Comput. Educ. **83**, 57–63 (2015). https://doi.org/10.1016/J.COMPEDU.2014.12.012

Aubé, C., Rousseau, V.: Team goal commitment and team effectiveness: the role of task interdependence and supportive behaviors (2005). https://doi.org/10.1037/1089-2699.9.3.189

Bass, B.M.: Leadership: good, better, best. Organ. Dyn. **13**(3), 26–40 (1985). https://doi.org/10.1016/0090-2616(85)90028-2

Bell, V.A., Cooper, S.Y.: Institutional knowledge: acquisition, assimilation and exploitation in internationalisation. Int. Mark. Rev. **35**(3), 475–497 (2018)

Bernstein, E.S., Shore, J.C., Jang, A.J.: Network centralization and collective adaptability to a shifting environment (2022). https://doi.org/10.1287/Orsc.2022.1584. https://doi.org/10.1287/ORSC.2022.1584

Bock, G.W., Zmud, R.W., Kim, Y.G., Lee, J.N.: Behavioral intention formation in knowledge sharing: examining the roles of extrinsic motivators, social-psychological forces, and organizational climate. MIS Q. Manag. Inf. Syst. **29**(1), 87–111 (2005). https://doi.org/10.2307/25148669

Bryman, A.: Social Research Methods, 4th edn. Oxford University Press, Cambridge (2012)

Burns, J.M.: Chapter 16: Toward a general theory. Leadership, 422–443 (1978). https://books.google.com/books/about/Leadership.html?id=DeztAAAAMAAJ

Butt, M.A., et al.: Individual knowledge management engagement, knowledge-worker productivity, and innovation performance in knowledge-based organizations: the implications for knowledge processes and knowledge-based systems. Comput. Math. Organ. Theory **25**(3), 336–356 (2019). https://doi.org/10.1007/s10588-018-9270-z

Cardador, M.T., Northcraft, G.B., Whicker, J.: A theory of work gamification: something old, something new, something borrowed, something cool? Hum. Resour. Manag. Rev. **27**(2), 353–365 (2017). https://doi.org/10.1016/j.hrmr.2016.09.014

Caruana, A., Morris, M.H., Vella, A.J.: The effect of centralization and formalization on entrepreneurship in exports firms. J. Small Bus. Manag. **36**(1), 16 (1998)

Chase, R.: The people factor. People Manag. **4**(2) (1998)

Chen, M.H., Huang, L.L., Wang, C.H.: Developing a digital game for stroke patients' upper extremity rehabilitation – design, usability and effectiveness assessment. Procedia Manuf. **3**(Ahfe), 6–12 (2015). https://doi.org/10.1016/j.promfg.2015.07.101

Cheong, R.K.F., Tsui, E.: From skills and competencies to outcome-based collaborative work: tracking a decade's development of personal knowledge management (PKM) models. Knowl. Process. Manag. **18**(3), 175–193 (2011). https://doi.org/10.1002/KPM.380

Choo, C.W.: The Knowing organization: how organizations use information to construct meaning, create knowledge and make decisions. Int. J. Inf. Manag. **16**(5), 329–340 (1996). https://doi.org/10.1007/BF01072060

Chou, Y.-K.: Actionable Gamification: Beyond Points, Badges, and Leaderboards, p. 502. Octalysis Media (2015)

Clark, M.C., Rossiter, M.: Narrative learning in adulthood. New Direct. Adult Contin. Educ. **2008**(119), 61–70 (2008). https://doi.org/10.1002/ACE.306

Cook, S.D., Brown, J.S.: Bridging epistemologies: the generative dance between organizational knowledge and organizational knowing. Organ. Sci. **10**(4), 381–400 (1999). https://doi.org/10.1287/Orsc.10.4.381

Costa, V., Monteiro, S.: Knowledge processes, absorptive capacity and innovation: a mediation analysis. Knowl. Process. Manag. **23**(3), 207–218 (2016)

Csikszentmihalyi, M.: Flow: the psychology of optimal experience. Harper & Row (1990)

Davenport, T.H., Prusak, L.:. Working knowledge: how organizations manage what they know. Knowl. Creat. Diffus. Utilizat. (1998). https://doi.org/10.1109/EMR.2003.1267012

Deci, E.L., Ryan, R.M.: Intrinsic Motivation and self-determination in human behavior. In: Intrinsic Motivation and Self-Determination in Human Behavior (1985). https://doi.org/10.1007/978-1-4899-2271-7

Deterding, S., Dixon, D., Khaled, R., Nacke, L.: From game design elements to gamefulness: defining "Gamification." In: Proceedings of the 2011 Annual Conference Extended Abstracts on Human Factors in Computing Systems - CHI EA '11 (2011). https://doi.org/10.1145/1979742.1979575

DiBella, A.J., Nevis, E.C., Gould, J.M.: Understanding organizational learning capability. J. Manag. Stud. **33**(3), 361–379 (1996). https://doi.org/10.1111/J.1467-6486.1996.TB00806.X

Donohoe, H.M., Needham, R.D.: Moving best practice forward: delphi characteristics, advantages, potential problems, and solutions. Int. J. Tour. Res. **11**(5), 415–437 (2009). https://doi.org/10.1002/JTR.709

Durst, S., Edvardsson, I., Foli, S.: Knowledge management in SMEs: a follow-up literature review. J. Knowl. Manag. **27**(11), 25–58 (2022). https://doi.org/10.1108/JKM-04-2022-0325

Ein-Dor, P., Segev, E.: Organizational context and MIS structure: some empirical evidence. MIS Q. Manag. Inf. Syst. **6**(3), 55–67 (1982). https://doi.org/10.2307/248656

Eppler, M.J., Sukowski, O.: Managing team knowledge: core processes, tools and enabling factors. Eur. Manag. J. **18**(3), 334–341 (2000). https://doi.org/10.1016/S0263-2373(00)00015-3

Eppmann, R., Bekk, M., Klein, K.: Gameful experience in gamification: construction and validation of a gameful experience scale [GAMEX]. J. Interact. Mark. **43**, 98–115 (2018). https://doi.org/10.1016/j.intmar.2018.03.002

Foss, N.J., Minbaeva, D.B., Pedersen, T., Reinholt, M.: Encouraging knowledge sharing among employees: how job design matters. Hum. Resour. Manag. **48**(6), 871–893 (2009). https://doi.org/10.1002/HRM.20320

Francisco-Aparicio, A., Gutiérrez-Vela, F.L., Isla-Montes, J.L., Sanchez, J.L.G.:. Gamification: Analysis and Application, pp. 113–126 (2013). https://doi.org/10.1007/978-1-4471-5445-7_9

Gaviria-Rivera, J.I., Zapata, E.L., López-Zapata, E., Gaviria-Rivera, J.I., López-Zapata, E.: Transformational leadership, organizational climate and job satisfaction in work teams. Eur. Res. Stud. J. **XXII**(3), 68–82 (2019). https://doi.org/10.11214/thalassinos.22.03.005

Gibson, D., Ostashewski, N., Flintoff, K., Grant, S., Knight, E.: Digital badges in education. Educ. Inf. Technol. **20**(2), 403–410 (2013). https://doi.org/10.1007/S10639-013-9291-7

Gloet, M., Terziovski, M.: Exploring the relationship between knowledge management practices and innovation performance. J. Manuf. Technol. Management, pp. 402–409 (2004). https://doi.org/10.1108/17410380410540390

Gold, A.H., Malhotra, A., Segars, A.H.: Knowledge management: an organizational capabilities perspective. J. Manag. Inf. Syst. **18**(1), 185–214 (2001). https://doi.org/10.1080/07421222.2001.11045669

Gottschalk, P.: Knowledge management in the professions: the case of IT support in law firms. In: Proceedings of the 33rd Annual Hawaii International Conference on System Sciences, vol. 2, p. 10 (2000)

Grisham, T.: The Delphi technique: a method for testing complex and multifaceted topics. Int. J. Manag. Proj. Bus. **2**(1), 112–130 (2009). https://doi.org/10.1108/17538370910930545

Gupta, A.K., Govindarajan, V.: Knowledge flows within multinational corporations. Strat. Manag. J. **21**(4), 473–496 (2000). https://doi.org/10.1002/(SICI)1097-0266(200004)21:4<473::AID-SMJ84>3.0.CO;2-I
Hadad, S.: Knowledge economy: characteristics and dimensions. Manag. Dyn. Knowl. Econ. **5**(2), 203–225 (2017). https://doi.org/10.25019/mdke/5.2.03
Hamari, J., Keronen, L.: Why do people play games? A meta-analysis. Int. J. Inf. Manage. **37**(3), 125–141 (2017). https://doi.org/10.1016/j.ijinfomgt.2017.01.006
Hamari, J., Koivisto, J.: Why do people use gamification services? Int. J. Inf. Manag. **35**, 419–431 (2015). http://10.0.3.248/j.ijinfomgt.2015.04.006
Hamari, J., Koivisto, J., Sarsa, H.: Does gamification work? - a literature review of empirical studies on gamification. In: Proceedings of the Annual Hawaii International Conference on System Sciences (2014). https://doi.org/10.1109/HICSS.2014.377
Hammedi, W., Leclerq, T., van Riel, A.C.R.: The use of gamification mechanics to increase employee and user engagement in participative healthcare services. J. Serv. Manag. **28**(4), 640–661 (2017). https://doi.org/10.1108/JOSM-04-2016-0116
Hansen, P., Järvelin, K.: Collaborative Information Retrieval in an information-intensive domain. Inf. Process. Manag. **41**(5), 1101–1119 (2005). https://doi.org/10.1016/J.IPM.2004.04.016
Harwood, T., Garry, T.: An investigation into gamification as a customer engagement experience environment. J. Serv. Mark. **29**(6–7), 533–546 (2015). https://doi.org/10.1108/JSM-01-2015-0045/FULL/XML
Hedlund, G.: A model of knowledge management and the N-form corporation. Strat. Manag. J. **15**(2 S), 73–90 (1994). https://doi.org/10.1002/SMJ.4250151006
Hidi, S.: Psychology of interest. In: International Encyclopedia of the Social & Behavioral Sciences, pp. 7712–7715 (2001). https://doi.org/10.1016/B0-08-043076-7/01535-7
Hill, M.M., Hill, A.: Investigação por Questioná¡rio. Silabo (2012)
Hollebeek, L.D.: Demystifying customer brand engagement: exploring the loyalty nexus. J. Mark. Manag. **27**(7–8), 785–807 (2010). https://doi.org/10.1080/0267257X.2010.500132
Hopper, M.D.: Rattling SABRE—new ways to compete on information. Harvard Bus. Rev. **68**(4), 118 (1990)
Huong, N. T., Katsuhiro, U., Chi, D.H.: Knowledge transfer in offshore outsourcing: a case study of Japanese and Vietnamese software companies. J. Glob. Inf. Manag. **19**(2), 27–44 (2011). https://Services.Igi-Global.Com/Resolvedoi/Resolve.Aspx?Doi=https://doi.org/10.4018/Jgim.2011040102. https://doi.org/10.4018/JGIM.2011040102
Hurley, R.F., Hult, G.T.M.: Innovation, market orientation, and organizational learning: an integration and empirical examination. J. Mark. **62**(3), 42–54 (1998). https://doi.org/10.2307/125 1742
Iansiti, M.. Real-World R&D: jumping the product generation gap. Harvard Bus. Rev., 138–147 (1993). https://www.hbs.edu/faculty/Pages/item.aspx?num=4299
Ichijo, K., von Krogh, G., Nonaka, I.: Knowledge enablers. In: von Krogh, G., Roos, J., Kleine, D. (eds.) Knowing in Firms: Understanding, Managing and Measuring Knowledge. SAGE Publications Ltd. (1998). https://doi.org/10.4135/9781446280256
McGonigal, J.: Reality is Broken - Why Games Make Us Better and How they Can Change the World. Vintage Publishing (2011)
Jarvenpaa, S.L., Staples, D.S.: The use of collaborative electronic media for information sharing: an exploratory study of determinants. J. Strat. Inf. Syst. **9**(2–3), 129–154 (2000). https://doi.org/10.1016/S0963-8687(00)00042-1
Jeng, J.-F., Dunk, N.: Knowledge management enablers and knowledge creation in ERP system success. Int. J. Electron. Bus. Manag. **11** (2013)
Johannessen, J.A., Olsen, B., Olaisen, J.: Aspects of innovation theory based on knowledge-management. Int. J. Inf. Manag. **19**(2), 121–139 (1999). https://doi.org/10.1016/S0268-401 2(99)00004-3

Jorge, C.F.B., Faléco, L.L.: A aplicação da gestão do conhecimento como estratégia de. Braz. J. Inf. Stud. Res. Trends **3**, 69–75 (2016)

Jorge, C.F.B., Sutton, M.J.D.: Games Como Estratégia na Construção e Gestão do Conhecimento no Contexto da Inteligência Organizacional. Perspectivas Em Gestão & Conhecimento **6**, 103–118 (2016)

Kakabadse, N.K., Kakabadse, A., Kouzmin, A.: Reviewing the knowledge management literature: towards a taxonomy. J. Knowl. Manag. (2003). https://doi.org/10.1108/13673270310492967

Keeney, S., Hasson, F., McKenna, H.P.: A critical review of the Delphi technique as a research methodology for nursing. Int. J. Nurs. Stud. **38**(2), 195–200 (2001). https://doi.org/10.1016/S0020-7489(00)00044-4

Kianto, A., Vanhala, M., Heilmann, P.: The impact of knowledge management on job satisfaction. J. Knowl. Manag. **20**(4), 621–636 (2016). https://doi.org/10.1108/JKM-10-2015-0398

von Krogh, G., Roos, J.: Organizational Epistemology. St Martin's Press (1995). https://doi.org/10.1002/smj.4250151005

Kumar, H., Raghavendran, S.: Gamification, the finer art: fostering creativity and employee engagement. J. Bus. Strateg. **36**(6), 3–12 (2015). https://doi.org/10.1108/JBS-10-2014-0119

Lee, H., Choi, B.: Knowledge management enablers, processes, and organizational performance: an integrative view and empirical examination. J. Manag. Inf. Syst. **20**(1), 179–228 (2003). https://doi.org/10.1080/07421222.2003.11045756

Leonard-Barton, D.: Wellsprings of Knowledge: Building and Sustaining the Sources of Innovation (Google eBoek) (1998). http://books.google.com/books?hl=nl&lr=&id=z2mQHT3PWWMC&pgis=1

Li, C., et al.: Influence of knowledge management practices on entrepreneurial and organizational performance: a mediated-moderation model. Front. Psychol. **11** (2020). https://doi.org/10.3389/fpsyg.2020.577106

Li, K.: Multi-context research on strategy characteristics of knowledge sharing in organization based on dynamic cooperative game perspective. J. Knowl. Manag. **22**(4), 850–866 (2018). http://10.0.4.84/JKM-09-2017-0420

Li, S., Martins, J.T., Vasconcelos, A.C., Peng, G.: Knowledge sharing in project work: the dynamic interplay of knowledge domains and skills. J. Knowl. Manag. (2022). https://doi.org/10.1108/JKM-06-2021-0455

Lin, S.W., Lo, L.Y.S.: Mechanisms to motivate knowledge sharing: integrating the reward systems and social network perspectives. J. Knowl. Manag. **19**(2), 212–235 (2015). https://doi.org/10.1108/JKM-05-2014-0209

Linstone, H.A., Turoff, M.: The Delphi Method Techniques and Applications. Addison-Wesley, Boston (1975)

Lithoxoidou, E., et al.: A novel social gamified collaboration platform enriched with shop-floor data and feedback for the improvement of the productivity, safety and engagement in factories. Comput. Ind. Eng. **xxxx**, 1–11 (2019). https://doi.org/10.1016/j.cie.2019.02.005

Maan, J.: Social business transformation through gamification. Int. J. Managing Inf. Technol. (IJMIT) **5**(3), 9–16 (2013). https://doi.org/10.5121/ijmit.2013.5302

Madhavan, R., Grover, R.: From embedded knowledge to embodied knowledge: new product development as knowledge management. J. Mark. **62**(4), 1–12 (1998). https://doi.org/10.2307/1252283

Mettler, T., Pinto, R.: Serious games as a means for scientific knowledge transfer—a case from engineering management education. IEEE Trans. Eng. Manag. **62**(2), 256–265 (2015). 10.0.4.85/TEM.2015.2413494

Mintzberg, H.: The structuring of organizations. In: Asch, D., Bowman, C. (eds.) Readings in Strategic Management, pp. 322–352. Macmillan Education UK (1989). https://doi.org/10.1007/978-1-349-20317-8_23

Moccozet, L., Tardy, C., Opprecht, W., Leonard, M.: Gamification-based assessment of group work. In: 2013 International Conference on Interactive Collaborative Learning, ICL 2013, pp. 171–179 (2013). https://doi.org/10.1109/ICL.2013.6644565

Mohajan, H.: The impact of knowledge management models for the development of organizations. J. Environ. Treat. Techn. **5**(1), 12–33 (2017). https://mpra.ub.uni-muenchen.de/83089/1/MPRA_paper_83089.pdf

Mubarak, K., Samantha, T.: The role of employee engagement on knowledge management and worker productivity: a case study in Sri Lanka. J. Asian Finan. Econ. Bus. **8**(4), 507–515 (2021). https://doi.org/10.13106/JAFEB.2021.VOL8.NO4.0507

Mullen, P.M.: Delphi: myths and reality. J. Health Organ. Manag. **17**(1), 37–52 (2003). https://doi.org/10.1108/14777260310469319

Nejatian, M., Nejati, M., Zarei, M.H., Soltani, S.: Critical enablers for knowledge creation process: synthesizing the literature. Glob. Bus. Manag. Res. (2013)

Nelson, K.M., Cooprider, J.G.: The contribution of shared knowledge to IS group performance. MIS Q. Manag. Inf. Syst. **20**(4), 409–429 (1996). https://doi.org/10.2307/249562

Nonaka, I., Takeuchi, H.: The Knowledge-Creating Company: How Japanese Companies Create the Dynamics of Innovation. Oxford University Press, Cambridge (1995)

Obeidat, B., et al.: The role of knowledge management infrastructure in enhancing innovation at mobile telecommunication companies in Jordan. Eacademic.Ju.Edu.Jo **50**, 313–330 (2015). http://eacademic.ju.edu.jo/m.aldalahmeh/Lists/Published%20Research/Attachments/16/Paper%20(%208%20)%20The%20Role%20of%20Knowledge%20Management.pdf

O'Dell, C., Grayson, C.J.: Knowledge transfer: discover your value proposition. Strat. Leadership **27**, 10–15 (1999)

Olander, H., Vanhala, M., Hurmelinna-Laukkanen, P., Blomqvist, K.: HR-related knowledge protection and innovation performance: the moderating effect of trust. Knowl. Process. Manag. **22**(3), 220–233 (2015). https://doi.org/10.1002/KPM.1476

Ologbo, A.C., Nor, K.: The 7-circle model: a practical and coherent km model for managing organizational knowledge. Mediterr. J. Soc. Sci. **6**(4), 120–128 (2015). https://doi.org/10.5901/mjss.2015.v6n4s1p120

Ozlati, S.: The moderating effect of trust on the relationship between autonomy and knowledge sharing: a national multi-industry survey of knowledge workers. Knowl. Process. Manag. **22**(3), 191–205 (2015). https://doi.org/10.1002/KPM.1474

Pansari, A., Kumar, V.: Customer engagement: the construct, antecedents, and consequences. J. Acad. Mark. Sci. **45**(3), 294–311 (2017). https://doi.org/10.1007/S11747-016-0485-6

Patricio, R., Moreira, A.C., Zurlo, F.: Gamification in innovation teams. Int. J. Innov. Stud. **6**(3), 156–168 (2022). https://doi.org/10.1016/J.IJIS.2022.05.003

Peng, W., Lin, J.H., Pfeiffer, K.A., Winn, B.: Need satisfaction supportive game features as motivational determinants: an experimental study of a self-determination theory guided exergame. Media Psychol. **15**(2), 175–196 (2012). https://doi.org/10.1080/15213269.2012.673850

Petelczyc, C.A., Capezio, A., Wang, L., Restubog, S.L.D., Aquino, K.: Play at work: an integrative review and agenda for future research. J. Manag. **44**(1), 161–190 (2018). https://doi.org/10.1177/0149206317731519

Petrov, Viktorija, Ćelić, Đorđe, Uzelac, Zorica, Drašković, Zoran: Three pillars of knowledge management in SMEs: evidence from Serbia. Int. Entrepre. Manag. J. **16**(2), 417–438 (2019). https://doi.org/10.1007/s11365-018-00557-2

Pe-Than, E.P.P., Goh, D.H.L., Lee, C.S.: Making work fun: Investigating antecedents of perceived enjoyment in human computation games for information sharing. Comput. Hum. Behav. **39**, 88–99 (2014). https://doi.org/10.1016/j.chb.2014.06.023

Polanyi, M.: The Tacit Dimension. Routledge & Legam Paul (1966)

Politis, J.: The relationship between team performance, authentic and servant leadership. In: European Conference on Management, Leadership & Governance. Eurcademic Conferences International Limited, pp. 237–244 (2013). https://www.proquest.com/docview/1467640113?pq-origsite=gscholar&fromopenview=true

Przybylski, A.K., Rigby,C.S., Ryan, R.M.: A Motivational Model of Video Game Engagement Motivation and Video Games (2010). https://doi.org/10.1037/a0019440

Quinn, J.B., Anderson, P., Finkelstein, S.: Leveraging intellect. the academy of management executive (1993–2005). Acad. Manag. Perspect. **19**(4), 78–94 (2005). http://www.jstor.org/stable/4166207

Raftopoulos, M., Walz, S., Greuter, S.: How enterprises play: Towards a taxonomy for enterprise gamification. In: DiGRA 2015: Diversity of Play: Games - Cultures - Identities (2015)

Rauch, M.: Best practices for using enterprise gamification to engage employees and customers. In: Kurosu, M. (ed.) HCI 2013. LNCS, vol. 8005, pp. 276–283. Springer, Heidelberg (2013). https://doi.org/10.1007/978-3-642-39262-7_31

Reese, D.D.: Dashboard effects challenge flow-learning assumption in digital instructional games. In: Emotions, Technology, and Digital Games, pp. 231–287 (2016). https://doi.org/10.1016/B978-0-12-801738-8.00011-7

Riege, A.: Three-dozen knowledge-sharing barriers managers must consider. J. Knowl. Manag. **9**(3), 18–35 (2005). https://doi.org/10.1108/13673270510602746/FULL/XML

Robson, K., Plangger, K., Kietzmann, J. H., McCarthy, I., Pitt, L.: Is it all a game? Understanding the principles of gamification. Bus. Horizons **58**, 411–420 (2015). https://doi.org/10.1016/j.bushor.2015.03.006

Rodrigues, L.F., Oliveira, A., Costa, C.J.: Playing seriously - how gamification and social cues influence bank customers to use gamified e-business applications. Comput. Hum. Behav. **63**, 392–407 (2016). https://doi.org/10.1016/J.CHB.2016.05.063

Ryan, R.M., Deci, E.L.: Self-determination theory and the facilitation of intrinsic motivation, social development, and well-being. Am. Psychol. **55**(1), 68–78 (2000). https://doi.org/10.1037/0003-066X.55.1.68

Ryan, R.M., Rigby, C.S., Przybylski, A.: The motivational pull of video games: a self-determination theory approach. Motivat. Emot. **30**(4), 344–360 (2006). https://doi.org/10.1007/S11031-006-9051-8

Salman, M., Becker, J., Fuchs, M., Heutelbeck, D., Hemmje, M.: Enhancing knowledge management and transfer in an applied gaming ecosystem. In: Proceedings of the European Conference on Knowledge Management, ECKM, 2016-Janua, pp. 786–795 (2016)

Sampaio, M.C., Sousa, M.J., Dionísio, A.: The use of gamification in knowledge management processes: a systematic literature review. J. Rev. Glob. Econ. **8**, 1662–1679 (2019). https://doi.org/10.6000/1929-7092.2019.08.150

Joaquim, S., António, C., Jorge, C.: As competências funcionais dos líderes e a eficácia das equipas. Revista Portuguesa e Brasileira de Gestão, Jul/Set, pp. 22–33 (2008). https://repositorio.iscte-iul.pt/bitstream/10071/1547/1/artigo.pdf

Santos, V., Almeida, N.: Ethical marketing model for luxury hotel chains: development and validation of a performance evaluation tool. Sustainability **14**(12), 7382 (2022). https://doi.org/10.3390/su14127382

Scott, J.E.: Organizational knowledge and the Intranet. Decis. Supp. Syst. **23**(1), 3–17 (1998). https://doi.org/10.1016/S0167-9236(98)00032-3

Seaborn, K., Fels, D.I.: Gamification in theory and action: a survey. Int. J. Hum. Comput. Stud. (2015). https://doi.org/10.1016/j.ijhcs.2014.09.006

Serrano, A., Fialho, C.J.F.: Gestão do Conhecimento - O novo paradigma das organizações, 2a Edição. FCA - Editora de Informática (2005)

Shpakova, A., Dörfler, V., MacBryde, J.: Changing the game: a case for gamifying knowledge management. World J. Sci. Technol. Sustainable Dev. **14**(2/3), 143–154 (2017). https://doi.org/10.1108/WJSTSD-01-2017-0002

Simões, J., Redondo, R.D., Vilas, A.F.: A social gamification framework for a K-6 learning platform. Comput. Hum. Behav. **29**(2), 345–353 (2013). https://doi.org/10.1016/J.CHB.2012.06.007

Singh, M.D., Kant, R.: Knowledge management barriers: an interpretive structural modeling approach. In: IEEM 2007: 2007 IEEE International Conference on Industrial Engineering and Engineering Management, pp. 2091–2095 (2007). https://doi.org/10.1109/IEEM.2007.4419560

Toma, S.-G., Catană, Ş, Grădinaru, C.: Leadership: an overview. Manager J. **32**(1), 51–59 (2020)

Stacey, R.D.: Complexity and creativity in organizations. Berrett-Koehler Publishers (1996)

do Rocio Strauhs, F.: Gestão do conhecimento em laboratório acadêmico: proposição de metodologia. Universidade Federal de Santa Catarina (2003)

Suh, A., Wagner, C.: How gamification of an enterprise collaboration system increases knowledge contribution: an affordance approach. J. Knowl. Manag. **21**(2), 416–431 (2017). https://doi.org/10.1108/JKM-10-2016-0429

Suh, A., Wagner, C., Liu, L.: Enhancing user engagement through gamification. J. Comput. Inf. Syst. **58**(3), 204–213 (2018). https://doi.org/10.1080/08874417.2016.1229143

Suorsa, A., Huotari, M.L.: Knowledge creation and the concept of a human being: a phenomenological approach. J. Am. Soc. Inf. Sci. **65**(5), 1042–1057 (2014). https://doi.org/10.1002/ASI.23035

Suorsa, A.R.: Knowledge creation and play – a phenomenological approach. J. Document. **71**(3), 503–525 (2015). https://doi.org/10.1108/JD-11-2013-0152

Swacha, J.: Gamification in knowledge management: motivating for knowledge sharing. Pol. J. Manage. Stud. **12**(2), 150–160. Czestochowa Technical University, Department of Management (2015)

Terra, J.C.C.: Implantando a Gestão do Conhecimento. Gestão de Empresas Na Era Do Conhecimento, 165–173 (2003)

Tondello, G.F., Elm, D., Kappen, D.L., Nacke, L.E.: Clever: gamification and enterprise knowledge learning. In: CHI PLAY'16 Extended Abstracts, Austin, TX, USA, 16–19 October 2016, vol. s5 XII, no. 301, p. 268 (2016). https://doi.org/10.1093/nq/s5-XII.301.268-d

Tseng, F.C., Fan, Y.J.: Exploring the influence of organizational ethical climate on knowledge management. J. Bus. Ethics **101**(2), 325–342 (2011). https://doi.org/10.1007/S10551-010-0725-5

Tsoukas, H.: A dialogical approach to the creation of new knowledge in organizations. Organ. Sci. **20**(6), 941–957 (2009). https://doi.org/10.1287/ORSC.1090.0435

Vallat, D., Bayart, C., Bertezene, S.: Serious games in favour of knowledge management and double-loop learning? Knowl. Manag. Res. Pract. **14**(4), 470–477 (2016). https://doi.org/10.1057/kmrp.2015.29

van den Hooff, B., Schouten, A.P., Simonovski, S.: What one feels and what one knows: the influence of emotions on attitudes and intentions towards knowledge sharing. J. Knowl. Manag. **16**(1), 148–158 (2012). https://doi.org/10.1108/13673271211198990

Velez-Pareja, I.: The Delphi Method (El Metodo Delphi). SSRN Electron. J. (2005). https://doi.org/10.2139/SSRN.420040

von der Gracht, H.A.: Consensus measurement in Delphi studies: review and implications for future quality assurance. Technol. Forecast. Soc. Chang. **79**(8), 1525–1536 (2012). https://doi.org/10.1016/J.TECHFORE.2012.04.013

Von Krogh, G.: Care in knowledge creation. Calif. Manag. Rev. **40**(3), 133–153 (1998). https://doi.org/10.2307/41165947

von Krogh, G., Roos, J., Kleine, D.: Knowing in Firms : Understanding, Managing, and Measuring Knowledge. Sage (1998)

Wenger, E.C., Synder, W.M.: Communities of practice: the organizational frontier. Harvard Bus. Rev. (2000). https://doi.org/10.1177/0170840603024003909

Werbach, K., Hunter, D.: For the Win: How Game Thinking Can Revolutionize Your Business. Wharton Digital Press (2012). https://doi.org/10.1017/CBO9781107415324.004

Weretecki, P., Greve, G., Bates, K., Henseler, J.: Information management can't be all fun and games, can it? How gamified experiences foster information exchange in multi-actor service ecosystems. Int. J. Inf. Manag. **61**, 102391 (2021). https://doi.org/10.1016/J.IJINFOMGT.2021.102391

Wu, W.L.: To share knowledge or not: dependence on knowledge-sharing satisfaction. Soc. Behav. Pers. **41**(1), 47–58 (2013). https://doi.org/10.2224/SBP.2013.41.1.47

Zichermann, G., Cunningham, C.: Gamification By Design - Implementing Game Mechanics in Web and Mobile Apps. Gabriel Z (2011). https://doi.org/10.1017/CBO9781107415324.004

Zucker, L.G., Brewer, M.B., Darby, M.R., Peng, Y.: Collaboration structure and information dilemmas in biotechnology: organizational boundaries as trust production. In: UCLA: Institute for Social Science Research, vol. 9999, pp. 1–13. Springer, Heidelberg (2016). https://escholarship.org/uc/item/0gd8j9k8

Author, F., Author, S., Author, T.: Book title, 2nd edn. Publisher, Location (1999)

Author, F.: Contribution title. In: 9th International Proceedings on Proceedings, pp. 1–2. Publisher, Location (2010)

LNCS Homepage. http://www.springer.com/lncs. Accessed 21 Nov 2016

Artificial Intelligence and New Pedagogies: Transforming Teaching and Learning Through AI-Driven Innovations

Maria José Sousa[1(✉)], Miguel Sousa[2], and Agnes Jenei[3]

[1] Instituto Universitário de Lisboa, Lisbon, Portugal
maria.jose.sousa@iscte-iul.pt
[2] University of Essex, Colchester, UK
[3] University of Public Services, Budapest, Hungary
jenei.agnes@uni-nke.hu

Abstract. In recent years, Artificial Intelligence (AI) has garnered significant attention across various sectors, particularly in higher education. The purpose of this chapter is to provide an in-depth analysis of AI technologies, including Machine Learning (ML), Deep Learning (DL), and Natural Language Processing (NLP), and their practical applications. Through a detailed review of the literature, the study aims to conceptualize AI, explore its categories and methodologies, and provide insights into its role in shaping education systems. AI's potential to enhance operational efficiencies, improve learning outcomes, and create a hybrid workforce is also examined. The paper concludes by offering recommendations for further research and policy considerations.

Keywords: Artificial Intelligence · Machine Learning · Deep Learning · Natural Language Processing · Higher Education

1 Introduction

Artificial Intelligence (AI) has become a foundation of modern technological advancements. In recent years, its integration into industries like manufacturing, healthcare, and education has opened up vast opportunities for innovation and optimization. The rise of Machine Learning (ML), Deep Learning (DL), and Natural Language Processing (NLP) has enhanced the capabilities of AI, making it a critical component of a wide range of applications. As a result, AI has sparked significant academic interest, prompting numerous studies to investigate its practical uses, theoretical frameworks, and potential future developments.

This article presents a comprehensive analysis of AI, its theoretical foundations, and its practical applications, with a particular focus on recent developments. The research is motivated by two main factors: first, to provide a conceptual overview of AI for newcomers in the field, and second, to respond to the growing interest in AI from scholars and industry practitioners alike. The first section explores the conceptualization of AI,

M. J. Sousa (Ed.): UFHRD 2024, ISEM 52, pp. 321–330, 2025.
https://doi.org/10.1007/978-3-031-92548-1_19

while the second section offers a literature review of its key categories and methodologies. The final section concludes by reflecting on the findings and challenges for future AI research and application.

2 Literature Review

AI encompasses a wide range of technologies and methodologies that enable machines to perform tasks that traditionally required human intelligence. Researchers have identified several dimensions of AI, each contributing to its growing capabilities. Among these are:

Case-Based Reasoning (CBR): CBR is a problem-solving methodology in AI where past cases or experiences are used to resolve new problems. The CBR cycle involves four steps: retrieval, reuse, revision, and retention. This approach is particularly useful in domains where solutions are based on specific experiences rather than general rules (Gent et al. 2008).

Genetic Algorithms (GA): GA is a search algorithm inspired by the mechanics of natural selection and genetics. It is used to solve optimization problems by evolving solutions through processes such as mutation, crossover, and selection (Gent et al. 2008).

Expert Systems: These are AI systems designed to emulate the decision-making abilities of human experts in specific fields. Expert systems use predefined rules and knowledge bases to provide solutions to complex problems (Heeks 2018).

Natural Language Processing (NLP): NLP involves the interaction between computers and human languages, enabling machines to understand, interpret, and generate human language. Recent advancements in NLP have improved applications such as sentiment analysis and chatbots (Sousa et al. 2019).

Machine Learning (ML) and Deep Learning (DL): ML allows machines to learn from data without explicit programming. Deep Learning, a subset of ML, mimics the human brain's neural networks, enabling more complex and autonomous learning from large datasets. Both technologies have revolutionized fields such as image recognition, autonomous driving, and natural language understanding (Sousa et al. 2019; Shalev-Shwartz and Ben-David 2014).

Cloud Computing and IoT: The convergence of AI with technologies like cloud computing and the Internet of Things (IoT) has further enhanced AI's capabilities. These technologies enable smarter machines and systems, contributing to the development of intelligent factories and ecosystems (Huawei 2017). Generative Artificial Intelligence (AI): Generative Artificial Intelligence refers to artificial intelligence systems with the capability to create text, images, or other forms of media through the utilization of generative models. These models acquire an understanding of patterns and structures within their training data, subsequently generating novel data with akin characteristics. Generative Artificial Intelligence encompasses various types, each tailored for specific tasks or forms of media generation Sengar et al. (2024).

In recent years, Artificial Intelligence has emerged as a transformative force in education, using those dimensions for the creation of AI-based tools offering new possibilities for pedagogical innovation. Traditional teaching models, which often rely on a one-size-fits-all approach, are being challenged by AI-powered technologies that can personalize learning experiences, adapt to individual needs, and optimize educational practices. AI systems, such as Intelligent Tutoring Systems (ITS), Virtual Reality (VR) and Mixed Reality (MR) powered by AI and Generative AI, Augmented Reality (AR), and learning analytics, are revolutionizing how students engage with learning materials and interact with educators.

AI-driven pedagogies are reshaping the roles of both teachers and learners, fostering a more interactive and data-informed learning environment. The application of AI in education is not only enhancing learning outcomes but also introducing new ways of thinking about teaching methodologies and educational assessment. This article examines the ways in which AI is influencing new pedagogical models, with a focus on personalization, data-driven decision-making, data-driven assessment, and the integration of AI tools in everyday teaching practices.

2.1 AI and Personalized Learning

Personalized learning refers to educational methods tailored to the individual needs, abilities, and interests of each learner. AI plays a crucial role in this process by analyzing large datasets about students' learning behaviors and performance to create customized learning experiences. By continuously tracking a learner's progress, AI can adjust the content, difficulty, and pace of instruction, ensuring that students receive the support they need at the right time.

A major example of AI in personalized learning is the Intelligent Tutoring System (ITS). ITS are designed to provide individualized instruction, often mimicking the role of a human tutor. These systems use AI to identify students' strengths and weaknesses, adapting instruction to address gaps in understanding. Research by VanLehn (2011) demonstrates that ITS can lead to better learning outcomes compared to traditional classroom instruction, particularly in subjects like mathematics and science. Additionally, systems like Carnegie Learning and Knewton have shown the potential of AI to personalize lessons, making the learning process more effective and engaging.

2.2 AI and Adaptive Learning Systems

Adaptive learning systems use AI algorithms to adjust the learning path based on real-time assessments of a student's knowledge. These systems can identify when a student is struggling with a concept and provide supplementary materials or alternative approaches to help them master the content. According to Baker and Siemens (2014), AI-powered adaptive learning systems can create a more flexible and responsive learning environment, improving educational outcomes by offering timely interventions.

Moreover, these systems are capable of providing feedback to students instantly, which enhances the learning process by reducing delays in correction and allowing students to learn from their mistakes more quickly. The work of Knewton and Smart Sparrow exemplifies how adaptive learning platforms can optimize instruction for both individual learners and entire classrooms, creating a more dynamic and personalized educational experience (Huang et al. 2020).

2.3 Natural Language Processing (NLP)

Natural Language Processing (NLP) is a subset of AI that enables machines to interpret, understand, and generate human language. In educational contexts, NLP is being used to support both students and educators. For example, AI systems powered by NLP can analyze student essays and provide feedback on grammar, spelling, and coherence, thereby assisting teachers in grading and offering timely comments. More advanced NLP applications can evaluate the content of essays, looking for critical thinking, argument structure, and originality, which goes beyond surface-level corrections (Russell et al. 2018).

Additionally, NLP is a key tool in facilitating more inclusive education. AI-driven language models can assist students with learning disabilities or those who speak different languages by providing real-time translation, reading assistance, and personalized explanations. Holmes et al. (2019) suggest that AI's role in NLP is not only about improving communication but also creating a more equitable and accessible educational environment for diverse learners.

2.4 Learning Analytics and Data-Driven Pedagogy

Learning analytics involves the use of data to inform educational decision-making and improve student outcomes. By analyzing data from various sources—such as student performance, attendance, and participation—AI can help educators identify trends and predict student outcomes. This allows for more proactive interventions and can support early warning systems for students at risk of falling behind.

Siemens (2013) argues that learning analytics can transform traditional teaching by making it more data-driven. Educators can tailor their strategies based on the insights provided by AI systems, resulting in a more efficient allocation of resources. Furthermore, AI-powered platforms can provide real-time feedback on student performance, allowing teachers to adjust instruction in response to individual needs (Ferguson 2012).

2.5 Virtual, Mixed and Augmented Reality in Pedagogy

Virtual Reality (VR) is "the use of computer modeling and simulation that enables a person to interact with an artificial three-dimensional (3-D) visual or other sensory environment" (Britannica 2025). While VR immerses the user in a completely virtual environment, AR only overlays digital content (e.g., images, text, or sounds) onto the real-world environment, enhancing it rather than replacing it. In MR, digital and physical objects can coexist and interact in real time. XR is used to refer to VR, AR and MR devices, however, VR is often used as umbrella term. All these technologies are revolutionizing pedagogy by offering immersive, interactive, experiential, collaborativelearning experiences that transcend traditional classroom boundaries.

Aczél (2017) defines educational VR as a subgroup of VR that require learning, knowledge extension, credible (specific, disciplinary) information, potential connection with other education methods, and target groups of teachers and learners. Aiello et al. (2012) highlights that the use of VR technologies in education not only simulate reality, but also actively contributes to the development of students' sensorimotor skills. VR is not seen as a mere technical innovation, but as a didactic tool capable of supporting the learning process.

VR creates entirely digital environments where students can explore simulations, conduct experiments, or visit historical landmarks, enabling experiential learning that enhances engagement and retention (Al-Azawi et al. 2019). AR, on the other hand, overlays digital content onto the physical world, blending real-world interactions with virtual elements. This approach enriches the learning experience by providing contextualized and dynamic visualizations of abstract concepts, such as complex biological processes or architectural designs (Akçayır and Akçayır 2017). Both technologies support differentiated instruction by catering to various learning styles—visual, auditory, and kinesthetic—while promoting active participation and collaboration.

In STEM education, VR, MR and AR enable students to perform virtual dissections, simulate chemical reactions, and explore the solar system without physical or safety limitations (Freina and Ott 2015). In Nursery Education, VR offers superior potential for advancing nursing students' theoretical knowledge, practice proficiencies, and overall satisfaction (Liu et al. 2023). In Built Environment (BE) education these technologies significantly enhance student engagement, bridging the gap between theoretical knowledge and practical application (Ibrahim et al. 2024).

Similarly, humanities and arts benefit from these tools, as students can virtually walk through historical events, practice languages with virtual avatars, or explore 3D art galleries. Moreover, these technologies support inclusivity by accommodating learners with disabilities through customized accessibility features, such as text-to-speech integration and gesture-based controls (Bacca et al. 2014).

VR promotes learning by experience, increasing motivation, engagement and understanding also in the field of soft skills development. For students enrolled in HRM or HRD programs (or in Business Schools) it can be particularly useful to have the opportunity to learn how to handle and how to conduct job interview or other types of difficult conversations, deliver public presentation or negotiate. They learn these skills in platforms like Bodyswap or Virtual Speech, learning-by-doing, in role play with a generative AI-powered avatars, in a safe, judgment-free place, repeating them as many times as possible. VR technology allows precise assessment of the performance. Learners receive immediate AI feedback and personalized learning instructions based on the collected and analyzed data. Recent research shows that VR simulation (VRS) has proved to be a powerful tool for developing students' communication skills (Cho and Kim 2024).

In addition, VR can help learners understand different perspectives and develop empathy toward others by the possibility of body swap and see himself/herself from the others's point of view. Finally, VR can be used for hybrid or multiplayer soft skills development exercises as well. In this case, the teacher /trainer can adopt dedicated VR experiences developed by psychologists (i.e. IDEGO) or other carefully selected software to fit the soft skills training.Furthermore, students and teachers can use Virtual Collaborative Platforms (Altspace VR, Engage, other specialized platforms), which are interactive, experiential, and collaborative learning environments, to realize complex projects, engineering experiments, or design projects in VR or Mixed Reality.

A recent review highlights the role of virtual reality in enhancing student engagement in educational settings (Lin et al. 2024). The article confirms that VR can enhance student engagement and learning outcomes, especially for those with learning disabilities. Despite these advantages, the adoption of VR, MR and AR in pedagogy faces challenges, including the education system's lag in adopting VR, the need for higher digital literacy among students, and teachers' limited VR proficiency (Lin et al. 2024). Celik and Baturay underscore the high implementation costs and the need for teacher training. Nevertheless, the potential of these technologies to transform education into an engaging, equitable, and future-ready domain continues to drive interest and innovation (Çelik, and Baturay 2024).

3 Methodology

This study employs a holistic literature review approach to explore the state-of-the-art developments in AI. The review draws on a wide range of academic articles, books, industry reports, and case studies to provide a comprehensive overview of AI's conceptualization, its core methodologies, and applications. Key sources include:

Scholarly articles on AI and its categories (Gent et al. 2008; Heeks, 2018).

Reports from industry leaders like IDC Technologies (2020) and Huawei (2017).

Research studies on AI applications in higher education and manufacturing (Sousa et al. 2019).

Government and policy reports on AI's role in economic development (OECD 2024).

The literature review focuses on analyzing the key factors driving AI advancements, such as the combination of vast computing power, data digitization, and breakthroughs in machine learning algorithms. Additionally, the review highlights the various AI applications in diverse sectors, with particular attention to the role of AI in education and manufacturing.

4 Findings

4.1 AI in Higher Education

AI has the potential to revolutionize higher education by improving learning opportunities, personalizing education, and supporting collaborative learning. AI applications in education can help ensure equitable access to learning resources, particularly for disadvantaged groups such as refugees, people with disabilities, and those living in remote areas (Holmes et al. 2019).

Intelligent Tutoring Systems (ITS) are a prime example of how AI can personalize learning experiences. These systems use algorithms to analyze student performance and provide tailored feedback and instructional materials. Additionally, AI can assist in grading essays and assignments, significantly reducing the time and effort spent on administrative tasks by educators (Holmes et al. 2019).

Moreover, AI can support collaborative learning by monitoring online discussions and providing teachers with insights into student engagement and participation. This is particularly valuable in distance learning environments, where students may not be physically present in the classroom (Holmes et al. 2019).

Generative AI has opened up remarkable possibilities for transforming education through innovative human-machine collaboration, such as personalized digital tutors for students, digital assistants for educators, and digital peers that enable new forms of collaborative learning. However, these advancements also bring significant ethical and regulatory challenges that must be addressed (Denny et al. 2024).

4.2 Enhanced Student Engagement and Retention

AI technologies have been shown to significantly enhance student engagement and retention by providing personalized learning experiences. Holmes et al. (2019) argue that AI allows educators to cater to diverse learning styles, ensuring that all students, regardless of their abilities, remain engaged with the material. By offering interactive, self-paced learning opportunities, AI tools help to keep students motivated and invested in their learning (VanLehn 2011). VR technology also boosts learner engagement and retention (Allcoat and von Mühlenen 2018 and Lin et al. 2024).

The generative AI models in education have boosted productivity, enhanced creativity, and personalized learning, providing timely feedback and support, and automating tasks (Farrokhnia et al. 2023 and Cotton et al. 2023).

Additionally, adaptive learning systems help identify students who may be at risk of disengaging or failing, allowing for timely interventions. This leads to improved retention rates and better learning outcomes.

4.3 AI's Role in Teacher Support

AI is not only beneficial to students but also plays a critical role in supporting teachers. AI systems can automate administrative tasks such as grading, lesson planning, and scheduling, freeing up time for educators to focus on teaching and student interaction. According to Baker (2015), AI can help teachers become more efficient by providing them with real-time data on student performance, enabling them to make more informed decisions about their teaching methods.

Regarding soft skills development in the most advanced VR platforms (combined with AI), teachers can help students to plan the training experience to their needs, based on the collected and analyzed data.AI also enables teachers to offer more individualized support to students VanLehn 2011). For example, ITS and learning analytics systems can identify students' struggles and provide teachers with actionable insights on how to help them improve.

Generative AI models are also valuable support for teachers as well, to deliver feedback more quickly and easily, and to create content for course materials (Mohammed 2023).

However, these technologies are useful only if teachers engage with them and adapt their methods in all subjects, not only in technological once (Alterares-Lopes 2024). Arndt (2023) introduces the concept of 'AI-learning-competence', emphasizing the need for formal training and pedagogical integration of AI tools.

5 Challenges and Ethical Considerations

While AI offers significant advantages in education, it also raises important challenges. One of the main concerns is the ethical implications of AI, particularly related to data privacy and algorithmic bias. AI systems require access to vast amounts of student data, and there is the risk of this data being misused or exploited. The researcher du Boulay (2023) highlights that privacy concerns must be addressed to ensure that AI tools do not inadvertently violate student rights or create inequalities in the classroom. Moreover, the use of AI in education could lead to an over-reliance on technology, potentially diminishing the role of human educators. There is a need for balance in integrating AI while maintaining the human element of teaching.

Challenges need to be tackled to guarantee a safe and responsible usage also of Generative AI (Su and Yang 2023) point out that the Cotton et al. (2023) highlight the relevance of academic integrity in this context.

6 Conclusions

Artificial Intelligence has the potential to revolutionize both industrial sectors and higher education systems. Its ability to enhance operational efficiencies, optimize production processes, and improve learning outcomes makes it a valuable tool across a range of applications. As AI technologies such as Machine Learning, Deep Learning, and Natural Language Processing continue to evolve, they promise even greater advancements in fields like manufacturing, education, and beyond.

For industries, the integration of AI with technologies like cloud computing and IoT is creating smarter, more efficient systems that can adapt to changing conditions and provide real-time insights. The hybrid workforce, where humans and machines collaborate, is set to become a key feature of the future labor market.

In higher education, AI's ability to personalize learning, support collaborative environments, and provide equitable access to educational resources is transforming the way students learn and interact with educators. The potential for AI to reduce administrative burdens, such as grading and assessment, further supports its integration into educational systems.

As AI continues to evolve, further research is needed to explore its ethical implications, data privacy concerns, and its impact on employment and labor markets. Policymakers must consider these factors to ensure that AI is developed and implemented in a way that benefits society as a whole.

References

Aczél, P.: Educational virtual reality: innovative methods for teaching and learning. In: Proceedings of the International Conference on Virtual Learning (ICVL) (2017)

Aiello, P., Di Tore, P.A., Sibilio, M.: Educational potentials of virtual reality: theoretical reflections and empirical evidences. J. e-Learn. Knowl. Soc. **8**(3), 55–65 (2012)

Akçayır, M., Akçayır, G.: Advantages and challenges associated with augmented reality for education: a systematic review of the literature. Educ. Res. Rev. **20**, 1–11 (2017). https://doi.org/10.1016/j.edurev.2016.11.002

Al-Azawei, A., Baiee, W.R., Mohammed, M.A.: Learners' experience towards e-assessment tools: a comparative study on virtual reality and moodle quiz. Int. J. Emerg. Technol. Learn. (iJET) **14**(05), 34–50 (2019). https://doi.org/10.3991/ijet.v14i05.9998

Allcoat, D., von Mühlenen, A.: Learning in virtual reality: effects on performance, emotion and engagement. Res. Learn. Technol. **26**, 1–13 (2018). https://doi.org/10.25304/rlt.v26.2140

Alterares-Lopes, R.: The transformation of teacher training: Integrating AI in pedagogical practice. J. Dig. Educ. Learn. Technol. **12**(1), 22–38 (2024)

Arndt, S.: Developing AI-learning competence in teacher education. Int. J. Educ. Technol. **18**(3), 112–128 (2023)

Bacca, J., Baldiris, S., Fabregat, R., Graf, S., Kinshuk.: Augmented reality trends in education: a systematic review of research and applications. Educ. Technol. Soc. **17**(4), 133–149 (2014)

Baker, R.S.: Big data and education. In: Fox, M.F., Koller, S.G., Guzdial, M.D. (eds.) Learning Analytics in Education, pp. 23–40. Springer, Heidelberg (2015)

Baker, R.S., Siemens, G.: Educational data mining and learning analytics. In: Sawyer, K. (ed.) The Cambridge Handbook of the Learning Sciences, 2nd edn, pp. 253–272). Cambridge University Press (2014)

Britannica. Virtual reality. In: Encyclopædia Britannica (2025). https://www.britannica.com/technology/virtual-reality

Çelik, H.C., Baturay, M.H.: Challenges and prospects of virtual reality integration in education: teachers' readiness and resource constraints. Technol. Pedagog. Educ. **33**(1), 67–83 (2024)

Cho, M., Kim, D.: Enhancing communication skills through virtual reality simulation: evidence from higher education. J. Educ. Comput. Res. **62**(2), 345–370 (2024)

Cotton, D.R.E., Cotton, P.A., Kneale, P.E.: Generative AI in education: benefits and dilemmas. Teach. High. Educ. (2023). https://doi.org/10.1080/13562517.2023.2264919

Denny, P., Luxton-Reilly, A., Simon, B.: Generative AI in education: opportunities, risks, and regulatory frameworks. Comput. Educ. Artif. Intell. **5**, 100153 (2024). https://doi.org/10.1016/j.caeai.2024.100153

du Boulay, B.: Ethics in AI-based education: a call for privacy and transparency. AI Soc. **38**(1), 55–70 (2023)

Farrokhnia, M.R., Esfijani, A., Rostami, M.: The impact of generative AI on learners' engagement and creativity in higher education. Educ. Inf. Technol. **28**, 1123–1145 (2023)

Ferguson, R.: The state of learning analytics in 2012: a review and future challenges. Technical Report KMI-12-01, Knowledge Media Institute. The Open University (2012)

Freina, L., & Ott, M. (2015). A literature review on immersive virtual reality in education: State of the art and perspectives. In *The International Scientific Conference eLearning and Software for Education* (Vol. 1, pp. 133–141). https://doi.org/10.12753/2066-026X-15-020

Gent, I.P., Jefferson, C., Kotthoff, L., Miguel, I., Nightingale, P.: AI Methodologies: Case-Based Reasoning and Genetic Algorithms in Practice. Springer, Heidelberg (2008)

Heeks, R.: Information and Communication Technology for Development (ICT4D), 2nd edn.. Routledge, Abingdon (2018)

Holmes, W., Bialik, M., Fadel, C.: Artificial intelligence in education: promises and implications for teaching and learning. Center for Curriculum Redesign (2019)

Huawei. The Intelligent World 2030 Report. Huawei Technologies Co., Ltd. (2017). https://www-file.huawei.com/Intelligent-World-2030-Report.pdf

Huang, R., Spector, J.M., Yang, J.: Educational TECHNOLOGY: A PRIMER for the 21st Century. Springer, Heidelberg (2020)

Ibrahim, R., Ahmed, V., Waseem, M.: Virtual reality in built environment education: bridging the gap between theory and practice. J. Eng. Educ. **113**(1), 88–106 (2024)

IDC Technologies. AI adoption in manufacturing and education: A global overview. IDC Research (2020)

Lin, H., Zhang, Y., Wu, F.: Virtual reality in education: enhancing engagement and learning outcomes—a review. Educ. Inf. Technol. **29**, 311–330 (2024)

Liu, Q., He, W., Wang, Y.: Using virtual reality to improve nursing education: a systematic review. Nurse Educ. Today **124**, 105805 (2023). https://doi.org/10.1016/j.nedt.2023.105805

Mohammed, A.-S.: Using generative AI to support teaching and learning: applications, implications, and challenges. Education Tech. Res. Dev. **71**, 995–1010 (2023)

OECD. AI and the future of skills: Insights from national strategies. Organisation for Economic Co-operation and Development (2024). https://www.oecd.org

Russell, S., Norvig, P., Davis, E.: Artificial Intelligence: A Modern Approach, 4th edn. Pearson Education (2018)

Sengar, V., Joshi, P., Ahuja, A.: Generative AI: foundations, applications, and challenges. Artif. Intell. Rev. **57**(1), 45–72 (2024). https://doi.org/10.1007/s10462-024-10412-3

Shalev-Shwartz, S., Ben-David, S.: Understanding Machine Learning: From Theory to Algorithms. Cambridge University Press, Cambridge (2014)

Siemens, G.: Learning analytics: the emergence of a discipline. Am. Behav. Sci. **57**(10), 1380–1400 (2013). https://doi.org/10.1177/0002764213498851

Sousa, M.J., Cruz, R., Martins, F.: Digital learning: developing skills for digital transformation in the context of industry 4.0. Education Sciences **9**(2), 140 (2019). https://doi.org/10.3390/educsci9020140

Staff, G.S.: Knewton brings adaptive and personalized learning to the masses. Getting Smart (2015). http://www.gettingsmart.com/2015/09/knewton-brings-adaptive-and-personalized-learning-to-the-masses/

UNCTAD. Technology and innovation in manufacturing: The role of artificial intelligence. United Nations Conference on Trade and Development (2017)

VanLehn, K.: The relative effectiveness of human tutoring, intelligent tutoring systems, and other tutoring systems. Educ. Psychol. **46**(4), 197–221 (2011). https://doi.org/10.1080/00461520.2011.611369

Support and Enhance Learning Communities AI Entrepreneurship Education Service Through the Focus Group

Candida Parlato(✉)

Insubria University, Varese, Italy
candi.parlato@gmail.com

Abstract. The concept of learning community refers to the process that brings together (into a community) otherwise individual actions of self-training for common learning goals, Wenger in the book dedicated to processing of the concept of community of practice (Wenger, 1998) describes the different characteristics of the learning that takes place there context and more generally in the interaction with the world. In the learning community, the key word is "Collaborate (colabore) means working together, which implies a sharing of tasks, and an explicit intention to "add value". Entrepreneurship education, is part of the theme, and has the aim of developing attitudes, knowledge, skills and competences in the participants, useful not only for their possible commitment in the entrepreneurial field, but in every working context and in every experience of active citizenship. The aim is to create a "transformative" tool to educate learning communities about entrepreneurship, the methodology will consist of a contextual and stateof-the-art analysis, implementing a SWOT Analysis (analysis of strengths), weaknesses, opportunities (opportunities) and threats (threats)) in progress and post intervention, using the focus group as a tool. The focus group can be a suitable tool to facilitate and build a common and effective language; in which a group of people is invited to talk, discuss and discuss attitudes personal to a topic. This dynamic of learning and change can be an effective way to connect and mediate between often contradictory demands regarding issues of environmental sustainability and companies.

Keywords: Learning community · focus group · SWOT Analysis

1 First Section

1.1 Nurturing Learning Communities

Entrepreneurship education extends beyond imparting technical skills; it aims to cultivate a mindset of innovation, adaptability, and collaboration. Learning communities play a pivotal role in achieving these goals. They provide a space for shared learning experiences, peer support, and collective growth. However, merely assembling individuals does not guarantee effective collaboration. Enter focus groups—a dynamic tool that

M. J. Sousa (Ed.): UFHRD 2024, ISEM 52, pp. 331–340, 2025.
https://doi.org/10.1007/978-3-031-92548-1_20

bridges theory and practice.The concept of a learning community refers to the process that brings together individual actions of self-training for common learning goals. In his book on the concept of community of practice, Wenger (1998) describes various characteristics of learning within such communities. These include:

Learning as Meaning Creation: Learning involves the construction of meaning. Learning as Identity Development: Learning contributes to the development of one's identity.

Learning as Belonging to a Community: Learning is intertwined with a sense of belonging to a community.

Learning as a Result of Practice within a Community: Learning emerges through active participation and practice within a community.

In the context of entrepreneurship education, the goal is to develop attitudes, knowledge, skills, and competences that are valuable not only in entrepreneurial contexts but also in other work settings and active citizenship experiences. These skills are considered transversal and applicable to life beyond entrepreneurship.

To create an effective educational tool for learning communities focused on entrepreneurship, the proposed methodology involves:

Contextual and State-of-the-Art Analysis: An analysis of the current context and existing knowledge.

SWOT Analysis: Assessing strengths, weaknesses, opportunities, and threats.

Focus Groups: Using qualitative discussions within interdisciplinary groups to build a common language and facilitate learning and change.

This approach aims to bridge conflicting demands related to environmental sustainability and entrepreneurship. By collaborating and adding value, learning communities can contribute to positive transformations in both fields.

1.2 Contextual and State-of-the-Art Analysis

Contextual analysis involves examining a subject or artifact within its broader context. It seeks to understand the surrounding circumstances, influences, and background; contextual analysis looks beyond the immediate object of study. It considers historical, cultural, social, and environmental factors that shape the subject. The goal is to gain deeper insights into the meaning, purpose, and significance of the subject. Contextual analysis answers questions like: Why was this created? What societal factors influenced it?

Instead, State-of-the-art analysis focuses on the current pinnacle of knowledge, technology, or practice in a specific field. It assesses the latest advancements and best practices, it examines the cutting-edge methods, tools, and theories that researchers and practitioners use. State-of-the-art analysis identifies trends and gaps and the researchers use it to stay informed, evaluate existing approaches, and identify areas for improvement.

In detail:

Contextual Analysis

Importance of Context: Understanding the context is crucial in various domains, including art, literature, and technology. It allows us to appreciate and interpret information more deeply.

External Factors: Context provides external factors that may have influenced a particular situation, artwork, or phenomenon. For instance, when analyzing an artwork, considering the historical, cultural, and social context can reveal hidden layers of meaning.

Enhanced Understanding: Context enhances our understanding by shedding light on the circumstances, motivations, and intentions behind a creation. It answers questions like: Why was this created? What was happening during that time?

Multiple Contexts: Artworks exist within multiple contexts simultaneously. These can include the artist's personal context, the cultural milieu, and the broader historical context.

State-of-the-Art Analysis

Defining State-of-the-Art: In the realm of research and technology, the term "state-ofthe-art" refers to the current highest level of achievement or advancement in a specific field. It represents the cutting-edge knowledge, techniques, and methodologies. Applications and Trends: State-of-the-art analysis involves examining the latest advancements, tools, and practices. Researchers explore what's currently considered the best practice and identify emerging trends.

Challenges and Gaps: It also highlights challenges and gaps in existing knowledge. By assessing the state-of-the-art, researchers can identify areas where further research is needed.

Evaluation Metrics: Researchers use evaluation metrics to measure the effectiveness of new approaches. These metrics help compare different methods objectively.

In summary, Contextual and State-of-the-Art Analysis are essential tools for under standing and advancing knowledge. Contextual analysis enriches our perception, while state-of-the-art analysis guides progress in research and innovation.

1.3 Swot Analysis

A SWOT analysis is a strategic planning tool used to evaluate the Strengths, Weaknesses, Opportunities, and Threats associated with a project, organization, or situation. Let's break down each component:

Strengths (S): These are internal factors that give an advantage. They could be skills, resources, or unique capabilities that contribute positively to the objective.

Weaknesses (W): These are internal factors that hinder progress. Identifying weaknesses helps in addressing them effectively.

Opportunities (O): These are external factors that can be leveraged to achieve the goal. Recognizing opportunities allows for strategic planning.

Threats (T): These are external factors that pose risks or challenges. Being aware of threats helps in risk mitigation.

In summary, a SWOT analysis provides a comprehensive view of the current situation, enabling informed decision-making and effective planning.

Let's explore some examples of opportunities and threats in the context of business:

Opportunities

Rapid Technological Advancements: Innovations in technology can create new markets, improve efficiency, and enhance products or services.

Market Expansion: A growing total addressable market (TAM) presents opportunities for businesses to reach more customers.

Changing Social Norms: Shifts in societal behavior can lead to the emergence of new markets or sub-segments.

Economic Trends: Positive economic changes can boost consumer spending and business growth.

Collaboration and Partnerships: Forming strategic alliances with other organizations can open up new possibilities.

Threats

Competition: Rival companies vying for the same market share can pose a threat.

Regulatory Changes: New laws or regulations may impact business operations.

Economic Downturns: Recession or economic instability can affect demand and revenue.

Changing Consumer Preferences: Shifts in consumer tastes can render existing products or services obsolete.

Supply Chain Disruptions: Dependence on specific suppliers or logistics disruptions can be risky.

Identifying opportunities and threats in your business involves a systematic analysis of both internal and external factors.

SWOT Analysis:

- Conduct a SWOT analysis (Strengths, Weaknesses, Opportunities, and Threats). Assess internal strengths and weaknesses, as well as external opportunities and threats.
- Strengths: Identify what your business does well, such as unique capabilities, resources, or competitive advantages.
- Weaknesses: Recognize areas where your business may be lacking, such as skill gaps or operational inefficiencies.
- Opportunities: Look externally for growth prospects, market trends, or emerging technologies.
- Threats: Consider external risks like competition, regulatory changes, or economic fluctuations.
- Market Research:

  Understand your industry, target market, and customer needs.
  Analyze market trends, customer preferences, and demand patterns.
  Investigate competitors and their strategies.

- PESTEL Analysis:

Evaluate external factors using a PESTEL analysis (Political, Economic, Social, Technological, Environmental, and Legal).A PESTEL analysis is a methodology for analyzing the external context of a business. It considers macrofactors from the macroenvironment and their impact on the company. The acronym PESTEL represents the following factors:

Political: Includes laws, government policies, political stability, and international relations.

Economic: Relates to interest rates, inflation, economic growth, and currency fluctuations.

Social: Considers cultural trends, demographics, social norms, and lifestyle changes.

Technological: Examines technological advancements, innovation, and digital transformation.

Environmental: Addresses environmental regulations, sustainability, and climate change.

Legal: Involves legal frameworks, compliance, and industry-specific regulations.

Consider how each factor impacts your business environment.

- Competitor Analysis:

  Study your competitors' strengths, weaknesses, and market positioning.
  Identify gaps where your business can excel.

- Customer Feedback:

  Gather feedback from existing customers.
  Understand their pain points, expectations, and satisfaction levels.

- Scenario Planning:

  Anticipate potential scenarios (positive and negative).
  Plan for contingencies based on different outcomes.

- Collaborate and Brainstorm:

  Involve your team in discussions.
  Brainstorm ideas, explore possibilities, and challenge assumptions.

1.4 Focus Group

The use of focus groups can significantly enhance communication and understanding within interdisciplinary groups. As a qualitative language-based technique, focus groups bring together individuals to discuss and explore personal attitudes related to specific topics. This dynamic process of learning and change serves as an effective bridge, mediating between conflicting demands concerning environmental sustainability and corporate interests.In today's complex and interconnected world, addressing environmental challenges requires collaboration across disciplines. The concept of a learning community—where individuals come together to share knowledge, experiences, and perspectives—plays a crucial role in fostering sustainable practices. This paper aims to explore the transformative potential of focus groups within learning communities, specifically in the context of environmental sustainability and business.Learning communities serve as incubators for collective knowledge creation. They go beyond individual learning by emphasizing shared goals, mutual support, and collaborative problem-solving. Within such communities, participants engage in meaningmaking—constructing understanding through dialogue, reflection, and interaction. Learning becomes a social process, shaped by context and community dynamics.A focus group is a valuable research method that brings together a small group of participants to discuss a specific topic or product. Here's a step-by-step guide:

Set Clear Goals

Define the purpose of your focus group. What specific insights or information are you seeking? Clarify your objectives.

Choose Relevant Participants

Select participants who represent your target audience. Include a mix of current customers and potential future customers.

Ensure diversity in demographics (age, gender, background) to capture varied perspectives.

Create an Agenda

Develop an agenda to keep the discussion focused. Key points to include:

Welcome: Introduce yourself and other researchers.

Participant Introductions: Allow participants to introduce themselves.

Ground Rules: Set guidelines for respectful communication.

Discussion Questions: Prepare open-ended questions related to your topic.

Wrap-Up: Summarize and thank participants.

Facilitate the Focus Group

Welcome participants warmly and establish rapport.

Encourage open dialogue and active listening.

Ask questions methodically, allowing everyone to share their insights.

Take notes or record the session (with participant consent).

Sample Questions

Here are some sample questions for your focus group:

"What do you think about our new product concept?"

"How would you use this service in your daily life?"

"What features are most important to you?"

"What concerns or reservations do you have?"

Analyze the Results

After the focus group, review your notes or recordings.

Look for patterns, common themes, and divergent opinions.

Use this qualitative data to inform your business strategies.

Focus groups offer a structured platform for dialogue. Here's how they work:

Participant Selection: A diverse group of stakeholders—academics, practitioners, policymakers, and community members—come together.

Facilitated Discussions:
A skilled moderator guides discussions around a central theme. Participants express their views, share experiences, and explore nuances.

Attitudes and Perspectives:
Focus groups delve into personal attitudes, beliefs, and values. They uncover hidden assumptions and reveal underlying motivations.

Common Language:
By engaging in dialogue, participants develop a common language. This shared vocabulary facilitates communication across disciplinary boundaries.

Environmental sustainability often clashes with corporate interests. Focus groups can mediate these tensions:

Identifying Trade-offs: Participants discuss trade-offs between profit-driven decisions and ecological well-being. They explore win-win solutions.

Stakeholder Perspectives: Focus groups include diverse stakeholders—environmentalists, business leaders, consumers, and policymakers. Their perspectives inform holistic strategies.

Innovative Solutions: By fostering creativity and critical thinking, focus groups generate innovative approaches. They challenge conventional wisdom and propose sustainable alternatives.

Consider a focus group comprising supply chain professionals. Their task: aligning business practices with environmental goals. Through facilitated discussions, they:

Map the Supply Chain: Identify environmental hotspots, from raw materials extraction to product disposal.

Collaborate on Solutions: Brainstorm ways to reduce waste, optimize logistics, and promote circular economy principles.

Negotiate Priorities: Balancing profit margins and ecological impact, they negotiate priorities and set actionable targets.

Focus groups serve as catalysts for change. By fostering dialogue, building common ground, and mediating conflicting demands, they empower learning communities to drive sustainable practices. As we navigate the intricate web of environmental challenges, let us embrace the power of collaborative conversations—one focus group at a time.

2 The Power of Focus Groups

2.1 Methodology: Implementing Effective Focus Groups

The Power of Focus Groups is:

- Contextual Understanding: Focus groups delve into contextual nuances. Participants discuss real-world challenges, drawing from their diverse backgrounds.
- Shared Language: By engaging in dialogue, participants develop a shared language. This linguistic bridge connects disciplines and fosters mutual comprehension.
- Attitudes and Beliefs: Focus groups uncover hidden attitudes and beliefs. These personal insights inform decision-making and strategy.

- Mediating Contradictions: Entrepreneurship often faces contradictory demands—profitability versus sustainability. Focus groups facilitate negotiation and creative problem-solving.

Implementing Effective Focus Groups is :

- Participant Selection: Curate a diverse group—students, entrepreneurs, educators, and industry experts.
- Topic Definition: Define the focus—whether it's sustainable business models, ethical practices, or innovation.
- Facilitator Skills: A skilled moderator guides discussions, ensuring active participation and respectful dialogue.
- Structured Sessions: Organize multiple sessions to explore different facets of entrepreneurship.
- Data Collection: Record discussions, capture key insights, and identify recurring themes.

Integrating Sustainability in Entrepreneurship Education:

- Scenario: A focus group of aspiring entrepreneurs explores sustainable practices.
- Discussion Points:
- Circular Economy: How can businesses adopt circular economy principles?
- Triple Bottom Line: Balancing profit, people, and planet.
- Eco-Innovation: Identifying eco-friendly product/service ideas.
- Collaboration: Opportunities for cross-disciplinary collaboration.
- 5. Outcomes and Recommendations:
- Shared Vision: Focus groups create a shared vision for sustainable entrepreneurship.
- Curriculum Enhancement: Use insights to enrich entrepreneurship courses.
- Policy Advocacy: Advocate for policies that promote responsible business practices.
- Long-Term Impact: Monitor participants' entrepreneurial journeys post-focus group.

2.2 Empowering Learning Communities

Focus groups are not mere discussions; they are catalysts for change. As we nurture learning communities, let us harness the power of dialogue, collaboration, and transformative thinking. By supporting entrepreneurship education through focus groups, we pave the way for a more sustainable and innovative future.

The triple bottom line (TBL) is a business concept that emphasizes measuring not only financial performance (the traditional "bottom line") but also social impact and environmental responsibility. Let's break it down:

Profit: This represents the financial aspect of a business. Maximizing profits has traditionally been the primary goal for companies in a capitalist economy.

People: Refers to the social dimension. It involves considering the well-being of employees, customers, communities, and other stakeholders. How does the business positively impact people's lives?

Planet: Focuses on environmental sustainability. Businesses should assess their ecological footprint, resource usage, and contributions to climate change. How can the company operate in an environmentally responsible manner?

By adopting the triple bottom line framework, organizations aim to create greater business value while addressing societal and environmental challenges.Environmental sustainability is the responsibility to conserve natural resources and protect global ecosystems to support health and well-being, both now and in the future. It encompasses several key aspects:

Conserving Resources: This involves using resources efficiently and minimizing waste. It includes practices like recycling, reducing energy consumption, and preserving biodiversity.

Protecting Ecosystems: Ecosystems provide essential services such as clean air, water, and soil. Environmental sustainability aims to prevent habitat destruction, pollution, and over exploitation.

Forward-Looking Approach: Because the impact of environmental decisions may not be immediately felt, sustainability considers the long-term effects. It's about meeting today's needs without compromising the ability of future generations to meet their needs1.

Regulations and Standards: Governments set environmental regulations to ensure responsible practices. These cover areas like air and water quality, waste management, and carbon emissions. Compliance with these standards is crucial for sustainability.

Balancing Profit and Planet: Businesses can adopt sustainable practices without compromising profitability. In fact, aligning profits with people and the planet is a win-win approach.Environmental sustainability focuses specifically on practices and actions that ensure the long-term health of our planet. It encompasses efforts to conserve natural resources, reduce pollution, protect ecosystems, and address climate change. Essentially, it's about balancing human needs with the well-being of the environment.

On the other hand, corporate social responsibility (CSR) is a broader concept. It refers to a company's commitment to operate ethically and contribute positively to society. CSR includes not only environmental aspects but also social and economic dimensions:

Environmental Responsibility: This aligns with environmental sustainability. Companies engage in practices like reducing their carbon footprint, using renewable energy, and minimizing waste.

Social Responsibility: CSR extends beyond the environment. It involves actions that benefit communities, employees, and other stakeholders. Examples include fair labor practices, philanthropy, and community development.

Economic Responsibility: Companies must remain financially viable. CSR encourages ethical business practices, transparency, and fair competition.

In summary, environmental sustainability is a subset of CSR, focusing specifically on ecological well-being. CSR encompasses a broader range of responsibilities, including social and economic aspects. Both are essential for responsible business practices.

3 Conclusion

Entrepreneurship education goes beyond traditional classroom learning. It emphasizes problem-solving, critical thinking, creativity, and collaboration. EE is relevant at all educational levels, from primary schools to higher education institutions. However, to truly support and enhance EE, we need to build strong learning communities that foster entrepreneurial mindsets. The Role of Focus Groups Focus groups provide a structured platform for dialogue and idea exchange. Here's how they can contribute to entrepreneurship education: Needs Assessment: Focus groups allow educators, students, and industry professionals to identify specific needs within the EE ecosystem. By discussing challenges, gaps, and opportunities, we can tailor our services effectively. Curriculum Development: Through focus groups, we can co-create relevant and engaging EE curricula. Participants can share insights on what topics, skills, and experiences should be included to prepare future entrepreneurs. Feedback and Improvement: Regular focus group sessions enable continuous improvement. Educators can receive feedback on teaching methods, resources, and program effectiveness. This iterative process ensures that EE remains dynamic and responsive. Networking and Collaboration: Focus groups bring together diverse stakeholders—teachers, students, entrepreneurs, policymakers, and industry representatives. These interactions foster networking, collaboration, and knowledge sharing. When organizing focus groups for entrepreneurship education, consider the following best practices: Diverse Representation: Invite participants from various backgrounds— students, educators, practitioners, and policymakers. Their diverse perspectives enrich discussions. Clear Objectives: Define the purpose of each focus group session. Are you exploring curriculum needs, evaluating a program, or brainstorming new initiatives? Facilitated Discussions: Skilled facilitators guide discussions, ensuring everyone has a chance to contribute. Encourage open dialogue and active listening. Actionable Outcomes: Focus groups should lead to actionable outcomes. Use the insights gathered to improve EE services, develop resources, or shape policies. Learning communities thrive when stakeholders collaborate. Focus groups provide a valuable space for dialogue, idea exchange, and collective problem-solving. By leveraging focus groups effectively, we can enhance entrepreneurship education and empower future innovators.

References

1. Bangalore, S., Rambow, O., Whittaker, S.: Evaluation metrics for generation. In: Proceedings of the First International Conference on Natural Language Generation, vol. 14, pp. 1–8. Assoc Comput Lingui, (2000)
2. Barry, E.S., Merkebu, J., Varpio, L.: State-of-the-art literature review methodology: a six-step approach for knowledge synthesis (2022)
3. Baud, R.H., Rassinoux, A.M., Scherrer, J.R.: ,Knowledge representation of discharge summaries. In: AIME 91, pp. 173–182. Springer, Heidelberg (1991)
4. Chomsky, N.: Aspects of the Theory of Syntax. MIT Press, Cambridge (1965)
5. Hill, T.R.: SWOT Analysis: It's Time for a Product Recall. Long Range Planning (1997)
6. Kolios, A., Read, G.: A political, economic, social, technology, legal and environmental (PESTLE) approach for risk identification of the tidalindustry in the united kingdom. Energies **6**(10), 5023–5045 (2013)
7. Makos, J.:, What is PESTLE Analysis? A Tool for Business Analysis. Weberience LLC (2011). http://PESTLEanalysis.com/what-is-PESTLE-analysis
8. Lewin, K.: Theory and experimentation in social psychology, Il Mulino (1951)
9. Maresca, S., Rizzello, S.: Insights for shaping entrepreneurship education: evidence from the European context. Sustainability **10**(11), 43233 (2018)
10. Wenger, E.: Community of Practice. Cambridge University Press, Cambridge (1998)

Becoming Change Agents at University: Challenges and Opportunity

Concetta Tino(✉) and Monica Fedeli

FISPPA Department, University of Padova, 35137 Padova, Italy
concetta.tino@unipd.it

Abstract. T4L for Change Agents (CAs) is a professional development project for volunteer faculty at University of Padova who want to play the role of facilitators and motivators of teaching innovation within their departments. Since 2018 two professional development paths for CAs have been realized, and two groups of faculties have been trained. This qualitative study was carried out during last wintertime to understand the transformation that the training experience have generated in terms of teachers' perspectives and teaching practices, and to know what challenges as CAs can be faced within academic contexts. In this study nine CAs were involved in a face-to-face interview. They belonged to different areas of knowledge. Findings show that training experiences offered them elements of dissonance that questioned their teaching and learning perspectives. The discoveries that faculties mentioned were related to the variety of methods, the role of students in the process of learning, the potentiality of feedback, the teacher-student relationship. The experience was impactful on their teaching practices. Concerning their role of CA all of them have recognized an opportunity in terms of personal and professional development, the possibility to be part of a valuable community at meso level. However, most of them have faced some challenges because of colleagues' resistance. Participants identified support and formal recognition of their role as conditions for a sustainable professional learning community.

Keywords: change agents · perspectives' transformation · professional learning community

1 Introduction

In 2016 at University of Padova (Unipd) developed a faculty professional development programme known as 'Teaching4Learning'(T4L), and consequently a process of change began, detecting significant transformations within teaching processes. Contributing to this process of change were those teachers who have voluntarily chosen to follow a dedicated training path to be able to play the role of Change Agent (CA) within their departments and the University Community. Exercising the role of CA in the university

Both the authors contributed to the development of the project design. Concetta Tino wrote the abstract; paragraphs: 2; 3; 4.2. Monica Fedeli wrote paragraphs: 1; 4.1. Both the authors wrote paragraph 5.

M. J. Sousa (Ed.): UFHRD 2024, ISEM 52, pp. 341–352, 2025.
https://doi.org/10.1007/978-3-031-92548-1_21

of Padova reality implies intersecting two trajectories and crossing the boundaries of two contexts: the internal reality of the departments and the broader "external" reality, which is the University Community.

In this context acting the role of CA means operating in two ways: on the one hand, as facilitators, mentors and animators of innovation for colleagues in their own departments, generating comparison and discussion on innovation of teaching/learning processes; on the other hand, as boundary spanners [1], as they are called to cross the boundaries of their department and their fields of knowledge to co-construct new sharing objects, made of practices, experiences, experimentation of methods and initiatives, in an attempt to contaminate the methods and processes of teaching and learning, as well as to open new perspectives within each Professional Learning Communities (PLCs). In fact, their role foresees the promotion of continuous professional development (CPD) among the teaching staff of their department through the care and growth of PLCs who can facilitate the transition from individual to collaborative professional development; to create a space where everyone can experience processes of sharing and mutual support [2], and recognize critical reflection as a determining factor that fuels and makes innovation and change sustainable, as well as the professional growth of teachers [3, 4]. The construction of PLCs is recognized as one of the most effective strategies to promote continuous and collaborative professional development [5, 6], where the role of facilitators becomes crucial for the growth and evolution of communities.

Therefore, the most important task for change agents (CAs) is to promote the development of PLCs as the key for sustainable change within their institutions. This task requires them to be aware of their own teaching/learning perspectives, and of their role within the community. Through the perspective of Transformative Learning [7, 8], and a qualitative approach, the study sought to explore: (i) What changes in learning and teaching they have experienced after training as CAs, (ii) What challenges they have faced within their departments, (iii) What elements need to be considered for the sustainability of a Professional Learning Community.

2 The Role of Change Agent in Promoting Continuing Professional Development (CPD) and Professional Learning Communities (PLCs)

In the current climate of change, which involves higher education institutions (HEIs), many studies analyze the role of faculty as Change Agents. They are considered as developers of innovative initiative to enhance teaching and learning processes [9]. They can act on three dimensions: (i) faculty development, when they provide resources and promote the growth of a faculty member such as a teacher and professional at an individual level; (ii) instructional development, when they focus on the quality of students' learning, and become supporters for curriculum definition, course design, and teaching; (iii) organizational development, when they implement effective strategies or programmes to engage students. [10]. According to this vision, effective faculty developers play a balanced role in offering pedagogy, learning, assessment, curricular programming, and participating in work groups, or projects to achieve organizational goals. O' Connel, Eddy, Iverson, and Mcdonald [11] consider change agents as catalysts of change, because they implement pedagogical practices to shape effective students' learning experiences and create inclusive learning environments; through their practices they influence other faculty and the culture of their department. Change Agents are also thought of as adult learners [12, 13]. There is a strong connection between adult situated learning, professional development and innovation and the sharing of teaching strategies with colleagues. Adult learners build new concepts based on their life experiences and recognize new learning as a tool for facing present issues [14]. According to Rogers [15] adults promote the process of innovation on the basis of four aspects: (i) the advantage that individuals can perceive from implementing new ideas; (ii) the level of compatibility with the values and culture already adopted; (iii) the level of complexity caused by the adoption of the new ideas; (iv) the testability of strategies, practices for a while before changing the entire course; (v) the observability that gives faculty the possibility to register successful change in students' learning. This typology of professional development experience allows change agents to adopt innovation and sharing approaches which suit their context, and needs. Based on these considerations the main role of T4L change agents is to enhance CPD through the development of PLCs to achieve one of the most important institutional goals: the excellence in teaching and learning.

PLC is not a professional development pathway guided by experts, but it is based on lifelong professional learning within a faculty work context, where change agents have the task of facilitating the sharing process of their expertise [16–18] and where a community for teachers' change and CPD can be built.

Effective change agents must investigate and understand the institutional pressures, cultures, and factors that create barriers between them and teachers [19]. Exercising the role of facilitators implies understanding teachers' contexts, enhancing their self-esteem, listening to them, implementing observation during the practice, and building a community of teachers. These are all essential elements for teachers' development [20]. Therefore, the most important task for change agents is to promote the development of PLCs as key of real change within their institutions and provide support for faculty as they work to transform their teaching pedagogy and practices.

Richardson and Placier [21] describe teacher change as a process that involves cognition and behaviors. Teacher cognition include all those beliefs, assumptions, knowledge, and perspective that the teachers nurture around teaching, learning, students, themselves, curricula, teaching activities and materials. Therefore, a process of teachers' change should promote reflection on teachers' beliefs [22] as they impact on teaching practices more than resources or policies [23]. While teachers' beliefs are implicit, teachers' behaviors are observable. In fact, they mirror the teaching practices implemented for supporting students' learning. The process of change around teachers' beliefs and behaviors can face resistance, and only a supportive PLC can motivate faculty to move from individual professionalism to collaborative professionalism, and facilitate them to work interdependently, exploring new curriculum, sharing new teaching strategies [24–26], and collectively experimenting with the way to face teaching innovation challenges. All this shows how the role, and the commitment of change agents are closely connected to the dimensions of CPD and PLC.

3 Transformative Learning

The potential of Transformative learning (TL) consists of questioning adults' rooted assumptions, known as 'frames of references and meaning perspectives' (beliefs and values) [27] that define individuals' identity and actions. In fact, the purpose of TL is to promote adult learners' critical reflection on their frame of references to recognize how they determine their world's interpretation and their decision-making process that influences their subsequent actions. Reflective practice acts as a powerful tool for highlighting the different aspects of being and allows for the consideration of alternative visions of individuals' identities; it is a process that enhances individuals' discovery of predominant perspectives. [28].

In precisely this way, when learners have an experience such as a disorienting dilemma, it allows them to critically assess their assumptions, and to develop a new open and perspective on the world. [7, 29]. This means that when learners recognize the validity of the new perspective, they start to plan their actions and to change their behaviors according to the new vision of the world. This change drives individuals' transformative learning process that usually happens in a safe and free environment, where people can exercise democratic thinking. [30] This aspect is strongly connected to the idea that the learning process is promoted by some contextual conditions where individuals can participate in a situated community and in the development of meaningful practices [31]. This means that the process of transformative learning is supported by the relationship. In fact, according to Taylor [32]"through trustful relationships that allow individuals to have questioning discussions, share information openly and achieve mutual and consensual understandings" (p.179) a significant process of transformation can happen. Therefore, a learning environment should provide conditions to allow learners to experience a critical incident [33] often identified in education as challenges that provoke teachers' reflections on their teaching practices [34, 35]. This reflective phenomenon happens during T4L path for CAs at Unipd. It stimulated the identification and questioning of teachers' assumptions and beliefs about teaching and learning, and while discovering unknown aspects of themselves and of teaching and learning processes, they looked for proper actions to face the challenges. This means that the power of reflective discourse empowers individuals to become open to a new understanding of themselves and others, and this motivates them to identify the importance of new skills in an encouraging environment [36]. Therefore, the reflection practice improves self-awareness and allows individuals to better understand their identities [37] and their assumptions. This shows that critical reflection is the central element of TL because it allows learners to interpret actions, or professional practices, and examine their belief systems [38, 39].

4 The Study

4.1 Methods and Participants

A case study methodology was adopted to carry out this study. Specifically, an explanatory approach was applied in this case study because it is useful to assess how and why interventions are working [40].

"T4L for change agents" is a 40 h-training path that was delivered in two different editions for volunteer faculty belonging to different areas of knowledge. The first edition was delivered in 2018 and the second one in 2021. The training paths have been delivered according to the approaches of active and student-centered learning, and with the aim of promoting critical reflection on individuals' teaching and learning perspectives. During each session participants learnt about constructivism perspective and were involved in the experimentation of active teaching/learning practices. Discussions and sharing of strengths and challenges of teaching, personal teaching/learning perspectives, and micro-teaching sessions followed by colleagues' feedback have been the heart of the programme.

At Unipd this study was approved by the Rector's Delegate for teacher training and innovative teaching. Then, through an email, the change agents were invited to participate in the study. From the list of the 40 trained change agents 9 of them (Table 1) gave their availability for participating in this study through a face-to-face semi-structured interview.

Table 1. Characteristics of the participants.

Position	Department	Group edition (ed)	Gender
Full professor	Industrial engineering	1st ed	M
Full professor	Economic-managerial engineering	1st ed	M
Associate professor	Environmental engineering	1st ed	F
Full professor	Comparative biomedicine and nutrition	1st ed	M
Full professor	Biomedical science	1st ed	F
Researcher	Biomedical science	2nd ed	F
Full professor	Mathematics	1st ed	M
Associate professor	Education	2nd ed	F
Full professor	Law	2nd ed	F

The interview, which lasted from 1 h to 1 h and 15 min, was focused on the following dimensions: i) changes in terms of teaching/learning perspectives and teaching practices after CAs training; ii) challenges faced in the role of CAs; iii) factors that need to be considered for creating a sustainable community.

The 9 interviews were carried out from November 2023 to January 2024 by one of the researchers in order to maintain consistency in data collected. With the participants' consent the interviews were audio recorded and then transcribed verbatim. To familiarize with collected data the researchers read the transcripts several times. Then, one researcher completed content analysis through ATLAS.ti 23 software. Phrases that were related to the aim of the study were coded and organized into categories and sub-categories. To guarantee the rigor analysis, a second researcher checked the data coding process.

To guarantee the reliability of the data a preview presentation of the analysis was presented to the whole group of CAs and the Rector's Delegate for teacher training and innovative teaching, who gave positive feedback on the results.

4.2 Results

From the analysis process there are three emerging themes which are summarised in Table 2. The themes were presented in the following.

Table 2. The three themes which emerged.

	Emerging themes
Theme 1	Critical reflection has supported teaching/learning perspectives and practices changes
Theme 2	Colleagues' resistance is the most difficult challenge for CAs
Theme 3	Formal recognition and support are the factors for a sustainable PLC

Critical Reflection has Supported Teaching/Learning Perspectives and Practice Changes. The training path gave CAs the opportunity to compare their teaching perspectives with dissonant elements of learning creating at the beginning of the training a sense of uncertainty, but then a sense of surprise at the discovery of new perspectives. In fact, during the training path the stimulus provided by instructors enhanced a deep reflection on participants' learning/teaching perspectives. Several participants gained awareness of having used teaching approaches based on their experience as students.

Training was a process that brought out a teacher-centered perspective as dominant characteristic among participants. In fact, participants often presented themselves in the class as talkative lecturers and the students as passive listeners who had gone to the class to collect content to memorize at home.

The formative class gave participants the opportunity to reflect on their experience as students and as teachers, and to critically revise their own perspectives according to the new discovery about teaching and learning. The elements of discovery and changes to which participants referred to were different as the participants' statements explained:

- The understanding of the role of teaching, no longer considered as a process of knowledge transfer, but as supportive process for learners and based on a holistic and active approach. *[…] Learning is a process which must be facilitated but which does not involve the simple transfer of knowledge from those who know something to those who do not know it. […] Teaching does not end with the transmission of knowledge, on the contrary, it is a minimal aspect of teaching. It foresees interaction and consideration of a personal way to learn.*
- The role of the teacher, moving from provider of knowledge to facilitator of learning. *[…] In our work we do not have to worry so much about the transfer of knowledge but rather about creating the conditions to facilitate learning.*

- The learning process no longer considered simply as achievable by transfer of knowledge, but as a transformative experience which implies active and emotional students' engagement. *[…] For me the main discovery was that the emotional level matters enormously in learning. So, when I prepare a lesson, I also must think about that aspect if I want learning to work well in the class.*
- The assessment, no longer just as summative but as formative process, identifying the potentiality of feedback to monitor and support students' learning, and to test teaching practices. *[…] Feedback has a big potential for students and teachers; it improves students' learning and as a mutual tool it improves the structure of the course, the choice of contents, teaching methods.*
- The teacher-students relationship is no longer too formal, but intentionally developed by teachers through the active listening, the care of proxemics during the class, the students' encouragement to participate in class activities and discussions. *[…] Now, I look for a relationship with my students because I think that they, more than anyone, know what they need and therefore it is necessary to give them a voice.*
- The variety of interactive teaching methods as way of re-imagining and reflecting on a new way of teaching and of shifting from a teacher-centered to a student-centered perspective. This new perspective involved reconsidering not only the choice of content, but above all the planning of lessons, which needs to include lesson objectives, activities, monitoring and feedback, choice of appropriate materials, and interactive strategies for learning. *The training path allowed me to think about teaching as I had never done. […] T4L path taught me a lot: from the choice of language, how to create relationships with students, how to move around the classroom, for example, how to pace explanations, how to organize breaks, how to involve students.*

Colleagues' Resistance is the Most Difficult Challenge for CAs. For the majority of participants colleagues' resistance has represented a real barrier to the innovation and changes. They expressed a general sense of dissatisfaction due to the difficulties encountered in generating change within their contexts. In fact, after an initial enthusiasm of CAs to try to involve everyone in the initiatives undertaken for the teaching innovation, the complex reality of some departments, due to the demonstration of resistance to change by colleagues, has led some CAs to choose to collaborate only with motivated people eager to improve teaching.

The phenomenon of resistance has generated mistrust, leading some CAs to remain completely inactive until now, even though they are planning to use a different strategy: for example, to *use technology as vehicle of innovation*, to *engage the new faculty in innovative initiatives*, to *train themselves on communicative skills*, considered useful to promote colleagues' engagement.

The resistance was thus explained by some participants as following: *[…] I am a little discouraged because mathematicians are part of a very traditional environment, there are those who still think that teaching is done with chalk on the blackboard; therefore, it becomes a little difficult to fight certain taboos and traditional beliefs.*

[…] the barriers are those of change because it costs effort and clashes with the idea of traditional role of the university.

However, it is important to highlight the positive experience of two CA who work in collaboration with other CAs in their own department. One of them underlined that in his department's resistance is a minority phenomenon compared to colleagues' engagement. The collaborative strategy with the other two CAs in his department helped them to implemented different innovative initiatives such as *peer observation, monitoring and supporting for teaching of new faculty.* This CA many times repeated that the key was probably the participation in T4L programme of the Head of his department. This was considered as an element of motivation for the others.

Another participant considered her position as CA and as President of Teaching Commission of her department impactful on colleagues' motivation. She stated: *it makes it easy to raise awareness for teaching innovation, the participation in the university calls for teaching projects, as well as for training as a CA.*

Formal Recognition and Support are the Factors for a Sustainable PLC. Before everything, it is important to underline how participants feel about belonging to two different PLCs: the first is the university CAs community, perceived as a safe space without any judgment, and where everyone find answers to their doubts, resources in terms of professional development, and the opportunity to also share teaching practices; the second is the departmental community, where the quality of the experience depends on the presence or lack of colleagues' resistance.

Some factors that participants consider as relevant for the sustainability of a departmental PLC emerged from all transcripts, as the reported statements show:

- The formal recognition of the role of CAs within the departments. Participants underlined that the formal recognition of the position as CAs can be one of the ways of justifying the innovative initiatives that they present to their colleagues. It would mean believing in the value of change in a systemic way. Therefore, [...] *if it is not clearly defined what this figure is, what contribution it can give, it becomes very difficult to generate change of teaching/learning culture. [...] it would be necessary to explicitly state what the expectations of the University from the PLCs are. This vision should be communicated to all CAs and to the whole personnel of the University.*
- The increase in the number of CAs within the departments. Most of participants think that the desired change of teaching culture requires the engagement of a larger number of CAs in each department, because it would support the awareness process among resistant colleagues, and it would be easier to implement innovative initiatives and to share new visions. *[...] It's a challenge in a large department [...]. For the entire teaching staff, one person as CA is too few, not so much for the numerical issue but for creating "consent".*
- The support that representatives with high positions in the department can offer for CAs and teaching innovation. Most of the participants were convinced that the effective engagement of university representatives with high position in the process of teaching innovation would support CAs in facing the challenge of resistance. [...] it would be important to engage in this awareness process *Presidents of different Teaching Commissions, Heads of Departments, the Teaching Advisors, and other representative who can support the innovation of the vision.*

5 Discussion and Conclusions

This study reports the results of the analysis of CAs case of University of Padova, where two groups of CAs attended T4L CAs training with the aim of generating changes within their departmental communities. The main goal of these figures is to play the role of facilitators and motivators for their colleagues in the process of teaching innovation, and in the building of a PLC as a sharing space of teaching practices, research, and solutions for facing teaching and learning challenges. Three research questions led this study. Concerning the first research question "What changes in learning and teaching have they experienced after training as CAs?", we can say that the professional development path determined a generative impact, both on the culture at professional/personal and organizational level and on the teaching practices. After the first disorienting learning experience, due to the discovery of interactive teaching strategies, the student-centred perspective, and the potential of feedback in supporting student learning, teachers began to develop a new vision of teaching and learning, developing new actions and attitudes to thinking and teaching. Therefore, the training path provided new visions that allowed participants to become "critically aware of how and why our presuppositions have come to constrain the way we perceive, understand, and feel about our world" [41] [p.14). The change of perspective is evidence not only of a transformative process relating to one's own perspectives, but also a demonstration of the teachers' availability to reflect on their teaching actions to improve them, thus starting a path of learning and professional growth [42].

At an organizational level there is an objective interest of the University of Padova for the innovative teaching processes because there is an investment of 1million of euros on teaching innovation every year, and each department can have access through the development of innovative teaching projects.

With reference to the second research question "What challenges have they faced within their departments?", we have identified CAs' initiatives their colleagues' resistance towards teaching innovation as a barrier for many. This means that the traditional culture of learning/teaching according to which still the teacher is "the central figure, the "sage on the stage" the one who has and transmits knowledge to the students, who simply memorize the information and later reproduce it on an exam, often without even thinking about it" is still dominant [43] (p.30). This old culture of teaching impedes the functions of CAs who should develop a PLC within their department as resources of CPD for all faculty, strategies, knowledge, building of solutions for teaching challenges. There is a good level community's maturity in two in only departments, where, probably due to a larger number of CAs, the engagement in teaching innovation of full professors have allowed the implementation of systemic practice innovation such as: peer observation, monitoring, and feedback of new faculty teaching. This shows the implementation of a CPD process, where the new faculty can find support and facilitators of teaching and where there is a positive environment for learning [16, 17].Therefore, at a micro level, we registered different experiences: some positive, some still challenging. The positive atmosphere was perceived by CAs at meso level, where the community of CAs in the university represents a comfort zone for all of them. It is made of faculty belonging to different areas of knowledge, and different departments. These aspects have not created barriers. On the contrary, they have created opportunities for faculty facilitating the boundary crossing process [1] and making individuals feel free and safe to participate in the development of a new culture and in the building of a community of shared teaching practices and ideas of innovation.

Regarding the third research question "What elements need to be considered for the sustainability of a Professional Learning Community?", we have identified the formal/institutional recognition of CAs' role as key elements for the sustainability of the PLC, because this would clearly show the strong interest of the University and the whole academic staff would be informed. It would also appear as an incentive to become a CA. Two other factors are for PLC sustainability are the increase of the number of CAs within the departments and the involvement in the awareness process of teaching innovation of academic representatives with high positions. This means that the more people are involved, the easier the process of culture change and innovation will be. The important aspect that emerges is that a systemic intervention would have the potentiality to generate an unexpected change, improving the relevant results already achieved, and with inversion of perspectives of resistant cultures.

5.1 Limitations

There are two main limitations to this study: Firstly, we present the results of a single case in Italy, because it is the only existing one. This does not allow us to compare it within other contexts. Secondly, even if we included in the study members of the two groups of CAs and belonging to different areas of knowledge, the small size of sample did not give us the possibility of knowing the perspectives of all CAs and how they feel to be part of a community.

Acknowledgments. We thank all the participants for generously contributing to this study.

References

1. Akkerman, S.F., Bakker, A.: Boundary crossing and boundary objects. Rev. Educ. Res. **81**(2), 132–169 (2011)
2. Cordingley, P., et al.: Developing great teaching: Lessons from the international reviews into effective professional development (2015). https://durham-repository.worktribe.com/output/1631202. Accessed 28 Apr 2024
3. Fedeli, M., Taylor, E.W.: Exploring the impact of a teacher study group in an Italian university. Formazione Insegnamento **XIV**(3), 166–178 (2016)
4. Stanley, A.M.: Professional development within collaborative teacher study groups: pitfalls and promises. Arts Educ. Policy Rev. **112**(2), 71–78 (2011)
5. Fedeli, M.: Migliorare la didattica universitaria: Il cambiamento organizzativo e il ruolo del Change Agent. Formazione Insegnamento **1**(1), 267–282 (2019)
6. Vangrieken, K., Meredith, C., Packer, T., Kyndt, E.: Teacher communities as a context for professional development: a systematic review. Teach. Teach. Educ. **61**, 47–59 (2017)
7. Mezirow, J.: Learning as Transformation: Critical Perspectives on a Theory in Progress. Jossey-Bass, San Francisco (2000)
8. Taylor, E.W.: Transformative learning theory. New Direct. Adult Contin. Educ. **119**, 5–15 (2008)

9. Grupp, L.L.: Faculty developer as change agent: a conceptual model for small institutions and beyond. J. Centers Teach. Learn. **6**, 45–58 (2014)
10. Cook, C.E., Meizlish, D.S.: Forging relationships with faculty and academic administrators. In: Cook, C.E., Kaplan, M. (eds.) Advancing the Culture of Teaching on Campus: How a Teaching Center Can Make a Difference, pp. 50–64. Stylus, Sterling (2011)
11. O'Connell, K., Eddy, P. L., Iverson, E. R., Macdonald, R. H.: Faculty change agent model: cultivating faculty to catalyze change. Change Maga. High. Learn. **54**(3), 11–18 (2022)
12. Eddy, P.L., Hao, Y., Markiewicz, C., Iverson, E.: Faculty change agents as adult learners: the power of situated learning. Commun. Coll. J. Res. Pract. **43**(8), 539–555 (2019)
13. Macdonald, R.H., et al.: Accelerating change: the power of faculty change agents to promote diversity and inclusive teaching practices. J. Geosci. Educ. **67**(4), 330–339 (2019)
14. Merriam, S.B., Bierema, L.L.: Adult Learning: Linking Theory and Practice. John Wiley & Sons, Hoboken (2013)
15. Rogers, E.M.: Diffusion of Innovations, 4th edn. Simon and Schuster, New York (2010)
16. Grossman, P., Wineburg, S., Woolworth, S.: Toward a theory of teacher community. Teach. Coll. Rec. **103**(6), 942–1012 (2001)
17. Stoll, L., Bolam, R., McMahon, A., Wallace, M., Thomas, S.: Professional learning communities: a review of the literature. J. Educ. Change **7**(4), 221–258 (2006)
18. Wenger, E.: Communities of Practice: Learning, Meaning, and Identity. Cambridge University Press, Cambridge (1998)
19. Fletcher, T., Beckey, A., Larsson, H., MacPhail, A.: The research and development challenge: better aligning teachers' and teacher educators' needs, priorities and demands. In: MacPhail, A., Lawson, A.H. (eds.) School Physical Education and Teacher Education, pp. 116–127. Routledge, Abingdon (2020)
20. Patton, K., Parker, M., Neutzling, M.M.: Tennis shoes required: the role of the facilitator in professional development. Res. Q. Exerc. Sport **83**(4), 522–532 (2012)
21. Richardson, V., Placier, P.: Teacher change. In Richardson V. (Ed.), Handbook of Research on Teaching, 4th edn., pp. 905–947. American Educational Research Association, Washington, DC (2001)
22. Tam, C.F.: How teaching beliefs of chinese language teachers influence their teaching practice. J. Curric. Stud. **1**(2), 97–124 (2006)
23. Cronin-Jones, L.L.: Science teacher beliefs and their influence on curriculum implementation: two case studies. J. Res. Sci. Teach. **28**, 235–250 (1991)
24. Harris, A., Jones, M.: Professional learning communities and system improvement. Improv. Sch. **13**, 172–181 (2010)
25. Kaasila, R., Lauriala, A.: Towards a collaborative, interactionist model of teacher change. Teach. Teach. Educ. **26**, 854–862 (2010)
26. Tam, C.F.: The development of an effective teacher learning community: a case study of Hong Kong. J. Educ. Res. Dev. **7**(2), 213 (2011)
27. Mezirow, J.: Education for Perspective Transformation: Women's Re-entry Programs in Community Colleges. Teacher's College, Columbia University, New York (1978)
28. Harrison, R.T., Leitch, C.M.: The Dynamics of Identity, Identity Work and Identity Formation in the Family Business: Insights from Identity Process Theory and Transformative Learning. Springer, Heidelberg (2019)
29. Taylor, E.W.: Intercultural competency: a transformative learning process. Adult Educ. Q. **44**(3), 154–174 (1994)
30. Lake, D.: Community building in the classroom: teaching democratic thinking through practicing democratic thinking. Partnerships J. Serv.-Learn. Civic Engage. **6**(1), 5–24 (2015)
31. Lave, J., Wenger, E.: Situated Learning: Legitimate Peripheral Participation. Cambridge University Press, Cambridge (1991)

32. Taylor, E.W.: An update of transformative learning theory: a critical review of the empirical research (1999–2005). Int. J. Lifelong Educ. **26**(2), 173–191 (2007)
33. Mezirow, J.: Toward a learning theory of adult literacy. Adult Basic Educ. **6**(3), 115126 (1996)
34. Kissock, C., Richardson, P.: Calling for action within the teaching profession: it is time to internationalize teacher education. Teach. Educ. **21**(1), 89–101 (2010)
35. Leigh, L.: Of course I have changed!: a narrative inquiry of foreign teachers' professional identities in Shenzhen, China. Teach. Teach. Educ. **86**, 102905 (2019)
36. Eschenbacher, S.: Transformative learning and the hidden dynamics of transformation. Reflective Pract. **21**(6), 759–772 (2020)
37. Mau, S.T., Harkness, S.S.: The role of teacher educators and university supervisors to help student teachers reflect: from monological reflection toward dialogical conversation. Reflective Pract. **21**(2), 171–182 (2020)
38. Cranton, P.: Types of group learning. New Direct. Adult Contin. Educ. **71**, 25–32 (1996)
39. Mezirow, J.: On critical reflection. Adult Educ. Q. **48**(3), 185–198 (1998)
40. Yin, R.K.: Case Study Research: Design and Methods, 2nd edn. Sage Publishing, Thousand Oaks (1994)
41. Mezirow, J.: How critical reflection triggers transformative learning. In: Mezirow, J. (eds.) Fostering Critical Reflection in Adulthood, pp. 1–20. Jossey-Bass Publishers, San Francisco (1990)
42. Schön, D.: The Reflective Practitioner: How Professionals Think in Action. Basic Books (1983)
43. King, A.: From sage on the stage to guide on the side. Coll. Teach. **41**(1), 30–35 (1993)

Educating the Digital Native Generation of Physicians in American Medicine: Generation Z in Graduate Medical Education Training Programs

Jennifer L. Pomicter(✉)

The George Washington University, Washington, D.C., USA
jpomicter@gwu.edu

Abstract. This literature review investigates the intersection of digital learning and Graduate Medical Education (GME) training, focusing on Generation Z medical trainees. Generation Z, the first true digital native generation, exhibits distinct learning preferences, heavily reliant on digital modalities. The COVID-19 pandemic has accelerated this trend, necessitating adaptation within medical education. Three overarching themes emerge: utilization and attitudes towards digital tools, efficacy and impact of digital learning interventions, and factors influencing readiness and satisfaction in clinical learning environments. While many studies lack theoretical frameworks, they highlight the commonality of the benefit of tailored educational strategies and digital integration in medical training. Future research should focus on GME trainees in the American medical system to understand if preferences shown in nursing and medical students translate to residency. Understanding Generation Z's digital learning needs is vital to optimizing graduate medical education and preparing future healthcare leaders.

Keywords: Generation Z · Digital Learning · Graduate Medical Education · Clinical Learning Environment · Clinical Education

1 Introduction

As medical students graduate from medical schools in the United States of America, they prepare for the next step in their journey to becoming physicians. This transition is marked by the students becoming Graduate Medical Education (GME) Trainees, also known as Interns or Residents in American medical education (Physicians and Surgeons: Occupational Outlook Handbook, 2023). Their chosen specializations within medicine, such as Internal Medicine, Pediatrics, Dermatology, or Surgery, lead these trainees to their specialized training program or Residency (ACGME Institutional Requirements, 2022; Fairbrother, 2023). These residency training programs can range from three to five years or more, depending on the specialty of training (ACGME Specialties, no date). During these years of education and training new physicians not only gain foundational knowledge of their chosen specialty but also the clinical experience and familiarity

M. J. Sousa (Ed.): UFHRD 2024, ISEM 52, pp. 353–363, 2025.
https://doi.org/10.1007/978-3-031-92548-1_22

necessary to interact with patients and navigate the American medical system. This type of learning is rooted in progressive interaction with superiors, peers, and patients in the clinical setting (Ryan et al., 2022). Furthermore, because GME training takes place in a real-world clinical setting, it is impacted by current and historical events (Silkens et al., 2023). Thus, the years spent in Residency are vital to the maturation of autonomously functioning physicians, ready to practice independently (Ryan et al., 2022).

The current generation of medical students preparing to graduate medical school and enter Residency demonstrate distinct learning preferences (Eckleberry-Hunt, Lick and Hunt, 2018). As Generation Z (born 1995 – 2010) is known as the first truly digital native generation, they rely more heavily on digital learning modalities than previous generations in higher education settings (Eckleberry-Hunt, Lick and Hunt, 2018; Shorey et al., 2021). The recent global COVID-19 pandemic has also greatly influenced their medical school learning experiences (Gallagher et al., 2022). This paradigm shift underscores the need for GME programs to adapt and integrate digital learning into residency training to engage this cohort effectively.

The efficacy of GME programs in preparing these trainees for the rigors of clinical practice hinges not only on acquiring medical knowledge and technical skills but also on cultivating professional competencies and adaptive learning behaviors. While national accrediting bodies outline the basic requirements for GME training completion, individual programs are left to interpret the execution of training to meet these requirements and the needs of their trainees (ACGME Common Program Requirements - Residency, 2023). Against this backdrop, it becomes imperative to explore the intersection of digital learning and GME training to optimize educational outcomes and support the development of future healthcare leaders.

This review draws upon a diverse array of scholarly articles, empirical studies, and theoretical frameworks to provide a comprehensive understanding of the current landscape and identify future areas of research and practice.

2 Methods

This literature review seeks to inform the multifaceted relationship between digital learning and GME training among Generation Z medical residents. The search process was informed by "Writing Literature Reviews: A Reprise and Update" (Callahan, 2014).

2.1 Selection Criteria

Empirical articles written in, or available by translation to, English without restrictions on publication date are included. Keywords used to refine the search are combinations of the following key terms: "Generation Z" AND (Learning OR "Digital Learning") AND ("Clinical Education" OR "Graduate Medical Education" OR "Clinical Learning"). Theoretical, Conceptual, and Literature Review articles were excluded. Articles that were ultimately deemed unrelated were also excluded.

2.2 Searches

Organized searches were conducted in the following databases in April 2024: Medline, a database that focuses on life and health sciences scholarly work; ProQuest, a large set of 42 databases covering a wide variety of literature; ERIC, a database for education research, Web of Science, a large set of databases covering scholarly journals and conference proceedings, and Google Scholar.

All database searches used each combination of key terms. The search strategy was limited by not searching the full text for the identified key terms. The database-specific parameters are as follows: Medline: Title/Abstract ONLY, ProQuest (All): Not Full Text, ERIC: Abstract Only, and Web of Science: Abstract Only.

2.3 Data Collection and Data Analysis

These databases and iterative keyword searches returned 176 articles in total. Non-empirical articles, articles unrelated to medical education, and articles unrelated to Generation Z were discarded, leaving 34 articles. Once duplicates were removed, 17 articles remained and were chosen for full review. Ultimately, 12 articles were included in this literature review study.

3 Organization of the Literature Review Findings

Based on the review of articles chosen for this literature review, three themes emerged as significant when seeking to understand the intersectionality of digital learning and GME training when accounting for Generation Z. 1: Uses and perspectives toward digital learning tools. What digital tools is Generation Z using for learning and networking, and how does this compare to current use by other generations (Persada et al. 2019; Lin et al. 2021, 2022; Yurdal et al., 2021; Picton, 2023)? 2: Effectiveness and impact of digital learning interventions. How effective are these digital learning strategies within a clinical learning environment, and what are some of the challenges faced (Ho et al., 2021; Manik et al., 2022; Sedaghatkar et al., 2023; Jiang et al., 2024)? 3: Factors influencing readiness and satisfaction in the clinical learning environment. How does the presence or absence of digital learning opportunities within clinical learning affect Generation Z trainees (Lee, 2023; DiMattio and Hudacek, 2020; Vizcaya-Moreno and Pérez-Cañaveras, 2020)? Each of these themes will be expanded in detail in the following sections.

3.1 Generation Z Usage and Perspectives Toward Digital Learning Tools

In their 2021 study, Lin et al. took a closer look at digital learning tool utilization and digital health tool comfortability among speech-language professionals in German-speaking countries. The study used a cross-sectional survey design with a convenience sample to explore knowledge, attitudes, and preferences toward digital tools, considering factors such as profession, generation, and experience. The study revealed variations in familiarity with digital learning terms, with Generation X showing greater knowledge compared to Generation Z and Millennials. While over three-quarters of respondents were familiar with digital learning tools, attitudes toward them varied, with benefits including flexibility and accessibility and concerns related to limited interaction and effectiveness for complex topics. Ultimately, Generation Z showed higher usage of digital learning tools. This study provided insights into digital learning preferences, interdisciplinary collaboration, and addressing barriers to utilization and comfort. Understanding the digital competency of Generation Z can aid in developing tailored educational strategies to ensure proficiency in clinical practice. Future studies were recommended to measure the implementation and retention of digital learning and training of clinical instructors, as well as studies conducted in other medical specialties. Proactively addressing challenges identified in interdisciplinary contexts can better prepare physicians for collaborative, digitally driven healthcare education (Lin et al. 2021).

In a 2022 follow-up study, Lin et al. explored the perceptions of speech-language students in several specialties in German-speaking countries on a digital learning toolbox app. The study utilized a structured questionnaire to assess interest levels and concerns and identify desired features, functions, and content. The study revealed strong interest among participants in adopting a digital learning app tailored to clinical speech-language education, with 98.8% expressing openness to trialing the proposed app. Key features like tutorials and filters were well-received. However, concerns over data privacy were raised. Suggestions for future research included refining the app, addressing privacy concerns, and tailoring the content to ensure usability, effectiveness, and knowledge retention (Lin et al. 2022).

Research was also conducted by Persada et al. (2019) to investigate how Generation Z, specifically, perceives and engages with digital learning platforms. The Unified Theory of Acceptance and Use of Technology (UTAUT) (Xavier and Oliveira, 2016) was applied while examining factors influencing Generation Z's intention to adopt and utilize these technologies. Their behavior in digital learning environments was explored using confirmatory factor analysis of a Likert-scale questionnaire sent to a purposive sample. The study found varied motivations for using digital learning tools, but the strongest correlation was found between facilitating conditions and intention. This result highlights the potential for enhancing engagement through user-friendly resources. Future research recommendations included broadening the study participant population and exploring the impact of Generation Z's desire for work-life balance on their desire to utilize digital learning and how this ultimately impacts readiness for independent medical practice (Persada et al. 2019).

In their 2023 dissertation research, Picton investigates how Generation Z diagnostic radiography students utilize social media for professional learning, aiming to develop a substantive theory grounded in student perspectives. This study employed a constructivist grounded theory approach to explore the interactions, perceptions, navigation patterns, barriers, and facilitators related to social media engagement among these students. Picton's findings reveal that Generation Z radiography students extensively utilize social media for learning, networking, and professional development. Online social platforms provided access to diverse educational resources and fostered collaboration, knowledge sharing, and community. Integrating social media into their education bridged traditional methods with contemporary digital practices, enabling access to resources, collaboration, and professional development opportunities. Further study regarding additional medical specialties and faculty development was suggested, as well as investigating how to align educational program policies with positive, evidence-based approaches utilizing social media and digital learning (Picton, 2023).

Finally, Yurdal et al. conducted a study to create and validate the Medical School Students' Attitudes towards Distance Education/Online Learning Scale (MATDLS) and explore its association with e-learning styles among medical school students during the COVID-19 pandemic. A quantitative correlational approach was utilized, with data gathered from 815 medical students. The analysis showed the complexity of medical students' attitudes towards online learning and e-learning styles. While a general preference for e-learning was evident, attitudes towards online medical education, specifically, were more ambivalent. Independent learning emerged as a significant predictor of positive attitudes towards online learning, contrasting with audio-visual and independent learning styles, which correlated with negative attitudes towards medical education. These findings emphasize the importance of tailoring educational strategies to accommodate diverse learning styles and attitudes among Generation Z medical students. Further investigation into specific interventions and instructional designs and exploration of different modalities of digital medical education were suggested (Yurdal et al., 2021).

These studies illuminate Generation Z's usage and perspectives toward digital learning tools within medical education contexts. While the studies vary in methodological approach and context, each study found that Generation Z is highly familiar and comfortable with digital learning tools and platforms. Each group of Generation Z participants showed interest in incorporating digital learning into their education but with accommodations for diverse learning styles and preferences. Lin et al. (Lin et al. 2021, 2022) and Persada et al. (Persada et al. 2019) reveal that Generation Z's familiarity with digital tools showcases their higher usage compared to other generations. Additionally, Lin et al. (Lin et al. 2021, 2022) and Picton (Picton, 2023) highlight Generation Z's openness to and extensive use of digital platforms for learning and professional development. Yurdal et al. (Yurdal et al., 2021) demonstrate the complexity of attitudes toward online medical education among medical students. These findings collectively suggest that Generation Z relies significantly on and is comfortable with digital learning tools. This indicates the necessity for tailored educational strategies within medical training programs to align with the preferences of incoming GME trainees to enhance learning outcomes.

3.2 Efficacy and Impact of Digital Learning Interventions

In their 2021 study, Ho et al. investigated the impact of a teaching intervention program utilizing the iLearning app in enhancing clinical reasoning ability and selfdirected learning regarding chest tube care among nursing students. Using a quasi-experimental design with cluster sampling, the intervention group utilized the iLearning app, while the comparison group received traditional teaching only. Clinical reasoning and self-directed learning data were analyzed using Generalized Estimating Equations (GEE) and regression analysis. The study findings reveal significant enhancements in clinical reasoning and self-directed learning among nursing students who utilized the iLearning app compared to those who received only traditional instruction. Instructors' guidance outside class was vital in maximizing the app's effectiveness, highlighting the importance of ongoing support and mentorship in digital learning environments. Additionally, the flexibility offered by mobile learning improved self-directed learning scores, while the incorporation of gamification fostered engagement and collaboration among students. Recommendations for future research include improving Cloud servers to mitigate interference in the learning process (Ho et al., 2021).

Jiang et al. also studied Generation Z nursing students in an Asian country in their 2024 study, seeking to understand the nuanced experiences and perceptions of these students during their ICU practicum. By examining the intersection of Generation Z upbringing in China, societal influences, and the educational journey of nurses, their study illuminates the multifaceted dynamics shaping the learning experiences of Generation Z students in medical education in China. Employing a phenomenological approach, semi-structured interviews with 15 Generation Z nursing students are analyzed to capture their lived experiences. The findings of the study found that integrating digital learning enhanced efficiency and safety in critical care settings. Despite these benefits, these students still perceived challenges to be evident, such as gaps between theory and practice, as well as interpersonal issues. However, when online and offline learning was integrated with positive role modeling by instructors, these Generation Z nursing students had enhanced knowledge acquisition and performance compared to the traditionally taught cohort (Jiang et al., 2024).

In a 2022 study, Manik et al. sought to understand Indonesian nursing students' viewpoints regarding the efficacy of vSim for Nursing™, a virtual simulation tool, in facilitating nursing skill acquisition through online learning during the COVID. The experiences of these nursing students were captured through qualitative interviews, and then thematic analysis was performed to identify perceived benefits and challenges related to digital learning through the app. The findings show that these nursing students, part of Generation Z, viewed the vSim for Nursing™ app as a valuable tool for refining clinical skills despite encountering technical proficiency and cultural adaptation challenges. While students acknowledged its potential benefits, technical difficulties impeded their learning. Moreover, cultural disparities posed barriers to fully engaging with simulated scenarios. Future research recommendations included exploring strategies to improve the integration of digital tools as well as investigating culturally responsive simulation approaches to foster inclusivity and accessibility across diverse cultural backgrounds (Manik et al., 2022).

Finally, Sedaghatkar et al. published a 2023 study evaluating the impact of integrating microlearning (ML) and task-based learning (TBL) on fourth-year medical students' knowledge acquisition and performance during their otolaryngology rotation. Employing a quasi-experimental design with pre and post-tests, the study compares traditional teaching methods and TBL with a combination of ML and TBL. Direct Observation of Procedural Skills (DOPS) assessments were utilized to measure skill proficiency. Study results showed that combining ML and TBL demonstrated significant enhancements in both knowledge and performance within the intervention group, emphasizing ML's asynchronous nature for just-intime learning and TBL's effectiveness in engaging students in authentic tasks to improve problem-solving skills. Based on the small sample size and single hospital setting, suggestions for future research focus on diversifying the clinical settings to understand better the preferences, challenges, and perceptions of digital learning for Generation Z medical trainees (Sedaghatkar et al., 2023).

Together, these studies emphasize the efficacy and impact of digital learning interventions, particularly among Generation Z medical trainees. As mentioned above, these studies vary in study settings, intervention types, and methodology.

However, they each concluded that there is efficacy in incorporating digital learning interventions within medical education, particularly among Generation Z trainees. Ho et al. demonstrated significant enhancements in clinical reasoning and self-directed learning among nursing students utilizing the iLearning app (Ho et al., 2021). At the same time, Manik et al. found that a simulation app facilitated procedure skill learning (Manik et al., 2022). It was also found that digital learning improved efficiency and safety in critical care settings (Jiang et al., 2024). In contrast, combined with task-based learning, microlearning significantly improved knowledge retention during a clinical rotation (Sedaghatkar et al., 2023). These studies underscore the efficacy of digital learning modalities in the retention of skills and medical knowledge in a clinical setting.

3.3 Factors Influencing Readiness and Satisfaction in Clinical Learning Environments

In this 2020 study, DiMattio and Hudacek aim to assess how specific psychosocial dimensions of the hospital clinical learning environment influence satisfaction among Generation Z nursing students in the United States. Grounded Psychosocial Learning Environment Theory, this study explores the interplay between psychosocial dimensions and the clinical learning environment. The researchers utilized a revised version of the Clinical Learning Environment Inventory (CLEI) to understand these psychosocial dimensions and their impact on the satisfaction of Generation Z nursing students in the clinical learning environment. Utilizing correlational regression analysis of the CLEI data, the study identified three key aspects of the clinical learning environment that predicted nursing student satisfaction: student decision-making, organized effective teaching, and concern for student welfare. These findings show that supporting student autonomy, clarifying expectations, and cultivating positive relationships enhance satisfaction among Generation Z nursing students. A suggestion for future research is to incorporate the impact of psychosocial dimensions of the clinical learning environment on the clinical judgment skills of these students (DiMattio and Hudacek, 2020).

Social media usage can also influence satisfaction and readiness for clinical practice, as suggested in the 2020 study by Vizcaya-Moreno and Pérez-Cañaveras. Using a descriptive cross-sectional survey methodology involving 120 Generation Z Spanish nursing degree students, this study examines potential associations between students' academic year, social media usage, teaching preferences, and perceptions of Generation Z characteristics. The findings indicate that, on average, students spend more time on social media for personal reasons rather than for clinical learning. While Google Currents/Google + emerged as the preferred platform for clinical learning, a weak correlation between academic year and Twitter usage for clinical learning was found. This suggests a potential shift in social media preferences over the course of students' academic progression. Regarding teaching method preferences, participants favored linking mentorship, digitalization, and gamification. Most students identified as high consumers of technology while also acknowledging the increased risks to mental health associated with such technology use. Recommendations for future research centered around exploring the efficacy and consequences of preferred teaching methods and considering the effects on patient safety and data protection in using social media for clinical learning purposes (Vizcaya-Moreno and Pérez-Cañaveras, 2020).

In the 2023 study, Lee et al. investigated the impact of personal factors and anxiety during clinical practicum rotations on the readiness for independent practice of Generation Z Korean senior nursing students. Utilizing Astin's theory of student involvement, the research employs a cross-sectional questionnaire design with 838 participants and analysis by structural equation modeling to identify relationships between variables. The findings indicate a moderate level of readiness to practice influenced by personal factors such as self-esteem, depression, and satisfaction with work-life balance. Positive correlations were observed between readiness and self-esteem, satisfaction with work-life balance, and the clinical learning environment. In contrast, negative correlations were found with depression and anxiety during clinical practicum. Notably, the study emphasizes the significant impact of the clinical learning environment on readiness to practice. Recommendations for future research include longitudinal studies to explore how these factors influence readiness to practice over time and evaluate the readiness to practice of the senior nursing students as perceived by nursing faculty and nursing educators in practice settings (Lee, 2023).

These studies illustrate the multifaceted nature of factors influencing readiness and satisfaction in the clinical learning environment for Generation Z medical trainees. DiMattio and Hudacek's study underscores the importance of psychosocial dimensions such as personal decision-making, effective teaching, and concern for student welfare in fostering satisfaction (DiMattio and Hudacek, 2020). VizcayaMoreno and Pérez-Cañaveras highlight the role of social media usage and teaching preferences in shaping readiness and satisfaction (Vizcaya-Moreno and Pérez-Cañaveras, 2020). Lee et al.'s findings emphasize personal factors alongside perceptions of clinical learning in determining readiness (Lee, 2023). These studies highlight how various dimensions of the lived experiences of Generation Z medical trainees influence satisfaction in clinical learning and, ultimately, readiness to practice.

4 Discussion

The studies presented in this literature review highlight the desires of Generation Z medical profession students and trainees to be provided with digital learning opportunities and customizable learning programs (Persada et al. 2019; Ho et al., 2021; Lin et al. 2021, 2022; Yurdal et al., 2021; Picton, 2023; Sedaghatkar et al., 2023; Jiang et al., 2024). Tailoring educational offerings by incorporating gamification (Vizcaya-Moreno and PérezCañaveras, 2020; Ho et al., 2021), simulation (Manik et al., 2022), and mentorship (Vizcaya-Moreno and Pérez-Cañaveras, 2020) enhances the learning experiences of this generation.

Generation Z is familiar with and comfortable with digital learning tools (Persada et al. 2019; Lin et al. 2021, 2022; Picton, 2023), which is indicative of this digital native generation (EckleberryHunt et al. 2018; Shorey et al., 2021). When presented with opportunities to utilize clinical learning apps or social media, the study participants have shown that these new avenues of learning enhance rather than detract from traditional learning in the clinical environment (Ho et al., 2021; Manik et al., 2022). When combined with task-based learning, microlearning significantly increases the retention of clinical knowledge and skills (Sedaghatkar et al., 2023). Together, these findings suggest that Generation Z can understand and quickly absorb clinical knowledge and that they prefer to, at least, have the opportunity to learn in this manner if it suits their learning style (Persada et al. 2019; Ho et al., 2021; Lin et al. 2021, 2022; Manik et al., 2022; Picton, 2023).

Generation Z is also highly affected by psychosocial dimensions and personal factors such as work-life balance. Whether these factors are felt in the clinical setting or in their personal lives, they impact clinical learning satisfaction and readiness to practice independently (Lee, 2023; DiMattio and Hudacek, 2020; Jiang et al., 2024). This generation is also cognizant of their extensive use of technology and how this can affect their well-being (Vizcaya-Moreno and PérezCañaveras, 2020; Picton, 2023). Study participants were aware of and influenced by anxiety and depression, which could impact their clinical training (Lee, 2023; DiMattio and Hudacek, 2020).

Furthermore, while these studies were thoroughly planned and executed, many lacked the underpinnings of explicit theoretical frameworks. Only four of the 12 scholarly works explicitly identified a theoretical framework. This could be attributed to omission by the researchers or the publishing journals. One could wonder if medical/clinical education research, in general, is lagging behind the move to incorporate research anchored in established theory (Lee, 2023; Persada et al. 2019; DiMattio and Hudacek, 2020; Picton, 2023).

It was also noted that while this literature review aimed to understand the digital learning needs of American GME Residents, studies involving this specific group have yet to be published. This can be attributed to the fact that American Graduate Medical Education (Residency) is typically begun by trainees at least 26 years old (Physicians and Surgeons: Occupational Outlook Handbook, 2023). This would mean that Generation Z GME trainees are only beginning to enter Residency. At the same time, the Generation Z students discussed in this literature review come from nursing programs (Lee, 2023; DiMattio and Hudacek, 2020; Vizcaya-Moreno and Pérez-Cañaveras, 2020; Manik et al., 2022; Jiang et al., 2024), medical schools (Yurdal et al., 2021; Sedaghatkar et al., 2023), or other programs outside of the American GME system (Lee, 2023; Vizcaya-Moreno and Pérez-Cañaveras, 2020; Lin et al. 2021, 2022; Manik et al., 2022; Jiang et al., 2024). Because of this gap in scholarly work, future research should specifically involve American GME trainees as they navigate their Residency training. Studies discussed in this review involve Generation Z nursing students (Lee, 2023; DiMattio and Hudacek, 2020; Vizcaya-Moreno and Pérez-Cañaveras, 2020; Manik et al., 2022; Jiang et al., 2024), and medical students (Vizcaya-Moreno and PérezCañaveras, 2020; Yurdal et al., 2021) show commonalities among students across the medical professions, and it would be beneficial to the medical education community to understand if these preferences translate to GME trainees. Further study could also incorporate clinical educator perceptions (Lee, 2023) and practical implications for national policy and requirements (Picton, 2023; ACGME Common Program Requirements - Residency, 2023) for incorporating digital learning in American GME training.

References

ACGME Common Program Requirements - Residency Accreditation Council for Graduate Medical Education (2023). https://www.acgme.org/globalassets/pfassets/programrequirements/cprresidency_2023.pdf

ACGME Institutional Requirements. Accreditation Council for Graduate Medical Education (2022). https://www.acgme.org/globalassets/pfassets/programrequirements/800_institutionalrequirements2022.pdf

ACGME Specialties. https://www.acgme.org/specialties/externalContentView/8755402b-fd2f-4665-b15e-1e617c746362?preview_generated=True. Accessed 5 Nov 2023

Callahan, J.L.: Writing literature reviews: a reprise and update. Human Res. Dev. Rev. **13**(3), 271–275 (2014). https://doi.org/10.1177/1534484314536705

DiMattio, M.J.K., Hudacek, S.S.: Educating generation Z: Psychosocial dimensions of the clinical learning environment that predict student satisfaction. Nurse Educ. Pract. **49**, 102901 (2020). https://doi.org/10.1016/j.nepr.2020.102901

Eckleberry-Hunt, J., Lick, D., Hunt, R.: Is medical education ready for generation Z? J. Grad. Med. Educ. **10**(4), 378–381 (2018). https://doi.org/10.4300/JGME-D-18-00466.1

Fairbrother, H.: How to choose the best residency program for you with Hilary Fairbrother, MD, MPH, American Medical Association (2023). https://www.ama-assn.org/medical-students/preparing-residency/howchoose-best-residency-program-you-hilary-fairbrother-md. Accessed 5 Nov 2023

Gallagher, A., et al.: Views of generation Z regarding care and care careers: a four-country study. Int. J. Care Caring **6**(3), 355–377 (2022). https://doi.org/10.1332/239788221X16308608299691

Ho, C.-J., et al.: The effectiveness of the iLearning application on chest tube care education in nursing students. Nurse Educ. Today **101**, 104870 (2021). https://doi.org/10.1016/j.nedt.2021.104870

Jiang, J., et al.: Perceptions and experiences of generation Z nursing students during their practicum in an intensive care unit: a qualitative study. Heliyon **10**(4), e26205 (2024). https://doi.org/10.1016/j.heliyon.2024.e26205

Lee, T., et al.: Personal factors and clinical learning environment as predictors of nursing students readiness for practice: a structural equation modeling analysis. Asian Nurs. Res. **17**(1), 44–52 (2023). https://doi.org/10.1016/j.anr.2023.01.003

Lin, Y., Lemos, M., Neuschaefer-Rube, C.: Digital health and digital learning experiences across speech-language pathology, phoniatrics, and otolaryngology: Interdisciplinary survey study. JMIR Med. Educ. **7**(4), e30873 (2021). https://doi.org/10.2196/30873

Lin, Y., Lemos, M., Neuschaefer-Rube, C.: Digital health and learning in speech-language pathology, phoniatrics, and otolaryngology: Survey study for designing a digital learning toolbox app. JMIR Med. Educ. **8**(2), e34042 (2022). https://doi.org/10.2196/34042

Manik, M., et al.: Virtual simulation learning from Indonesian nursing students' perspectives. Maced. J. Med. Sci. **10**, 112–117 (2022). https://oamjms.eu/index.php/mjms/article/view/8239. Accessed 6 Apr 2024

Persada, S.F., Miraja, B.A., Nadlifatin, R.: Understanding the generation Z behavior on D-learning: a unified theory of acceptance and use of technology (UTAUT) approach. Int. J. Emerg. Technol. Learn. (Online) **14**(5), 20–33 (2019). https://doi.org/10.3991/ijet.v14i05.9993

Physicians and Surgeons: Occupational Outlook Handbook. U.S. Bureau of Labor Statistics (2023). https://www.bls.gov/ooh/healthcare/physiciansand-surgeons.htm#tab-1. Accessed 12 Nov 2023

Picton, R.B.: 'Mind the gap' a constructivist grounded theory exploring the role of social media in augmenting the professional learning journey of generation Z undergraduate student diagnostic radiographers. Professional Doctorate in Health and Social Care. London South Bank University, Institute of Health and Social Care (2023). https://openresearch.lsbu.ac.uk/download/20b52120200e01ca049711cdf627b00d5e2e11222e1dbb890b20f000321ad692/2718430/Mind%20the%20Gap%20Thesis%20submission%2015.09.23%20v2.pdf. Accessed 6 Apr 2024

Ryan, M.S., et al.: Competency-based medical education across the continuum: How well aligned are medical school EPAs to residency milestones? Med. Teach. **44**(5), 510–518 (2022). https://doi.org/10.1080/0142159X.2021.2004303

Sedaghatkar, F., et al.: Enhancing medical students' knowledge and performance in otolaryngology rotation through combining microlearning and taskbased learning strategies. Int. J. Environ. Res. Public Health **20**(5), 4489 (2023). https://doi.org/10.3390/ijerph20054489

Shorey, S., et al.: Learning styles, preferences and needs of generation Z healthcare students: Scoping review. Nurse Educ. Pract. **57**, 103247 (2021). https://doi.org/10.1016/j.nepr.2021.103247

Silkens, M.E.W.M., et al.: A national qualitative investigation of the impact of service change on doctors' training during Covid-19. BMC Med. Educ. **23**, 1–11 (2023). https://doi.org/10.1186/s12909-023-04143-1

Vizcaya-Moreno, M.F., Pérez-Cañaveras, R.M.: Social media used and teaching methods preferred by generation Z students in the nursing clinical learning environment: a cross-sectional research study. Int. J. Environ. Res. Public Health **17**(21), 8267 (2020). https://doi.org/10.3390/ijerph17218267

Xavier, N., Oliveira, T.: Factors affecting behavioural intention to adopt e-participation. In: Proceedings of the 10th European Conference on Information Systems Management, ECISM 2016, pp. 322–325 (2016). http://www.scopus.com/inward/record.url?scp=85016094892&partnerID=8YFLogxK. Accessed 11 May 2024

Yurdal, M.O., et al.: 'Development of medical school students' attitudes towards online learning scale and its relationship with E-learning styles. Turk. Online J. Dist. Educ. 310–325 (2021). https://doi.org/10.17718/tojde.961855

Tailoring and Transferability of Coaching and Coaching Supervision Models: Challenges and Considerations for Educational Application

Helen Smith(✉)

Manchester Metropolitan University, Manchester, UK
helen.smith@mmu.ac.uk

Abstract. This paper draws upon the professional experience of a group of five coaching supervision practitioners to inform and reflect upon practice in this field of supervision to compare and contrast with supervision in education. The author is one of a team of coach supervisors practicing across multiple sectors, with strong educational links providing a unique perspective of coaching supervision models and techniques that may be tailored and transferred to the educational sector. Models and approaches from coaching supervision practice are shared and assessed in terms of transferability of outcomes for potential application within educational supervision. Models and approaches traditionally used in coaching supervision are explored. Exploration of tools and methods that can be applied to education may enrich the experience for the academic supervisor and student thesis dissertation. The aim is to offer a sustainable and lasting experience for the student that can be carried through to their future employment and result in a rewarding and fulfilling journey for themselves and the academic supervisor.

Methodology applied is the analysis of practitioner case studies and applied models drawing upon the practitioner insight and experience to critique each model for academic 'fit.'

Keywords: coaching · coaching supervision · supervision · reflection · supervision models

1 Introduction

Educational supervisors, coaches and coach supervisors enact important roles in the change process through a supporting dialogic conversation (Risko et al. 2009). Thoughtful dialogue encourages supervisors and students to focus upon and analyse instructional decisions and thought processes underlying research design and methodology. As academic supervisors become acquainted with their student's research, their thinking process, and actions, they raise opportunities for growth and emergence of positive change. How supervisors interact with each student on their dissertation or thesis becomes a dance where we seek the perfect flow of steps and avoid standing on the students' toes and strive to become the perfect dance partner. Educational supervisors scaffold learning by mediating the student experiences by probing, prompting, and encouraging a

M. J. Sousa (Ed.): UFHRD 2024, ISEM 52, pp. 364–374, 2025.
https://doi.org/10.1007/978-3-031-92548-1_23

re-visioning of their work (Kibler et al. 2014) whilst aligning to the awarding criteria rubric and the student's research aim.

Following a review of reading teachers, Risko et al. (2009) recommended that the process of learning and doing be unpacked as facilitated by dialogue associated with purposeful conversation, facilitated through supervisor support and feedback. Supervision sessions provide a major place for active, reflective learning (Laster 2013, p. 4). Scaffolding in this context enhances learning (Collet 2015) with purposeful discussion in combination with reflection as in a supervision session supporting pedagogical change (Collet 2012; Edwards-Groves 2012; Whitcomb et al. 2009). Reflection involves consideration of one's beliefs and recognition of potential consequences and maximises the construction of meaning (Schön 1987). In addition, according to Lev Vygotsky (1896–1934), thoughts are "completed in the word" (Vygotsky 1987, p. 252), rendering the use of words, as in dialogic supervision, a compelling basis for guiding purposeful reflection and learning. Through such reflection we recognise areas that require strengthening, consider alternatives and potentially reconstruct as required. For example, this often happens when a student realises that the focus of their original aims and objectives may not be clear or concise enough to allow logical execution of their research focus. Achieving relevance in a supervision relationship relies upon more than simple dialogue, which we will *unpack* now through the coaching supervision lens.

2 Coaching and Coaching Supervision

As with any relationship, it takes both partners to make the association work. In coaching supervision, Hewson and Carroll (2016) advise the process begins with the basics, starting with the supervisor focusing upon the student's learning and progress. The supervisor needs to have an awareness of the psychological functioning of the student to support any issues that may impede their progress; being attentive to the student and informing them that their research matters to you as their supervisor. Establishing trust and rapport from the outset is crucial to this relationship. These responsibilities can be represented by pillars of supervision as in Fig. 1.

Student wellbeing	Professional standards	Student development
Regular Check-in	Academic framework / criteria	Observation of development
Contracting at each session	Focus upon standards / rubric	Co-create milestones to success

Fig. 1. Pillars of Supervision

Supervisors are responsible for providing equal attention to all three pillars, though this will likely oscillate depending upon the student needs and stage of their work. Thus, the role of a supervisor can be described as developing and maintaining the students' competence and wellbeing whilst upholding and monitoring professional standards. Bransford et al. (2000) state "the contemporary view of learning is that people construct new knowledge and understanding based on what they already know and believe" (p. 10). These authors argue that paying attention to what the student brings to supervision is a sensible first step. The observations from Hewson and Carroll and Bransford et al. infer that there are multiple layers to the supervision process. Once rapport is established between supervisor and student, the first step is to clarify what the goal (aim) of the research is and whether the student has any ideas on how they will achieve their research goal. At this point, a simple coaching framework such as GROW (Whitmore 1992) or OSCAR (McKergow and Jackson 2002) may be helpful in offering a structure to the exploratory phase of the supervision conversation, as illustrated in Fig. 2.

Goal – research aim/objective	**O**utcome – intended research aim
Reality – methodology, access to targets, time and competence considerations	**S**ituation – what is possible
Options – possible to achieve goal	**C**hoice/**C**onsequences – of various methods, approaches
Will/**W**ay forward/**W**rap up – what the student feels comfortable and able to do	**A**ction – which choices adopted
	Review

Fig. 2. GROW and OSCAR Frameworks

Each of these frameworks ends with a review element, providing an opportunity to establish a clear direction of next steps for the student and supervisor in stepwise progress with opportunities for constructive feedback.

The second stage of Hewson and Carroll's (2016) process is to create the correct space for dialogue to occur. Co-creating a space where the student feels **S**afe, with the application of a clear **P**rocess, and through the supervision sessions becoming **A**ware of what they need to do, maintaining **C**uriosity throughout the research journey and **E**mbracing (described as 'critically accepting' by Fook and Gardener 2007) the resultant outcome and learning. Within this *space,* there may be different roles that a supervisor adopts; namely directive, evaluative, passive, restorative (getting a student on track), active and reflective (Table 1). Whatever space is occupied, it is with the intention of monitoring, providing valuable feedback and supporting. Monitoring should be a transparent collaborative activity, allowing for any concerns or challenges to be shared.

In a coaching or coaching supervision context, we do not wish to create a reliant or dependent relationship but create and build sustainable skills in the supervisee by guiding them 'only in what they are unable to achieve themselves' (Hewson and Carroll 2016). The same can apply to academic supervision; to establish learning and appreciation of processes and models that can be drawn upon in the student's future career. Staying connected and supporting the student throughout the supervision process is key to providing valuable formative feedback.

Table 1. Supervision space/stance

Supervision space/stance	Educational benefit
Directive	Facilitating ethical practice, adherence to awarding criteria and university regulations
Evaluative	To monitor and provide summative feedback
Passive	Provide information or optional choices
Restorative	Listen and note progress, celebrate, recalibrate
Active	Interactive thinker and learner, teaching, learning, problem-solving
Reflective	Facilitative, attention to milestones and criteria required

Reflective space is not just physical, it is co-created between the student and supervisor, both between and in each of them. Within a coaching context, there are five separate areas of focus to achieve the correct reflective space as illustrated in Fig. 3.

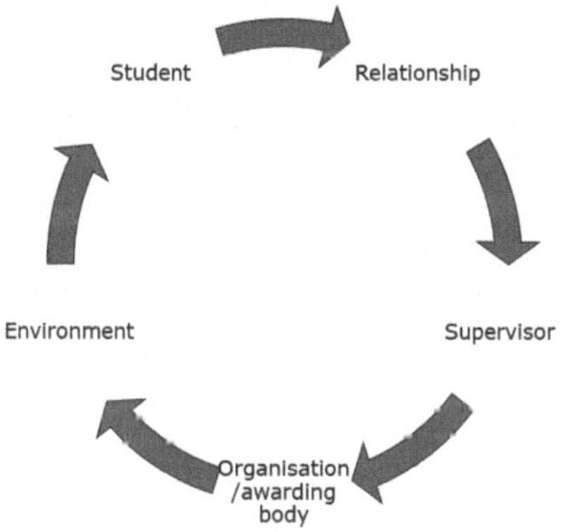

Fig. 3. Creating Reflective Space.

As a mindful consideration, reflection is critical to learning (Kolb 1984) and described by Jon Kabat-Zinn (2003) as "The awareness that emerges through paying attention on purpose, in the present moment, and nonjudgmentally to the unfolding of experience moment by moment". Noting the experience that the student is gaining for the purpose of completing their dissertation or thesis is valuable for supervisors to assess their competence, progress and support. This requires focus by the supervisor upon the student's competency and learning journey. As supervisors, we are employing Schon's (1983) reflective lenses (in and on action) to assist the student during their research journey, to understand its outcomes. The supervision relationship and process is dynamic and reinforces the need to contract during each supervision session to acknowledge where the student is, within their research journey. From our informal survey of ten UK Universities, no reports were received in the use of a contract within a dissertation or thesis context. A formal written contract in a one-to-one and group coaching supervision session is standard good practice and provides an opportunity to outline expectations and boundaries of the supervisory relationship. This initial written contract is routinely supported by ongoing verbal contracting at the outset of each coaching supervision session to calibrate, facilitate alignment, check progress, and offer an assessment opportunity to the supervisor and coachee. The value this may add to academic supervision is yet unknown, offering research focus going forward.

A further point that may add value in academia is that within coaching supervision, the coachee (student) makes the choice of 'matching' with the supervisor. Matching in academia is based upon topic 'fit' and available hours for the supervisor allocation. The impact of such matching in education is unknown and offers a potential topic for research. Time is a critical element in a learning journey and a potential concern for supervision allocation given that academics are becoming more time starved. In addition, to growing student numbers, the consideration of group supervision is becoming a reality. Group supervision in coaching is a commonplace occurrence offering deep and purposeful deliberation, allowing for group and individual enhanced learning through sharing, and network support. However, the challenges of orchestrating group supervision delivery is a skill demanding full attention and management of the group by the supervisor. Group supervision drop-in sessions may offer the choice to students as per their supervisor choice and goes some way to resolve the strict 'no-choice' matching allocation commonly applied in academia.

One group case study from the Coaching Supervision Practice group references supervision delivery to The Save the Children Coaches. These coaches are placed globally in locations where children are experiencing the outcomes of wars, famine, natural disasters, and other stressful situations. The Save the Children coaches draw many parallels with student group supervision in coming together to share a trusted relationship to empathise and understand the ongoing challenges. Students from different backgrounds, cultures, values, with different research topics all share the same goal; namely to complete and finalise their research focus.

As a member of a practice driven group, we observe one another's coaching supervision and use this association to test models and vary approaches. In group supervision, the practice group observations are to allow sufficient time for each participant trust the group supervision process (Fig. 4), establish relationships, be attentive to the whole group, ask questions to stimulate reflection, illicit comment from group members, listen, encourage them to draw upon their own knowledge and end each group session with a reflection and individual clear actions.

Step	Description
Start with a check-in	Allows for an individual assessment/observation within the group.
Equal time allocation	Rotate around group and involve group in supportive dialogue, dealing with challenges, resources etc builds supportive network.
Review and Actions	Allow reflective time for each group member. Establish actions to achieve before next supervision session.

Fig. 4. Group Supervision Process

There is a visible power difference in group supervision which devolves some of the power beyond the process to the students and coachees. This is akin to Bruner's handover principles (1983, p.60), where student learning is encouraged through guided inquiry and open-ended questions. This approach empowers learners to seek answers, make connections, develop their cognitive abilities, encourage participation, and foster network support and social learning. Guided discussions such as in group supervision illustrates reflectivity and an enhanced learning gain (Zahid and Khanam 2019). A model that is applicable in one-to-one and group coaching is the Newcastle Cake Stand Model, (NCSM, Armstrong and Freeston 2006); so, named due to its four tiers: primary inputs (from each participant), parameter setting (clarifying goals, outputs, roles, structures - potentially through contracting), dynamic forces (individual research topic and rubric foci), and learning process, as per Kolb's reflective cycle. From the practitioner's perspective, this model provides a logical and flexible structure for a supervision session.

The Process Evaluation of Training and Supervision (PETS) model is based upon a review of leadership techniques in teaching (Milne and James 2000) and identifies 13 activities of supervision. These are listening, observing, managing, supporting, questioning, formulating, informing, feeding back, challenging, disagreeing, evaluating, guiding experiential learning and *other* such as social chat that occurs in group supervision with a focus upon feeding back. Hewson and Carroll (2016) suggest "feeding back is throwing catchable feedback balls to someone who is ready, willing and able to catch them", suggesting a certain preparation is required; namely preparation of the feedback and of the student to facilitate their readiness to catch the ball and receive feedback. How feedback is interpreted and received, is reliant upon culture. Erin Meyer highlighted the cultural differences of positive and negative feedback in her book *The Culture Map*. Ellis et al. reaffirm the cultural importance when noting that 'human development relies upon the appropriation of pre-existing cultural tools and that appropriation occurs through social interchange' (Ellis et al. 2010, 4). This is possibly the reason why feedback conversations are reported as being the most difficult conversations to have. In our multi-cultural education environment, this is a further consideration for supervisors which can be addressed *in part* through thorough contracting in addition to individual attentiveness, empathy, respect for cultural differences and acceptance of this role model. An awareness and cognition on behalf of the supervisor should be part of the preparation for supervision, i.e., knowing yourself and your student.

Hawkins et al. (2012) summarised the guidelines for feedback in coaching supervision as CORBS – clear, owned, regular, balanced, and specific. CORBS transfers well to the academic context, with reinforcement of ownership being especially useful and important to pursue with the student within the academic context. Students need to take charge of their own research and learning.

While CORBS is one potential reflective feedback framework, there are others that may be applicable to academic scenarios as illustrated below in Table 2.

Table 2. Reflective Feedback Frameworks

Authors	Reflective feedback elements	Potential academic application
Claiborn and Goodyear (2005)	Corrective, Descriptive, Emotional disclosure, Evaluative (+ve & −ve), Interpretive	Rarely required in academia, but useful in cases of student trauma during supervision
Falender and Shafranske (2004)	Analytical, Empowering, Evaluative (+ve), Confrontative, Concerned, Reflective, Responsive	Framework aligns strongly with student wellbeing, especially the Concern element
Heron (1999)	Catalytic, Cathartic, Confrontative, Perspective; Informative; Supportive	Confrontation is not usual in academic supervision, but may prompt student action when needed
Loganbill et al. (1982)	Catalytic; Conceptual; Confrontative; Supportive	Similar to Heron's approach
Stone and Heen (2014)	Appreciation; Coaching; Evaluative (+ve & −ve), Interpretive	Roots in appreciative enquiry, applicable in academic settings

Familiarity with reflective feedback frameworks by academic supervisors may add to their competence in supporting their student supervisees. These models further support a positive feedback mechanism that steers us (as supervisors) away from mistakes, and reinforce opportunities for formative feedback; namely.

- Start with positive feedback
- Offer ideas with consent (and as part of the contracted expectations), do not prescribe what the students should do
- Notice their response to ideas offered – back off if required
- Work with your student to get them ready for feedback (part of contract)
- Avoid confrontation – unless all other feedback has failed
- Ask before you tell, asking is far less threatening and ensures they are still in control of their research.

Academic rubrics often seek evidence of critical reflection from students. This is certainly true of dissertation and thesis work in supporting the transformation that often occurs due to curiosity and critical assessment of the student's chosen topic. Critical reflection can unearth new insights and lead to a rethinking or refinement of our approach (both of supervision and research approach for the student). As supervisors, we need to facilitate space for students to experience this reflectivity whenever possible.

Tom Christensen (Professor of Public Administration, Oslo) stated, "if we cannot question the way we are doing things and thinking about things in the present, it will not occur to us that they could be thought of or done differently".

This brings into question academic supervision, the process, the underpinning, and the training of higher education institutions' supervisors. With increasing focus being placed upon research in education, perhaps our approach in academic supervision needs a rethink.

3 Why Change Academic Supervision?

In the last decade the focus has been upon The Research Excellence Framework (REF), which is the UK's system for assessing the excellence of research in UK higher education institutions (HEIs). REF outcomes are used to facilitate an annual allocation of ca. £2bn of public funding for university research via. a rigorous research process. The demands being made of academics of their research have escalated to meet the criteria of the REF, impact outcomes and a future revised focus upon People, Culture and Environment (PCE) thereby offering the possibility for associated community engagement opportunities.

Community engagement opportunities can certainly involve dissertation and thesis students as community stakeholders and contributors to the PCE agenda and activities. This was recently illustrated through a Chartered Institute of Personnel Development (CIPD) collaborative research opportunity, where students were offered the opportunity to gain publication exposure for their research work in specific areas sponsored by the CIPD. Such activities and focus draws a spotlight upon research rigour, including that of supervision as a part of the process.

Coaching including supervision in a clinical setting is often combined with professional development methods such as discussion and self-reflection as in the case of supervision. The process of engaging dialogically, as in a supervision conversation, encourages consideration of new experiential perspectives. When teachers discuss and critically consider their own assumptions about teaching and learning, they adapt their teaching in powerful, positive ways (Collet 2012). Extending the clinical teaching experience through dialogue encourages students to voice their new understandings and adapt them to other pedagogical contexts (Zwart et al. 2009). These observations from Collet and Zwart would suggest that those who supervise *in context* should engage with training and reflection upon their practice to add greater value to those they supervise.

4 The Skills of Supervisors

The role, tasks, functions, and work of university supervisors in the professional development of Master's and Doctoral students, has attracted little research interest to date, but with the high stakes driven by the REF and other impact frameworks, more rapid focus is being attributed to this area. Understanding the role, functions, and work of supervisors is crucial in establishing more effective methods to contribute to impactful learning within the increasing context of REF and PCE. The need to envisage the potential of future supervisory roles and their functions with the capacity to transform HEI education is paramount. Supervision in the teacher training context reveals considerable variation in the adopted roles and tasks undertaken by supervisors during the practicum (Correa 2014), including those characteristic of a facilitator, mediator, role-model, critical friend, assessor, and as a supervisor who checks the policy of the awarding body and adherence to the required ethical standards.

It has been suggested that the lack of certainty and clarity about the nature of effective supervisory practice has influenced the quality of learning in both from a positive and negative perspective, thereby affecting the learning trajectories of students and their levels of achievement. In the context of teacher training, supervisors often represent the sole source in the university of those who are explicitly responsible for bridging the gap between theory and practice, making this negotiated presence critical to the effectiveness of preservice teacher learning (Montecinos et al. 2015). While we can posit if this is similar for masters and doctoral student outcomes and the supervision experience, the supervisor must be aware of bridging the theory and practice gap in addition to the other role definitions that are attributed to supervision. In essence, there are many unknown variables in academic supervision. However, a good starting place is to ensure clarity and certainty as to the status of supervision in the UK's HEIs.

This practitioner group aims to understand the academic supervision landscape more fully by:

- Conducting a rigorous audit of the supervision status in UK HEI's
- Applying for appropriate funding to develop a supervision training programme and tool kit
- Conducting a multi-centre trial (of demographically similar universities) of the training and tool kit with embedded impact outcome measures
- Reporting outcomes and amending training and tool kit as required
- Launching and sharing training and the tool kit

This paper represents the starting point for this journey to better understand supervision within academia. The Coaching Supervision Practice feels they have the necessary credentials to analyse and evolve a robust model that can be tailored and transferred within the education sector. Their experience and insights is deeply rooted in the foundation of coaching.

References

1. Armstrong, P., Freeston, M.: Conceptualising and formulating cognitive therapy supervision. In: Tarrier, N. (ed.) Case formulation in Cognitive Behaviour Therapy. Routledge, London (2006)
2. Bransford, J.D., Brown, A.L., Cocking, R.R.: How People Learn. National Academy Press, Washington, DC (2000)
3. Claiborn, C., Goodyear, R.: Feedback in Psychotherapy. J. Clin. Psychol. **61**, 209–217 (2005)
4. Collet, V.S.: The gradual increase of responsibility model: Coaching for teacher change. Lit. Res. Instr. **51**(1), 27–47 (2012)

5. Collet, V.S.: The Gradual Increase of Responsibility Model for coaching teachers: Scaffolds for change. Int. J. Mentor. Coach. Educ. **4**(4), 269–292 (2015)
6. Correa, E.: Las Prácticas: Primer Espacio De Profesionalización Docente (The Practicum: First Spache of Professionalisation). In: Cortés, I., Hirmas, C. (ed.) Primer Seminario sobre Formación Práctica Docente: Vinculación entre el sistema universitario y el sistema escolar, pp. 25–40. Organización de Estados Iberoamericanos para la Educación de la Ciencia y la Cultura (2014)
7. Edwards-Groves, C.: Interactive creative technologies: changing learning practices and pedagogies in the writing classroom. Aust. J. Lang. Lit. **35**(1), 99–113 (2012)
8. Ellis, V., Edwards, A., Smagorinsky, P.: Cultural-Historical Perspectives on Teacher Education and Development: Learning Teaching. Routledge, London (2010)
9. Falender, C., Shafranske, E.: Clinical Supervision. American Psychological Society (2004)
10. Fook, J., Gardener, F.: Practicing Critical Reflection: A Resource Handbook. McGraw-Hill (2007)
11. Hawkins, P., et al.: Supervision in the Helping Professions, 4th ed. Open University Press, Maidenhead, Berkshire, England (2012)
12. Heron, J. The Complete Facilitators: The Modern Revolution in Learning. London Kogan (1999)
13. Hewson, D., Carroll, M.: Reflective Practice in Supervision. Moshpit Publishing (2016)
14. Kabat-Zinn, J.: Mindfulness based interventions in context: past present and future. Clin. Psychol.: Sci. Pract. **10**, 144–156 (2003). (quote 145)
15. Kibler, A., Valdes, G., Walqui, A.: Re-visioning pedagogies in response tostandards-based reform. TESOL Q. **48**(3), 615 (2014)
16. Kolb, D.A.: Experiential Learning. Prentice Hall, New Jersey (1984)
17. Laster, B.: A historical view of student learning and teacher development in reading clinics. In: Ortlieb, E., Cheek, E.H. (eds.) Advanced Literacy Practices (Literacy Research, Practice and Evaluation), vol. 2, pp. 3–20. Emerald Group Publishing, Bingley (2013)
18. Loganbill, C., Hardy, W., Delworth, U.: Supervision: a Conceptual Model. Couns. Psychol. **10**, 3–42 (1982)
19. McKergow, M., Jackson, P.: The Solutions Focus: Making Coaching and Change SIMPLE (2002)
20. Milne, D., James, I.A.: A systematic review of effective cognitive-behavioural supervision. Br. J. Clin. Psychol. **41**, 55–72 (2000)
21. Montecinos, C., Walker, H., Cortez, M.: Sugerencias De Docentes Directivos Para Mejorar La Formación Práctica En Las Carreras De Pedagogía: Transitando De Acciones Fragmentadas a Una Participación Legítima En Los Colegios (Suggestions of Principlas to Improve the Practicum: From Peripheral Participation to Legitimate School Members. Estudios pedagógicos (Valdivia) **41**, 157–176 (2015)
22. Risko, V.J., Roller, C., Cummins, C., Bean, R., Block, C.C., Anders, P.: Making sense of reading teacher education research and prospects for future research. In: Presentation at the Annual Conference of the International Reading Association, Minneapolis (2009)
23. Schön, D.A.: Educating the Reflective Practitioner. Jossey-Bass, San Francisco (1987)
24. Stone, D., Heen, S.: Thanks for the Feedback. Penguin (2014)
25. Vygotsky, L.S.: Thinking and speech. In: Rieber, R.W., Carton, A.S. (eds.) The Collected Works of L.S. Vygotsky, Volume 1: Problems of General Psychology, pp. 39–285. Plenum Press, New York (1987)
26. Whitcomb, J., Borko, H., Liston, D.: Growing talent: promising professional development models and practices. J. Teach. Educ. **60**(3), 27–212 (2009)
27. Whitmore, J.: Coaching for Performance. Brealey, London (1992)

28. Zahid, M., Khanam, A.: Effect of reflective teaching practices on the performance of prospective teachers. The Turkish Online Journal of Educational Technology **18**(1), 32–43 (2019)
29. Zwart, R.C., Wubbels, T., Bergen, T., Bolhuis, S.: Which characteristics of a reciprocal peer coaching context affect teacher learning as perceived by teachers and their students? J. Teach. Educ. **60**(3), 243–257 (2009)

AI as the New Antecedents of Technostress at Work

Aykut Arslan[1,2,3], Serdar Yener[1,2,3], and Helena Belchior-Rocha[1,2,3](✉)

[1] Piris Reis University, Istanbul, Turkey
Helena_rocha@iscte-iul.pt
[2] Sinop University, Sinop, Turkey
[3] Iscte-University Institute of Lisbon, Lisbon, Portugal

Abstract. This study explores the role of Artificial Intelligence (AI) across various business sectors and underscores the importance of the human element in harnessing AI's potential. It addresses the concept of "technostress" and aims to understand the implications of rapid AI proliferation on the workforce. The research seeks to understand the dimensions of technostress exacerbated by AI adoption and elucidate the factors that followers deem crucial in navigating and adapting to the transformative trajectory of AI technologies. A significant proportion of observed efficiency improvements can be attributed to the integration of systems involving collaborative efforts between human workers and AI. However, the realization of these efficiency gains is contingent upon the workforce's acceptance of these systems. Resistance to technological adoption due to technostress could potentially hinder the full exploitation of the envisioned efficiency enhancements associated with Human-AI Collaboration systems. Most literature has focused on the perception of AI adoption, and studies investigating the effect of human-AI collaboration through actual co-working with AI are scarce. This study aims to fill this gap in the literature, offering valuable insights into the interplay between human cognition, technology-induced stress, and its broader impact on individual well being and organizational dynamics.

Keywords: Artificial Intelligence (AI) · Technostress · Human-AI Collaboration systems

1 Introduction

Artificial Intelligence (AI) is acknowledged as the primary force instigating disruption across diverse industries and enterprises at both national and international levels [1]. Its envisaged impact encompasses substantial transformations in various facets of human existence, with the potential to augment organizational efficacy, streamline operational processes, and optimize resource utilization [2]. Artificial Intelligence (AI) Integration refers to the incorporation of AI technologies into various aspects of the workplace, ranging from routine tasks to complex decision-making processes. Sectors such as manufacturing, law, broadcasting, and banking have already borne witness to noteworthy shifts attributable to the integration of AI [3]. Businesses incorporating AI applications

M. J. Sousa (Ed.): UFHRD 2024, ISEM 52, pp. 375–391, 2025.
https://doi.org/10.1007/978-3-031-92548-1_24

are poised to realize enhancements in terms of increased business value, encompassing heightened revenue, diminished costs, and heightened operational efficiency [4].

Beyond serving as a tool for operational automation and cost reduction, the strategic asset status of AI is underscored by its capacity to facilitate organizational adaptation and repositioning within competitive markets [5], rendering it an indispensable component of novel product development [6]. According to a recent report [7], which surveyed over 3000 managers globally, a substantial proportion of companies are actively piloting or implementing AI initiatives (57%), possess a dedicated AI strategy (59%), and comprehend the potential for AI to generate business value (70%). Nevertheless, despite the considerable benefits that AI promises, a mere 8% of companies employing strategic AI practices achieve comprehensive organizational adoption.

Furthermore, the study revealed that more than 90% of businesses attaining successful scaling have allocated over half of their analytics expenditures to initiatives fostering adoption, such as workflow redesign, communication, and training. Despite these increased investments and activities, only 10% of organizations report significant financial gains from the incorporation of artificial intelligence. The study underscores the frequently underestimated role of mutual learning between humans and machines in extracting value from AI, emphasizing that companies incorporating diverse forms of interaction and feedback between humans and AI stand six times more likely to enhance their AI-related successes. The report accentuates the paramount importance of the human element in the equation, asserting that "The single most critical driver of value from AI is not algorithms, nor technology—it is the human in the equation." Consequently, the rapid proliferation of AI across various business domains underscores the imperative of scrutinizing its implications on the workforce [8].

While existing literature on AI implementation provides guidance on how leaders should address and mitigate technostress factors, an equally crucial endeavor involves examining the considerations followers deem essential for anticipating AI's transformative trajectory. Gaining insights from followers can furnish leaders with a more nuanced comprehension of effective human resource management during this transitional phase. Furthermore, foreknowledge of followers' perceptions and expectations can play a pivotal role in bolstering their engagement, alleviating tensions associated with the transformation process, and mitigating subsequent technostress.

Technostress is a psychological phenomenon arising from the use of technology, characterized by feelings of overload, anxiety, and dissatisfaction [9]. In the context of AI integration, it encompasses stressors unique to the adoption of AI technologies in the work environment. This study explores dimensions of technostress exacerbated by AI adoption. This includes factors such as information overload, techno-invasion, role ambiguity, techno-uncertainty, and techno-inefficiency. Each dimension contributes to the overall experience of technostress in employees. Examining the factors that employees consider crucial for navigating and adapting to the transformative trajectory of AI technologies requires understanding how individuals perceive, cope with, and integrate AI into their daily work routines. This encompasses both individual and organizational strategies for managing the challenges posed by AI adoption.

Technostress is defined as a modern disease of adaptation caused by an inability to cope with the new computer technologies in a healthy manner [10]. It manifests in various dimensions, including techno-overload, where individuals feel compelled to work faster and for longer hours due to technology [11]. Techno-invasion refers to the encroachment of technology into personal life, eroding the boundary between work and personal time [12]. Techno-complexity arises from the overwhelming complexity of technology and the constant need to learn and understand new systems [13]. Technoinsecurity is the fear of job loss or obsolescence due to the perceived inability to adapt to technological advancements [14]. Techno-uncertainty is the feeling of uncertainty due to frequent changes and updates in technology [15]. Lastly, techno-inefficiency is the perception that technology is creating more problems and inefficiencies rather than enhancing productivity [16]. As AI continues to advance, its integration into the workplace can potentially exacerbate these dimensions of technostress, necessitating careful consideration of its implementation and the provision of adequate support for employees.

The aim of this research is to comprehensively explore the implications of Artificial Intelligence (AI) integration in the workplace, with a particular focus on its impact on employees' technostress. The study seeks to understand the dimensions of technostress exacerbated by AI adoption and elucidate the factors that followers consider crucial in navigating and adapting to the transformative trajectory of AI technologies. By examining the perspectives of employees, the research aims to provide some insights that can guide leaders in effective human resource management strategies during the transitional phase of AI implementation.

2 AI Widening Its Practices in Everyday Life

The practical applicability of research findings in the field of Artificial Intelligence (AI) can be significantly enhanced by providing specific examples or case studies illustrating successful human-AI collaboration. In the field of healthcare, AI-powered diagnostic tools have collaborated with medical professionals to analyze medical images, such as X-rays and MRIs. For instance, the collaboration between radiologists and AI in medical imaging has led to improved diagnostic accuracy [17]. AI algorithms can analyze medical images and highlight potential areas of concern, assisting radiologists in making more accurate diagnoses, this collaboration has significantly improved the accuracy and speed of diagnoses, allowing for more timely and precise medical interventions. IBM Watson for Oncology a AI-powered system designed to assist oncologists in making treatment decisions analysing vast amounts of medical literature, clinical trial data, and patient records to provide evidence-based treatment options [17, 18] by inputting patient data and medical history. The AI analyzes this information to recommend personalized treatment plans based on the latest medical research and best practices. The collaboration between oncologists and IBM Watson for Oncology has led to more accurate and tailored treatment recommendations. The system's ability to process extensive data quickly has empowered healthcare professionals to stay updated on the latest advancements in oncology.

One more example is DeepMind, a subsidiary of Google, they developed AI algorithms for analyzing medical images, specifically focusing on ophthalmology which involved using AI to detect and diagnose eye conditions, such as diabetic retinopathy [19–22]. Ophthalmologists have worked alongside DeepMind's AI system by providing it with retinal images for analysis, in this case AI identifies potential signs of eye diseases, and ophthalmologists use this information to make informed diagnoses and treatment decisions [21, 22]. The collaboration between ophthalmologists and DeepMind's AI has improved the early detection of eye diseases and the system's ability to analyze large datasets of retinal scans has contributed to faster and more accurate diagnoses, enabling timely interventions to prevent vision loss [21, 22].

AI-powered diagnostic tools in healthcare can enhance the capabilities of medical professionals, leading to improved accuracy and speed in diagnosing medical conditions, making collaboration between human expertise and AI technologies are a promise for advancing healthcare outcomes.

Another example is the use of AI in supply chain management. Companies like Amazon have successfully integrated AI technologies to optimize their supply chain processes, resulting in increased efficiency and reduced costs [23]. Many companies have successfully implemented AI-powered chatbots to assist customer service representatives. By handling routine inquiries and providing quick responses, these chatbots free up human agents to focus on more complex and personalized customer interactions. This collaboration has led to improved efficiency and enhanced customer satisfaction.

In the field of education, intelligent tutoring systems have been developed to provide personalized learning experiences for students, demonstrating the potential of AI in enhancing educational outcomes [24], these platforms analyze student performance data to provide tailored recommendations for improvement. Teachers then use this information to adapt their teaching strategies, creating a more adaptive and effective learning environment [25]. The successful application of AI provides valuable insights into the factors that contribute to effective human-AI collaboration, serving as practical guides.

Concerning the financial sector, AI algorithms collaborate with human analysts to detect fraudulent activities. By continuously analyzing vast amounts of transaction data, AI can flag potentially suspicious patterns, allowing human experts to investigate and make informed decisions. This collaboration has proven effective in mitigating financial risks [26–29].

To create safer and more efficient transportation, the development of autonomous vehicles involves a collaboration between AI systems and human drivers. Advanced driver-assistance systems (ADAS) use AI to enhance safety and navigation, while humans remain responsible for complex decision-making and situational awareness [30].

AI-powered robotic systems collaborate with human workers in manufacturing processes. Robots handle repetitive and physically demanding tasks, while humans focus on complex problem-solving, creativity, and quality control, this improves overall production efficiency and product quality [31, 32].

Therefore, it becomes clear how human-AI collaboration has been successfully implemented to achieve specific goals, addressing challenges, and unlocking new possibilities. These can help stakeholders understand the practical implications and benefits of integrating AI technologies into various aspects of human activities. However, over time the introduction of technology has also had consequences, namely the processes of technostress and it is in this sense that this paper aims to contribute.

3 AI as a New Antecedent of Technostress

Technostress, as conceptualized by [10], is a phenomenon that arises from the challenges individuals face when adapting to new technologies. Traditional technostress, often associated with the introduction of computers and digital technologies in the workplace, has been extensively studied in various dimensions such as techno-overload, techno-invasion, techno-complexity, techno-insecurity, and techno-uncertainty [33].

As we transite into the era of Artificial Intelligence (AI), these dimensions of technostress acquire new nuances. For instance, techno-overload and techno-invasion can intensify as AI systems often require users to process vast amounts of information and may infringe on personal time. Techno-complexity can escalate when users find AI technologies challenging to understand and use. Techno-insecurity can heighten with fears of job loss due to AI automation, and techno-uncertainty can increase when users are unsure about the consequences of using AI [34]. Thus, the transition from traditional technostress to AI-related technostress is not merely a shift in technology but also a deepening and broadening of the stress dimensions. Recognizing and addressing these evolving stressors is crucial for successful AI integration and user well-being [35].

The integration of Artificial Intelligence (AI) into the workplace can significantly influence the dimensions of technostress. AI's capacity to optimize and automate tasks can lead to techno-overload, where employees may feel pressured to match the efficiency and speed of AI systems, potentially resulting in longer work hours and increased workload [11]. AI systems, characterized by their high efficiency and rapid task execution, set an accelerated pace for work, leading employees to feel pressured to match this speed and fostering heightened expectations for task completion. The continuous evolution of AI technologies necessitates ongoing adaptation and learning, creating a persistent demand for employees to acquire new skills to remain effective. The 24/7 availability of AI-driven systems contributes to an "always-on" culture, blurring the boundaries between work and personal life and intensifying work-related stress. The integration of AI driven tools and platforms introduces a multitude of interfaces and applications, adding to the cognitive load and contributing to techno-overload. Additionally, the fear of redundancy, stemming from AI automation, may drive employees to take on additional tasks and responsibilities to demonstrate ongoing value, thereby increasing workload and stress levels. The pressure to match or exceed AI efficiency, coupled with reduced perceived control over work processes due to AI autonomy, further contributes to feelings of overwhelm. The efficiency gains achieved by AI may lead to a lack of downtime, as the saved time is often redirected towards taking on additional responsibilities, leaving employees with limited opportunities for rest and recuperation. Addressing these challenges requires a comprehensive organizational strategy that prioritizes employee well-being and provides support for managing the psychological impact of AI integration into the modern workplace.

Techno-invasion: The phenomenon of techno-invasion is intensified by the proliferation of AI-driven tools that foster an "always-on" culture, contributing to the erosion of boundaries between work and personal life [12]. The constant availability and functionality of AI-driven technologies create an expectation of continuous engagement, blurring the traditional lines that delineate work-related activities from personal time.

This persistent connectivity, facilitated by AI, extends beyond conventional working hours, making it challenging for individuals to disengage and establish clear distinctions between professional responsibilities and personal life. The integration of AI-driven tools, with their inherent efficiency and accessibility, not only enables but reinforces the expectation of constant availability, exacerbating techno-invasion. As a consequence, individuals may experience heightened stress and difficulty in achieving a healthy work-life balance, underscoring the need for organizational policies and practices that address the impact of AI on boundary management and support the well-being of employees in the face of evolving technological landscapes.

The techno-complexity: The techno-complexity inherent in AI systems contributes to the overwhelming nature experienced by employees, stemming from the intricate algorithms and advanced data processing capabilities characteristic of these technologies. The sophisticated nature of AI algorithms, designed for intricate decision-making and task automation, demands a continuous commitment to learning and adaptation from employees [13]. The evolving landscape of AI technologies necessitates ongoing efforts by individuals to understand and navigate the intricate processes, resulting in a constant need for skill development and knowledge acquisition. The intricate nature of AI systems, encompassing complex algorithms and data analytics, can create a sense of cognitive overload as employees strive to comprehend and keep pace with the rapid advancements in technology. Consequently, this techno-complexity poses challenges to employee well-being, emphasizing the imperative for organizations to provide comprehensive training programs and supportive environments that facilitate continuous learning, thereby empowering employees to navigate the complexities of AI integration in the modern workplace.

Furthermore, the integration of AI into the workplace can heighten techno-insecurity, amplifying concerns related to job displacement and skill redundancy [14]. The advanced capabilities of AI, enabling it to perform tasks that were traditionally carried out by humans, instill a sense of unease among employees about the potential obsolescence of their skills. The fear of job displacement arises as AI's efficiency in executing certain functions may lead to the automation of roles previously performed by human workers. This perceived threat to job security can contribute to heightened stress and techno-insecurity among employees. The evolving nature of work roles due to AI implementation may also create anxieties about skill redundancy, prompting individuals to question the ongoing relevance of their expertise in the face of technological advancements. To address these concerns, organizations must implement strategies that emphasize the complementarity of AI and human skills, providing avenues for skill development and career adaptation, and fostering a supportive environment that mitigates the negative psychological impact of techno-insecurity on the workforce.

Moreover, the rapid evolution of AI technologies contributes to techno-uncertainty, engendering ambiguity for employees concerning their roles and the imperative to stay abreast of continuous technological advancements [15]. The swift pace of innovation in AI introduces a dynamic environment where the nature of tasks and job requirements may undergo frequent transformations. This constant state of flux can create uncertainty among employees about the future trajectory of their roles, generating concerns about the stability and continuity of their professional responsibilities. Additionally, the need to stay informed about ongoing technological advancements to remain effective in their positions can contribute to a sense of unease and ambiguity. Employees may grapple with the challenge of predicting the specific skills and knowledge that will be crucial for future success in the rapidly evolving technological landscape. Addressing technouncertainty necessitates organizational strategies that promote transparent communication, provide avenues for ongoing education and upskilling, and cultivate a workplace culture that embraces adaptability and resilience in the face of technological change.

Finally, techno-inefficiency may emerge when AI implementation lacks alignment with users' needs or when there exists a mismatch between the technology's capabilities and organizational processes, resulting in frustration and perceived inefficiency [16]. The successful integration of AI into the workplace hinges on the seamless synergy between technological solutions and the specific requirements of users. When this alignment is lacking, employees may encounter difficulties in effectively leveraging AI tools to fulfill their tasks, leading to frustration and a sense of inefficiency. A mismatch between the capabilities of AI and the intricacies of organizational processes can impede the intended benefits of automation and optimization, creating disruptions and hindering the overall workflow. Recognizing and addressing techno-inefficiency requires a meticulous evaluation of user needs, a strategic alignment of AI implementation with organizational processes, and a commitment to refining the integration over time. This proactive approach is crucial to optimizing the benefits of AI technology while minimizing the negative impact of techno-inefficiency on user experience and overall workplace effectiveness.

3.1 Managing Technostress Phenomenon

Kumar [36] aims to inquire into the technostress phenomenon at an organizational level from machine learning (ML) and artificial intelligence (AI) deployment. The findings of the investigation revealed that technostress arising from the implementation of Machine Learning (ML) and Artificial Intelligence (AI) technologies was influenced by factors such as role ambiguity, job insecurity, and the technological milieu. Specifically, stressors associated with the technology environment encompassed complexity, uncertainty, reliability, and utility, constituting primary sources of stress. The study also proposed a novel approach to address technostress at the organizational level. This involved the integration of ML and AI automation with a focus on augmentation interdependence. Additionally, the incorporation of socio-technical systems was identified as a crucial component in the effective management of technostress. This innovative integration was posited as a strategic mechanism for mitigating the challenges posed by the deployment of ML and AI technologies within organizational contexts. In another study, [37] synthesizes her findings to explore the intricate relationship between technostress and human-AI collaboration. Her analysis is structured around the categorization of technostress creators, distinguishing between challenge technostressors and hindrance technostressors. Furthermore, her study delves into the examination of positive and negative psychological responses that ensue from the interaction with AI technologies. The ultimate focus lies on understanding the impact of these dynamics on outcomes related to work related well being and the intention to use AI. Weber [38] stresses the critical significance of recognizing technostress as a constraining factor influencing the acceptance of new technologies, particularly evident in scenarios such as the introduction of the proposed machine feedback tool. This assertion underscores the imperative of considering the psychological impact of technology on individuals, wherein the manifestation of technostress becomes a pivotal determinant in shaping the reception and adoption of innovative tools. By acknowledging technostress as a significant variable in the equation of technological acceptance, Weber's [38] perspective underscores the necessity for a holistic understanding of the human experience within the realm of emerging technologies, thereby contributing to a more nuanced and comprehensive approach to technology adoption and implementation.

Organizations aspiring to maintain a competitive edge recognize the imperative of proactive engagement with the substantial impact anticipated from the widespread integration of Artificial Intelligence (AI) in the foreseeable future. The transformative potential of AI implementation can empower organizations, but this necessitates a strategic preparedness to harness its benefits [39]. However, akin to any paradigm shift, the extent of employee participation, whether supportive or resistant, can significantly influence an organization's readiness for AI adoption. In this context, organizational executives find themselves in a pivotal role, tasked with steering the trajectory of their organizations through the impending AI landscape. Organizational leaders with valuable insights and strategies to address technostress factors among employees that may arise in the context of future AI implementation within their respective organizations. By understanding and mitigating these technostress factors, leaders can foster a more conducive environment for AI integration, thereby maximizing the positive impact and minimizing potential disruptions associated with this transformative technological shift [40].

There are several real-world examples where organizations have successfully managed technostress related to AI integration. One such example is highlighted in a study by [36], who investigated the role of machine learning (ML) and AI in inducing technostress at an organizational level. They found that role ambiguity, job insecurity, and the technology environment contributed to technostress due to ML and AI deployment. However, they also discovered that the integration of ML and AI automation-augmentation interdependence, along with socio-technical systems, could be effectively used for technostress management at the organizational level. The study suggested that executives could follow the simultaneous deployment of ML and AI technologies-based automation-augmentation strategy along with socio-technical measures to cope with technostress [36]. Another comprehensive study integrated and synthesized empirical data from 102 articles, encompassing 113 independent studies with a total of 49,955 observations [40]. The analysis confirmed that technostress is meaningful in terms of its detrimental impact on both psychological and behavioral outcomes. However, it also revealed that in certain contexts, a disaggregated account of technostress creators can reveal meaningful empirical information [40].

These studies provide valuable insights into how organizations can manage technostress related to AI integration. They highlight the importance of understanding the specific stressors associated with AI and implementing effective strategies to manage these stressors. This includes providing comprehensive training programs, fostering a supportive work environment, ensuring work-life balance, reassuring employees about their job security, and implementing effective change management strategies [36, 40].

4 Match Between Employee Skills and AI Implementation to Mitigate Technostress

To mitigate technostress arising from the integration of Artificial Intelligence (AI) into the workplace, a pivotal strategy is to establish a harmonious match between employee skills and AI implementation. The rapid evolution of AI technologies introduces challenges related to techno-overload, techno-invasion, techno-insecurity, techno-uncertainty, and techno-inefficiency as discussed above. Aligning employee skills with AI implementation serves as a proactive measure to address these challenges and promote a conducive work environment. Firstly, fostering a match between employee skills and AI capabilities alleviates the pressures associated with techno-overload. By aligning the tasks automated by AI with the skills and expertise of employees, organizations can optimize efficiency without placing undue burdens on individuals to match the pace and proficiency of AI systems. This strategic alignment enables employees to focus on tasks that complement AI capabilities, fostering a collaborative work environment. Secondly, addressing techno-invasion requires a thoughtful consideration of how AI tools impact the work-life balance of employees. Ensuring that AI-driven technologies enhance, rather than encroach upon, personal time requires a careful calibration of tasks, recognizing the boundaries between work and non-work activities. This alignment mitigates the "always-on" culture promoted by certain AI applications, contributing to a healthier work-life balance for employees. Thirdly, the match between employee skills and AI implementation plays a crucial role in alleviating techno-insecurity. Providing employees with the necessary training and upskilling opportunities ensures they remain relevant and capable in a technologically evolving landscape. By demonstrating the complementarity of human skills with AI, organizations can mitigate fears of job displacement and skill redundancy, fostering a sense of security and professional confidence among employees. Moreover, addressing techno-uncertainty involves establishing clear communication channels regarding the evolving nature of job roles and the skills required for future success. By aligning employee skills development initiatives with the anticipated technological advancements, organizations can empower employees with the knowledge and adaptability needed to navigate a dynamically changing workplace. Lastly, to combat techno-inefficiency, organizations must carefully assess the skills of their workforce and tailor AI implementations to fit seamlessly within existing organizational processes. Understanding the specific needs and competencies of employees ensures that AI tools enhance workflow efficiency rather than causing frustration or perceived inefficiency. This alignment contributes to a more effective and streamlined integration of AI into the organizational framework.

Implement transparent communication about the goals, benefits, and potential changes associated with AI adoption, provide ongoing updates and address concerns proactively. Clear communication helps in managing expectations and reducing anxiety about the impact of AI on job roles, thus minimizing resistance. Introduce flexible work policies to accommodate the potential technostress-related challenges, allowing employees to manage their workloads effectively helps employees balance their professional and personal lives, mitigating concerns related to techno-invasion and work-life balance. Invest in developing intuitive and user-friendly interfaces for AI technologies to reduce the cognitive load on users. A simplified user experience minimizes the feeling of information overload and promotes a positive perception of technology. Establish feedback channels to gather insights from employees about their experiences with AI technologies and use this feedback to make continuous improvements, creates a sense of involvement and empowers employees to voice concerns, leading to more responsive adjustments. The implementation of wellness programs and support services to address the potential impact of technostress on employees' mental well-being can provide resources for stress management and mental health support contributing to a healthier work environment.

4.1 Adaptation to AI at Work

Adapting to AI at work involves a multi-faceted approach that draws upon various fields of psychology and behavior. From a cognitive perspective, employees need to develop a sound understanding of AI technologies and their implications [41]. This involves not only technical knowledge but also an understanding of how AI can be integrated into their work processes. From a behavioral perspective, employees need to develop new habits and routines that incorporate AI tools. This could involve using AI for data analysis, decision-making, or routine tasks [42]. Emotionally, employees may need to overcome fears and anxieties related to AI, such as fear of job loss or concerns about privacy [42].

This can be facilitated through emotional intelligence training and support from the organization. Socially, employees need to navigate the changing dynamics in the workplace brought about by AI, such as changes in power structures or team dynamics [41]. Finally, from an organizational perspective, companies need to provide the necessary training, support, and resources to help employees adapt to AI. Incuding not only technical training but also support for the emotional and social aspects of AI adaptation [42].

4.2 What Could HR Do to Smoothen the Process?

In the context of implementing Artificial Intelligence (AI) in the workplace, Human Resources (HR) serves a critical role in managing the transition and alleviating technostress among employees. Human Resources (HR) plays a pivotal role in facilitating the smooth transition of AI implementation and mitigating technostress among employees. HR can take measures to support this process such as we describe in Table 1, below:

The role of Psychological safety contributing into the smoothening of the process is essencial in facilitating the smooth transition of AI implementation in the workplace and mitigating technostress among employees. The ways in which psychological safety contributes are showed in Table 2, below:

Table 1. Mitigating Technostress

Dimensions	Examples
Open Communication	Psychological safety fosters an environment where employees feel comfortable expressing their thoughts, concerns, and questions related to AI implementation. Open communication channels enable the free exchange of ideas, allowing employees to voice their opinions without fear of reprisal
Risk-Taking and Innovation	A psychologically safe environment encourages employees to take risks and experiment with new approaches, fostering innovation. This mindset is essential during AI implementation, where experimentation and adaptation are integral to finding optimal ways to integrate technology into existing workflows
Learning and Adaptation	Psychological safety promotes a culture of continuous learning and adaptation. Employees are more likely to embrace the learning curve associated with AI technologies when they feel psychologically safe, knowing that mistakes will be viewed as opportunities for growth rather than punitive measures
Collaboration and Teamwork	Teams that operate in a psychologically safe environment are more likely to collaborate effectively. In the context of AI implementation, collaboration is essential for sharing knowledge, insights, and best practices, leading to a collective understanding of how to leverage AI for mutual benefit
Reduced Fear of Job Insecurity	A psychologically safe environment promotes inclusive decision-making processes. In the context of AI integration, involving employees in decision-making regarding the selection, implementation, and use of AI tools ensures that diverse perspectives are considered, enhancing the effectiveness of the transition
Inclusive Decision-Making	A psychologically safe environment promotes inclusive decision-making processes. In the context of AI integration, involving employees in decision-making regarding the selection, implementation, and use of AI tools ensures that diverse perspectives are considered, enhancing the effectiveness of the transition
Support for Well-being	Psychological safety contributes to an organizational culture that values employee well-being. This support is crucial during times of change, such as AI implementation, as it helps alleviate stress and anxiety, fostering an environment where employees can adapt more seamlessly
Feedback Loops	Psychological safety encourages the establishment of feedback loops. Employees feel more comfortable providing feedback on their experiences with AI tools, enabling organizations to make continuous improvements and refine the integration process based on real-time insights
Trust in Leadership	A psychologically safe workplace fosters trust in leadership. Trust is essential during times of change, and employees are more likely to trust that leadership has their best interests in mind when psychological safety is prioritized

Source: Authors elaboration.

In this sense, psychological safety contributes significantly to the smoothening process of AI implementation by creating an environment where employees feel comfortable, supported, and empowered. This, in turn, enables organizations to harness the full potential of AI technologies while minimizing the negative psychological impacts associated with change and technological advancements.

Table 2. Psychological Safety to facilitate transition to AI

Dimensions	Example
Open Communication	Psychological safety fosters an environment where employees feel comfortable expressing their thoughts, concerns, and questions related to AI implementation. Open communication channels enable the free exchange of ideas, allowing employees to voice their opinions without fear of reprisal
Risk-Taking and Innovation	A psychologically safe environment encourages employees to take risks and experiment with new approaches, fostering innovation. This mindset is essential during AI implementation, where experimentation and adaptation are integral to finding optimal ways to integrate technology into existing workflows
Learning and Adaptation	Psychological safety promotes a culture of continuous learning and adaptation. Employees are more likely to embrace the learning curve associated with AI technologies when they feel psychologically safe, knowing that mistakes will be viewed as opportunities for growth rather than punitive measures
Collaboration and Teamwork	Teams that operate in a psychologically safe environment are more likely to collaborate effectively. In the context of AI implementation, collaboration is essential for sharing knowledge, insights, and best practices, leading to a collective understanding of how to leverage AI for mutual benefit
Reduced Fear of Job Insecurity	A psychologically safe environment promotes inclusive decision-making processes. In the context of AI integration, involving employees in decision-making regarding the selection, implementation, and use of AI tools ensures that diverse perspectives are considered, enhancing the effectiveness of the transition
Inclusive Decision-Making	A psychologically safe environment promotes inclusive decision-making processes. In the context of AI integration, involving employees in decision-making regarding the selection, implementation, and use of AI tools ensures that diverse perspectives are considered, enhancing the effectiveness of the transition
Support for Well-being	Psychological safety contributes to an organizational culture that values employee well-being. This support is crucial during times of change, such as AI implementation, as it helps alleviate stress and anxiety, fostering an environment where employees can adapt more seamlessly
Feedback Loops	Psychological safety encourages the establishment of feedback loops. Employees feel more comfortable providing feedback on their experiences with AI tools, enabling organizations to make continuous improvements and refine the integration process based on real-time insights
Trust in Leadership	A psychologically safe workplace fosters trust in leadership. Trust is essential during times of change, and employees are more likely to trust that leadership has their best interests in mind when psychological safety is prioritized

Source: Authors elaboration.

5 Conclusion

The long-term effects of AI adoption can be profound, impacting both individual well-being and organizational dynamics. On an individual level, AI can enhance well-being by automating routine tasks, freeing up time for more meaningful and creative work [44]. However, it can also lead to job displacement and increased stress due to the need to constantly update skills [44]. On an organizational level, AI can lead to increased efficiency and productivity, and open new opportunities for innovation [45]. However, it can also disrupt existing business models and require significant changes in organizational structures and processes [46]. In the long term, successful AI adoption requires a holistic approach that considers not only the technical aspects of AI but also its social, psychological, and organizational implications. This includes providing support and training for employees, fostering a culture of continuous learning, and implementing ethical guidelines for AI use [41].

A substantial proportion of the efficiency improvements observed can be attributed to the integration of systems that involve collaborative efforts between human workers and artificial intelligence (AI). The application of Human-AI Collaboration systems holds promise across various facets of daily work, facilitating heightened precision and reliability in output generation while concurrently diminishing the time and effort invested in such endeavors [46]. However, the actualization of these efficiency gains is contingent upon the acceptance of these systems by the workforce.

The successful implementation of Human-AI Collaboration systems hinges crucially upon the extent to which employees embrace these technological advancements. A seminal work by [48] posited that the lack of acceptance for new information technology systems jeopardizes their successful integration. In this context, the realization of optimal efficiency gains may encounter constraints linked to the levels of technostress experienced by employees in their workplace. The reluctance or resistance towards technological adoption due to technostress could potentially impede the complete exploitation of the envisioned efficiency enhancements associated with Human-AI Collaboration systems.

It is interesting that the majority of literature focused on the perception of AI adoption [37] and studies investigating the effect of human-AI collaboration through actual co-working with AI are scarce. The surge in interest surrounding human-AI collaboration has not been proportionately matched by extensive research investigating the implications of working with AI on technostress and its subsequent effects on psychological well-being and organizational behavior [37]. This dearth of scholarly inquiry underscores a notable gap in the existing literature, as the intricate dynamics and consequences of human-AI interaction remain inadequately explored. Addressing this research gap is crucial for a comprehensive understanding of the implications associated with the integration of AI technologies into the workplace, offering valuable insights into the nuanced interplay between human cognition, technology-induced stress, and its broader impact on both individual well-being and organizational dynamics.

It may be concluded that technostress manifests through dimensions such as information overload, techno-invasion, role ambiguity, and techno-uncertainty. These dimensions contribute to employees' stress levels and impact their overall well-being in the workplace. Perceived usefulness and ease of use play pivotal roles in follower's acceptance of AI technologies. Adequate training, leadership support, and organizational culture significantly influence the successful adaptation of AI. Establish clear communication channels regarding AI's purpose, benefits, and its impact on roles to reduce role ambiguity. Provide user-friendly interfaces and comprehensive training programs to enhance perceived ease of use. Cultivate a supportive leadership culture that endorses AI adoption and encourages ongoing learning. Continuous monitoring and addressing of technostress dimensions are crucial for sustaining individual wellbeing. Positive adaptation to AI can contribute to a culture of innovation, increased productivity, and improved organizational dynamics.

The global shift towards remote work, accelerated by the COVID 19 pandemic, has emphasized the importance of AI-driven collaboration tools. While these tools enhance virtual teamwork, the constant connectivity and reliance on AI for communication may introduce new dimensions of technostress related to digital overload, virtual presence fatigue, and maintaining work-life boundaries.

Growing awareness of ethical considerations and biases in AI algorithms has gained prominence. Employees may experience technostress related to concerns about the ethical implications of AI applications, especially if they perceive biases in decision-making processes or worry about the impact of AI on privacy and fairness.

The increased integration of AI in decision-making processes, especially in sectors like finance and healthcare, might introduce concerns about transparency and accountability. Employees may experience technostress if they feel uncertain about how AI algorithms make critical decisions and the potential consequences of those decisions on their roles.

The development and use of AI-powered mental health monitoring applications have gained attention. While these tools aim to support well-being, concerns about the privacy of mental health data and the reliance on AI for emotional analysis could introduce unique dimensions of technostress in the context of personal well-being.

It's crucial to stay updated and from these key findings is perceived the need to create to develop comprehensive training programs focusing on both technical skills and the psychological aspects of AI adaptation. Ensure ongoing learning opportunities to keep employees updated on AI advancements. Foster transparent communication about AI implementations, addressing concerns and emphasizing the positive impacts on individual and organizational goals. Cultivate a culture that values innovation, learning, and adaptability. Recognize and reward employees who effectively integrate AI into their work processes and invest in Well-being Initiatives.

To contribute to filling this void this research provides a comprehensive exploration of the implications of AI integration, focusing specifically on the dimensions of technostress. By dissecting these dimensions, the study uncovers the nuanced ways in which AI adoption influences employees' psychological well-being.

In developing effective human resource management strategies during the transitional phase of AI implementation, were included recommendations for mitigating technostress, fostering a positive organizational culture, and ensuring a smooth adaptation to AI technologies.

References

1. Davenport, T.H., Ronanki, R.: Artificial intelligence for the real world. Harv. Bus. Rev. **96**(1), 108–116 (2018)
2. Stahl, B.C., et al.: A systematic review of artificial intelligence impact assessments. Artif. Intell. Rev. [Preprint] (2023). https://doi.org/10.1007/s10462-02310420-8
3. Bollier, D.: Artificial intelligence comes of age: the promise and challenge of integrating AI Into cars, healthcare and journalism (2017). https://www.aspeninstitute.org/wpcontent/uploads/2017/01/2017-Artificial-In telligence-REPFINAL.pdf
4. AlSheibani, S., Messom, C., Cheung, Y.: Re-thinking the competitive landscape of artificial intelligence. In: Proceedings of the 53rd Hawaii International Conference on System Sciences (2020)
5. Jarrahi, M.H., Kenyon, S., Brown, A., Donahue, C., Wicher, C.: Artificial intelligence: a strategy to harness its power through organizational learning. J. Bus. Strategy (2022)

6. Elliot, B., Andrews, W.: A framework for applying AI in the enterprise. Gartner 1–38 (2017)
7. Ransbotham, S., Khodabandeh, S., Kiron, D., Candelon, F., Chu, M., LaFountain, B.: Expanding AI's impact with organizational learning. MIT Sloan Review and Boston Consulting Group (2020)
8. Selenko, E., et al.: Artificial intelligence and the future of work: a functional-identity perspective. Curr. Dir. Psychol. Sci. **31**(3), 272–279 (2022). https://doi.org/10.1177/09637214221091823
9. Bahr, T.J., Ginsburg, S., Wright, J.G., Shachak, A.: Technostress as source of physician burnout: An exploration of the associations between technology usage and physician burnout. Int. J. Med. Inform. **177**, 105147 (2023). https://doi.org/10.1016/j.ijmedinf.2023.105147. Epub 2023 Jul 8. PMID: 37517300
10. Brod, C.: Technostress: The Human Cost of the Computer Revolution. Addison Wesley, Boston (1984)
11. Ayyagari, R., Grover, V., Purvis, R.: Technostress: technological antecedents and implications. MIS Q. **35**(4), 831–858 (2011)
12. Dery, K., Kolb, D.G., MacCormick, J.: Working with connective flow: how smartphone use is evolving in practice. Eur. J. Inf. Syst. **23**(5), 558–570 (2014)
13. Maier, C., Laumer, S., Eckhardt, A., Weitzel, T.: Giving too much social support: social overload on social networking sites. Eur. J. Inf. Syst. **24**(5), 447464 (2015)
14. Vakola, M., Wilson, I.E.: The challenge of virtual organisation: critical success factors in dealing with constant change. Team Perform. Manag.: Int. J. **10**(5/6), 112–120 (2004)
15. Spreitzer, G., Cameron, L., Garrett, L.: Alternative work arrangements: two images of the new world of work. Ann. Rev. Organ. Psychol. Organ. Behav. **4**, 473–499 (2017). Author, F.: Article title. Journal 2(5), 99–110 (2016)
16. Parasuraman, A., Colby, C.L.: An updated and streamlined technology readiness index: TRI 2.0. J. Serv. Res. **18**(1), 59–74 (2015)
17. Liu, X., et al.: A comparison of deep learning performance against health-care professionals in detecting diseases from medical imaging: a systematic review and meta-analysis. Lancet Digit. Health **1**(6), e271–e297 (2019)
18. Jie, Z., Zhiying, Z., Li., L.: A meta-analysis of watson for oncology in clinical application. Sci. Rep. **11**, 5792 (2021). https://doi.org/10.1038/s41598-021-84973-5
19. Wolf, R.M., Channa, R., Liu, T.Y.A., et al.: Autonomous artificial intelligence increases screening and follow-up for diabetic retinopathy in youth: the ACCESS randomized control trial. Nat. Commun. **15**, 421 (2024). https://doi.org/10.1038/s41467-023-44676-z
20. Geada, N., Jamil, G.: Perspectives on Artificial Intelligence in Times of Turbulence: Theoretical Background to Applications. IGI Global (2023). https://doi.org/10.4018/978-1-6684-9814-9
21. Wang, C., et al.: The impact of COVID-19 pandemic on physical and mental health of Asians: a study of seven middleincome countries in Asia. PLoS ONE **16**(2), e0246824 (2021). https://doi.org/10.1371/journal.pone.0246824
22. Grzybowski, J.: Separatists, state subjectivity, and fundamental ontological (in)security in international relations. Int. Relat. (2021). https://doi.org/10.1177/00471178211045619
23. Chen, K., Wang, Y., Nie, K., Zhang, Z., Wang, Q., Tan, T.: Deep learning for smart manufacturing: methods and applications. J. Manuf. Syst. **54**, 124–138 (2020)
24. Zawacki-Richter, O., Marín, V.I., Bond, M., Gouverneur, F.: Systematic review of research on artificial intelligence applications in higher education–where are the educators? Int. J. Educ. Technol. High. Educ. **16**(1), 1–27 (2019)
25. Kamalov, F., Santandreu Calonge, D., Gurrib, I.: New era of artificial intelligence in education: towards a sustainable multifaceted revolution. Sustainability **15**(16), 12451 (2023)

26. Bao, Y., Hilary, G., Ke, B.: Artificial intelligence and fraud detection. In: Babich, V., Birge, J.R., Hilary, G. (eds.) Innovative Technology at the Interface of Finance and Operations, vol. I, pp. 223–247. Springer International Publishing, Berlin (2022)
27. Priya, G. Saradha, S.: Fraud detection and prevention using machine learning algorithms: a review, pp. 564–568 (2021). https://doi.org/10.1109/ICEES51510.2021.9383631
28. Jha, N., et al.: Psychosocial and stress-related risk factors for abnormal menstrual cycle pattern among adolescent girls: a case-control study. J. Educ. Health Promot. **9**, 313 (2020). https://doi.org/10.4103/jehp.jehp_419_20. PMID: 33426117; PMCID: PMC7774621
29. Donning, H., Eriksson, M., Martikainen, M., Lehner, O.M.: Prevention and detection for risk and fraud in the digital age-the current situation. ACRN Oxford J. Financ. Risk Perspect. **8**, 86–97 (2019)
30. Giannaros, T.M., Dafis, S., Kotroni, V., Lagouvardos, K.: The extreme heat wave of late July/early August 2021 in Greece under the context of the direct effect of anthropogenic greenhouse gases. Atmos. Sci. Lett. **25**(1) (2023). https://doi.org/10.1002/asl.1191
31. Othman, U., Yang, E.: Human–robot collaborations in smart manufacturing environments: review and outlook. Sensors **23**, 5663 (2023). https://doi.org/10.3390/s23125663
32. Borboni, A., Reddy, K.V.V., Elamvazuthi, I., AL-Quraishi, M.S., Natarajan, E., Azhar Ali, S.S.: The expanding role of artificial intelligence in collaborative robots for industrial applications: a systematic review of recent works. Machines **11**, 111 (2023). https://doi.org/10.3390/machines11010111
33. Tarafdar, M., Tu, Q., Ragu-Nathan, B.S., Ragu-Nathan, T.S.: Crossing to the dark side: examining creators, outcomes, and inhibitors of technostress. Commun. ACM **54**(9), 113–120 (2015)
34. Ragu-Nathan, T.S., Tarafdar, M., Ragu-Nathan, B.S., Tu, Q.: The consequences of technostress for end users in organizations: Conceptual development and empirical validation. Inf. Syst. Res. **19**(4), 417–433 (2008)
35. Weil, M.M., Rosen, L.D.: Technostress: Coping with Technology @Work @Home @Play. Wiley, New York (1997)
36. Kumar, A., Krishnamoorthy, B., Bhattacharyya, S.S.: Machine learning and artificial intelligence-induced technostress in organizations: a study on automation-augmentation paradox with socio-technical systems as coping mechanisms. Int. J. Organ. Anal. **32**, 681–701 (2023)
37. Xia, M.: Co-working with AI is a double-sword in technostress? an integrative review of human-AI collaboration from a holistic process of technostress. In: SHS Web of Conferences, vol. 155, p. 03022. EDP Sciences (2023)
38. Weber, P.: The effect of technostress on the acceptance of artificial intelligence-enabled machine feedback systems (Master Thesis) (2020)
39. Htahet, H., Johansson, E.: Leaders, followers, and AI technostress: a study on how leaders can mitigate AI implementation related technostress through transformational leadership and employees' engagement (2023)
40. Salanova, M., Llorens, S., Cifre, E., Martínez, I.M., Schaufeli, W.B.: From "technostress" to "technohealth": theoretical and methodological implications. In: Perrewé, M., Halbesleben, J., Rosen, C. (eds.), Research in Occupational Stress and Well-Being, vol. 14, pp. 1–33. Emerald Group Publishing Limited (2016)
41. Davenport, T.H., Ronanki, R.: Artificial Intelligence for the Real World. Harvard Business Review (HBR) (2018). https://www.bizjournals.com/boston/news/2018/01/09/hbr-artificial-intelligence-for-the-real-world.html
42. Brougham, D., Haar, J.: Smart Technology, Artificial Intelligence, Robotics, and Algorithms (STARA): employees' perceptions of our future workplace. J. Manag. Organ. **24**(2), 239–257 (2018)

43. Bessen, J.E.: AI and Jobs: The Role of Demand. NBER Working Paper No. 24235 (2019)
44. Arntz, M., Gregory, T., Zierahn, U.: The Risk of Automation for Jobs in OECD Countries: A Comparative Analysis. OECD Social, Employment and Migration Working Papers, No. 189. OECD Publishing, Paris (2016)
45. Brynjolfsson, E., McAfee, A.: The Second Machine Age: Work, Progress, and Prosperity in a Time of Brilliant Technologies. W. W. Norton & Company, New York City (2014)
46. Ford, M.: Rise of the Robots: Technology and the Threat of a Jobless Future. Basic Books, New York City (2015)
47. Amershi, S., Cakmak, M., Knox, W.B., Kulesza, T.: Power to the people: the role of humans in interactive machine learning. AI Mag. **35**(4), 105–120 (2014). https://doi.org/10.1609/aimag.v35i4.2513
48. Davis, F.D.: Perceived usefulness, perceived ease of use, and user acceptance of information technology. MIS Q.: Manag. Inf. Syst. **13**(3), 319–339, (1989). https://doi.org/10.2307/249008

Author Index

M. J. Sousa (Ed.): UFHRD 2024, ISEM 52, p. 393, 2025.
https://doi.org/10.1007/978-3-031-92548-1

The manufacturer's authorised representative in the EU is Springer Nature Customer Service Centre GmbH, Europaplatz 3, 69115 Heidelberg, Germany. If you have any concerns regarding our products, please contact ProductSafety@springernature.com

Printed and bound by CPI Group (UK) Ltd, Croydon, CR0 4YY
30/09/2026
02225588-0016